APPROXIMATION THEORY VII

Proceedings of the Seventh International Symposium on Approximation Theory hosted by the University of Texas at Austin, Texas on January 3-7, 1992.

Previously published:

Approximation Theory, edited by G.G. Lorenz, 1973
Approximation Theory II, edited by G.G. Lorenz, C.K. Chui, L.L. Schumaker, 1976
Approximation Theory III, edited by E.W. Cheney, 1980
Approximation Theory IV, edited by C.K. Chui, L.L. Schumaker, J.D. Ward, 1983
Approximation Theory V, edited by C.K. Chui, L.L. Schumaker, J.D. Ward, 1986
Approximation Theory VI, Volume I, edited by C.K. Chui, L.L. Schumaker, J.D. Ward, 1989
Approximation Theory VI, Volume II, edited by C.K. Chui, L.L. Schumaker, J.D. Ward, 1989

APPROXIMATION THEORY VII

Edited by E.W. Cheney, C.K. Chui, and L.L. Schumaker
Department of Mathematics
University of Texas - Austin
Austin, Texas

ACADEMIC PRESS, INC.
Harcourt Brace Jovanovich, Publishers

Boston San Diego New York
London Sydney Tokyo Toronto

This book is printed on acid-free paper. ∞

Copyright © 1993 by Academic Press, Inc.

All rights reserved.
No part of this publication may be reproduced or
transmitted in any form or by any means, electronic
or mechanical, including photocopy, recording, or
any information storage and retrieval system, without
permission in writing from the publisher.

ACADEMIC PRESS, INC.
1250 Sixth Avenue, San Diego, CA 92101-4311

United Kingdom Edition published by
ACADEMIC PRESS LIMITED
24-28 Oval Road, London NW1 7DX

Approximation theory VII / edited by E.W. Cheney, C.K.,Chui, L.L.
  Schumaker.
    p. cm.
  "Proceedings of the Seventh International Symposium on
Approximation Theory, held in Austin, Texas, on January 3-7, 1992"-
-CIP galley.
   Includes bibliographical references.
   ISBN 0-12-174589-9
   1. Approximation theory--Congresses. I. Cheney, E. W. (Elliott
Ward) . 1929-   . II. Chui, C. K.   III. Schumaker, Larry L., 1939-
. IV. International Symposium on Approximation Theory (7th : 1992 :
Austin, Tex.)   V. Title: Approximation theory 7.
QA221.A6448   1993
511'.4--dc20   92-33039
                                                                    CIP

Printed in the United States of America

92 93 94 95 96 97   QW   9 8 7 6 5 4 3 2 1

CONTENTS

Preface vii

Participants ix

Approximation Order without Quasi-Interpolants
   Carl de Boor 1

Wavelets and Signal Analysis
   Charles K. Chui 19

Wavelet Bases, Approximation Theory, and Subdivision Schemes
   Albert Cohen 63

Approximation with Convex Rational Functions
   Bo Gao, Donald J. Newman, and Vasil Popov 87

Block Structure and Recursiveness in Rational Interpolation
   Martin Gutknecht 93

Multivariate Approximation from the Cardinal Interpolation
Point of View
   Kurt Jetter 131

Ridge Functions, Sigmoidal Functions, and Neural Networks
   Will Light 163

Knot Removal for Spline Curves and Surfaces
   Tom Lyche 207

Approximation by Algebraic Polynomials
   Vilmos Totik 227

PREFACE

During the week of January 3–7, 1992, the Seventh International Symposium on Approximation Theory was held in Austin, Texas, at the Hyatt Regency Hotel. The University of Texas at Austin was the host. Previous conferences in this series had been held in 1973, 1976 and 1980 in Austin, and in 1983, 1986, and 1989 in College Station, where Texas A&M University was the host. The proceedings of those conferences were published by Academic Press, and we are pleased that they are continuing that tradition.

This conference was attended by over 185 mathematicians from 23 different countries. The survey lectures dealt with some particularly active subareas of approximation theory, including quasi-interpolants, wavelets, subdivision algorithms, rational approximation and interpolation, cardinal interpolation, neural networks, knot removal for spline curves and surfaces, and approximation by algebraic polynomials. These Proceedings contain the nine survey papers from the conference.

We are indebted to the National Science Foundation and the U.S. Army Research Office for financial support of the conference. In addition, we would like to extend our appreciation to Texas A&M University and The University of Texas at Austin for providing secretarial assistance, and for helping to defray many incidental costs. In particular, the Center for Numerical Analysis at the University of Texas, under the direction of Professor David M. Young, Jr, helped with logistic support. The Mathematics Department of the University of Texas, under the chairmanship of Professor Efraim Armendariz, underwrote the expenses of one of our invited speakers, and made technical help available to us.

Several of our colleagues and students helped with the conference in various ways. In particular, we thank Drs. Joachim Stöckler, N. Sivakumar, and Ewald Quak for their assistance in preparing the conference program. Our students Kuei-Fang Chang, Debao Chen, Greg Fasshauer, Dong Hong, Jianao Lian, Jeff Marasovich, and Valdir Menegatto, were very helpful in the day-to-day running of the conference.

We want to express our special appreciation to Stephanie Sellers, Margaret Chui, and Katy Burrell who served as conference secretaries. Also, we thank Jean Johnson and Sheri Brice for handling the accounting and other technical details. Finally, we are grateful to Margaret Combs of the University of Texas Mathematics Department for editing the TeX files, instilling some stylistic conformity, and preparing the final camera-ready masterscript.

Austin, Texas
August 1, 1992

PARTICIPANTS

G. DONALD ALLEN, *Department of Mathematics, Texas A&M University, College Station, TX 77843-3368* [dallen@hilbert.tamu.edu]

RADWAN AL-JARRAH, *Department of Mathematics, Southwestern Oklahoma State University, 100 Campus Dr., Weatherford, OK 73096*

DAN AMIR, *Tel Aviv University, Rector's Office, Ramat Aviv, Tel Aviv, Israel* [amird@taunivm.bitnet]

GEORGE ANASTASSIOU, *Dept. of Mathematical Sciences, Memphis State University, Memphis, Tennessee 38152* [anastasg@hermes.msci.memst.edu]

ERLEND ARGE, *Senter for Industriforskning, Box 124 Blindern, N-0314 Oslo, Norway* [arge@bivrost.cs.si.no]

AHMED AYED, *Department of Mathematics, Texas A&M University, College Station, TX 77843-3368*

CATALIN BADEA, *Université de Paris-Sud, Mathématique, Bat. 425, 91405 Orsay Cedex, France*

CHANDERJIT BAJAJ, *Dept. of Computer Science, Purdue University, West Lafayette, IN 47907* [bajaj@cs.purdue.edu]

BELA BAJNOK, *Dept. of Applied Mathematical Sciences, University of Houston - Downtown, One Main Street, Houston, TX 77002* [bajnik@dt3.dt.uh.edu]

IAN BARRODALE, *Barrodale Computing Services, Ltd., 8-1560 Church Avenue, Victoria, B.C., Canada V8P 2H1* [ibarroda@csr.uvic.ca]

MARTIN W. BARTELT, *Department of Mathematics, Christopher Newport College, Newport News, VA 23606*

B. J. C. BAXTER, *DAMTP, Silver Street, Cambridge CB3 9EW, England*

R. K. BEATSON, *Department of Mathematics, University of Canterbury, Christchurch, New Zealand* [rkb@maths.canterbury.ac.nz]

HUBERT BERENS, *Mathematical Institute, Univ. of Erlangen-Nüremberg, Bismarckstr. 1 1/2, D-8520 Erlangen, Germany* [berens@cnve.rrze.uni-erlangen.dbp.de]

HANS PETER BLATT, Katholische Universität Eichstätt, Ostenstrasse 26-28, D-8078 Eichstätt, Germany [blatt@kueichstaett.dbp.de]

THOMAS BLOOM, Department of Mathematics, University of Toronto, Toronto, Canada M5S 1A1 [bloom@math.toronto.edu]

CARL DE BOOR, Center for Mathematical Sciences, University of Wisconsin-Madison, 610 Walnut Street, Madison, WI 53706 [deboor@cs.wisc.edu]

PETER BORWEIN, Dept. of Math., Stats, and C.S., Dalhousie University, Halifax, Nova Scotia, Canada B3H 3J5 [pborwein@cs.dal.ca]

L. BOS, Department of Mathematics & Statistics, University of Calgary, Calgary, Alberta, Canada T2N 1N4 [lpbos@uncamult.bitnet]

KEN W. BOSWORTH, Chesapeake Biological Lab, Solomons, MD 20688-0038 [boz@cbl.umd.edu]

ELSBETH BREDENDIECK, Mathematics Seminar der Universität, Bundesstrasse 55, D-2000 Hamburg 13, Germany

HERMANN BURCHARD, Department of Mathematics, Oklahoma State University, Stillwater, OK 74078 [u2927aa@vms.ucc.okstate.edu]

JOHN C. BURKETT, 999 S.W. 16th Ave. #3, Gainesville, FL 32601

MICHELE CAMPITI, Dept. of Mathematics, University of Bari, Via Edoardo Orabona, 4, 70125 Bari, Italy

RALPH E. CARLSON, L-316 Lawrence Livermore National Lab, Livermore, CA 94550 [carlson@lll-crg.llnl.gov]

B. L. CHALMERS, Department of Mathematics, University of California - Riverside, Riverside, CA 92521

KUEI-FANG CHANG, Department of Mathematics, University of Texas at Austin, Austin, TX 78712 [kuei@math.utexas.edu]

DEBAO CHEN, Department of Mathematics, University of Texas at Austin, Austin, TX 78712 [chen@math.utexas.edu]

GUANRONG CHEN, Dept. of Electrical Engr., University of Houston, Houston, TX 77204-4793 [gchen@uh.edu]

TIANPING CHEN, Dept. of Mathematics, Fudan University, Shanghai, China

WEIYU CHEN, Dept. of Mathematics, University of Alberta, Edmonton, Alberta, Canada T6G 2G1 [userchwy@mts.ucs.ualberta.ca]

WARD CHENEY, Dept. of Mathematics, University of Texas, Austin, TX 78712 [cheney@math.utexas.edu]

ZHENG-XING CHENG, Dept. of Mathematics, Xi'an Jiatong University, Xi'an, China

CHARLES K. CHUI, Center for Approximation Theory, Texas A&M University, College Station, TX 77843-3368 [cchui@tamu.edu]

Participants xi

ALBERT COHEN, *University of Paris - Dauphine, CEREMADE, Place du Marechal de Lattre de Tassigny, 75775 Paris Cedex 16 France* [chavent@frulm63.bitnet]

LISA OSTERMAN COULTER, *Dept. of Mathematics and Computer Sciences, Stetson University, DeLand, FL 32720* [coulter@stetson.bitnet]

ANNIE CUYT, *Dept. of Mathematics and Computer Sciences, University of Antwerp (UIA), Universiteitsplein 1, B-2610 Wilrijk-Antwerp (Belgium)* [cuyt@ccu.uia.ac.be]

MORTEN DAEHLEN, *Senter for Industriforskning, Box 124 Blindern, N-0314, Oslo, Norway* [mortend@ifi.vio.no]

WOLFGANG DAHMEN, *Institut für Mathematik I, Freie Universität Berlin, Arnimallee 2-6, 1000 Berlin 33, Germany* [dahmen@math.fu.berlin.de]

LUBOMIR T. DECHEVSKI, *Dept. of Mathematics, Stats, and C.S., University of Chicago, Box 4248, Chicago, Illinois 60680* [u54502@uicvm.bitnet]

CATHERINE DETAILLE, *FNDP, Dept. de Mathématique, rempart de la Vièrge, 8, B-5000, Namur, Belgium* [cdetaille@bnandp51.bitnet]

FRANK DEUTSCH, *Dept. of Mathematics, Penn State University, University Park, PA 16802* [deutsch@math.psu.edu]

HARVEY DIAMOND, *Dept. of Mathematics, West Virginia University, Morgantown, WV 26506* [diamond@cerc.wvu.wvnet.edu]

ZEEV DITZIAN, *Dept. of Mathematics, University of Alberta, Edmonton, Alberta, Canada T6G 2G1* [ditz@ualtavm.bitnet]

KATHY DRIVER, *Dept. of Mathematics, Witwatersrand University, Private Bag 3, P.O. Wits 2050, Johannesburg, South Africa* [036kad@witsvma.wits.ac.za]

CHARLES B. DUNHAM, *Computer Science Dept., University Western Ontario, London, Ontario, Canada N6A 587* [czhu@uwovax.uwo.ca]

NIRA DYN, *School of Mathematical Science, Tel-Aviv University, Tel-Aviv 69978 Israel* [niradyn@taurus.bitnet]

KEVIN ENGELBRECHT, *226 West 9th Avenue, E., Columbus, Ohio 43201*

TAMAS ERDELYI, *Dept. of Mathematics, OSU, 231 West 18th Ave., Columbus, Ohio 43210* [terdelyi@function.mps.ohio.state.edu]

DAVID EUBANKS, *P.O. Box 1505, Hartsville, SC 29550*

MARTINA FINZEL, *Mathematisches Institut, Universität Erlangen-Nürnberg, Birmarckstrasse 1 1/2, W-8520 Erlangen, Germany* [finzel@cnve.rrze.uni-erlangen.dbp.de]

BERND FISCHER, *Institute of Applied Mathematics, University of Hamburg, Bundesstrasse 55 D-W-2000, Hamburg 13, Germany* [na.fischer@na-net.ornl.gov]

THOMAS A. FOLEY, *Computer Science Dept., Arizona State University, Tempe, AZ 85287* [foley@asuvax.eas.asu.edu]

ROLAND FREUND, *RIACS, Mail Stop Ellis Street, NASA Ames Research Center, Moffett Field, CA 94035* [freund@riacs.edu]

MYRON GOLDSTEIN, *5914 S. Newberry Road, Tempe, Arizona 85283*

MANFRED VON GOLITSCHEK, *Institut für Angewandte Mathematik und Statistik, Universität Würzburg, 8700 Würzburg, Germany* [golitschek@vax.rz.uni-wuerzburg.dbp.de]

HEINZ H. GONSKA, *Dept. of Computer Science, European Business School, D-W-6227 Oestrich-Winkel, Germany* [gonska@ebs.uucp]

C. W. GROETSCH, *Dept. of Mathematics, University of Cincinnati, Cincinnati, OH 45221-0025* [groetsch@ucbeh.san.uc.edu]

RENE GROTHMANN, *Inst. Mathematik, Kath. University, Eichstätt, D-8078 Eichstätt, Germany* [mga010@ku-eichstaett.dbp.de]

MARTIN GUTKNECHT, *IPS, ETH-Zentrum, CH-8092 Zürich, Switzerland* [mhg@ips.ethz.ch]

D. C. HANDSCOMB, *Oxford University Computing Lab, 11 Keble Road, Oxford OX1 3QD England* [dch@comlab.oxford.ac.uk]

WERNER HAUSSMANN, *Department of Mathematics, University of Duisburg, Lotharstr. 65, D-4100, Germany* [hn277ha@unidui.uni-duisburg.de]

MATTHEW HE, *Math and Science Dept., Nova University, 3301 College Ave., Ft. Lauderdale, FL 33314* [hem@polaris.nova.edu]

MYRON S. HENRY, *Office of Academic Affairs, Old Dominion University, Norfolk, VA 23508*

DARVIS HOLLAND, *2500 University Drive NW, Calgary, Alberta, Canada*

DOUG HONG, *Dept. of Mathematics, Texas A&M University, College Station, TX 77843-3368* [doh4737@tamvenus.bitnet]

RIZHUANG HU, *Zhongshan University, Guangzhou, China*

YINGKANG HU, *Dept. of Mathematics/CSC, Georgia Southern University, Statesboro, GA 30460-8093* [ykh@uscn.uscn.uga.edu]

KURT JETTER, *Department of Mathematics, University of Duisburg, Lotharstr. 65, D-4100, Germany* [hn277je@unidui-uni-duisburg.de]

SHING-WHU JHA, *National Central University, Chung-Li, Taiwan 320, R.O.C.* [t212345@twncu865.bitnet]

RONG-QING JIA, *Dept. of Mathematics, University of Alberta, Edmonton, Canada T6G 2G1* [rjia@vega.math.ualberta.ca]

DECHANG JIANG, *Dept. of Mathematics, University of Alberta, Edmonton, Canada T6G 2G1* [userdjng@vega.math.ualberta.ca]

Participants xiii

EDWIN H. KAUFMAN, JR., *Dept. of Mathematics, Central Michigan University, Mount Pleasant, Michigan 48899* [32zej7n@cmuvm.bitnet]

KATHERINE BALAZS KILGORE, *309 Payne Street, Auburn, AL 36849* [kilgore@audncvax.bitnet]

THEODORE KILGORE, *Department of Mathematics, Auburn, AL 36849* [kilgore@audncvax.bitnet]

PER ERIK KOCH, *University of Trondheim, Dept. of Mathematics, 7034 Trondheim-NTH, Norway* [pek@imf.unit.no]

RALITZA KOVACHEVA, *Institute of Mathematics, Bulgaria Academy of Sciences, 1090 Sofia, Bulgaria*

ANDRAS KROÓ, *Math. Inst. of the Hungarian Academy of Sciences, Budapest, Realtanoda u. 13-15, H-1053 Hungary* [h1467kro@ella.hu]

MING-JUN LAI, *Dept. of Mathematics, University of Utah, Salt Lake City, UT 84112* [lai@math.utah.edu]

JUNJIANG LEI, *Dept. of Mathematics, Oklahoma State University, Stillwater, OK 74078* [jlei@hardy.math.okstate.edu]

DAVID J. LEEMING, *Dept. of Mathematics and Statistics, University of Victoria, P.O. Box 3045, Victoria, B.C. Canada V8W 3P4* [leeming@uvvm.uvic.ca]

DAVID LEGG, *Dept. of Mathematics, Indiana-Purdue University at Ft. Wayne, Ft. Wayne, IN 46805* [legg@cuax.ipfw.indiana.edu]

ALAIN LE MÉHAUTÉ, *Dept. de Mathématique, Université de Nantes, 2 rue de la Houssinière, 44072 Nantes Cedex, France* [alm@cicb.fr]

INDY LAGU, *Dept. of Mathematics, University of Calgary, Calgary, Canada* [islagu@uncamult.bitnet]

EITAN LAPIDOT, *P.O. Box 2250, Haifa, Israel*

PIERRE-JEAN LAURENT, *LMC-IMAG, Université Joseph Fourier, BP 53X, 38041 Grenoble, France* [pjl@imag.fr]

NORMAN LEVENBERG, *Department of Mathematics, Wellesley College, Wellesley, MA 02181*

DANY LEVIATAN, *Dept. of Mathematics, University of California, Riverside, CA 92521* [leviatan@math.tau.ac.il]

A. L. (ELI) LEVIN, *Dept. of Mathematics, The Open University of Israel, 16 Klausner St. POB 39328, Tel-Aviv 61392 Israel* [elevin@taunos]

CHUN LI, *Center for Approximation Theory, Department of Mathematics, Texas A&M University, College Station, TX 77843-3368*

WU LI, *Dept. of Mathematics, Old Dominion University, Norfolk, VA 23529* [li@xanth.cs.odu.edu]

XIN LI, Department of Mathematics, Texas A&M University, College Station, TX 77843-3368

XIN LI, Department of Mathematics, University of Central Florida, Orlando, FL 32816

J. A. LIAN, Department of Mathematics, Texas A&M University, College Station, TX 77843-3368

W. A. LIGHT, Department of Mathematics, University of Leicester, University Road, Leicester LE1 7RH, England [maa007@cent1.lancs.ac.uk]

XIAOYAN LIU, Dept. of Mathematics, University of South Florida, Tampa, FL 33620-5700 [xliu@gauss.math.usf.edu]

GEORGE G. LORENTZ, Department of Mathematics, University of Texas, Austin, TX 78712

RUDOLPH LORENTZ, G.M.D., Schloss Birlinghoven, D-5205 St. Augustin 1, Germany [gmap27@dbngmd21.bitnet]

D. S. LUBINSKY, Dept. of Mathematics, Witwatersrand University, P.O. Wits 2050, South Africa [036DOR@witsvma.wits.ac.za]

GUAN LUTAI, Dept. of Mathematics, Texas A&M University, College Station, TX 77843-3368 [lutai@triton.tamu.edu]

TOM LYCHE, Institutt for Informatikk, P.O. Box 1080, Blindern 0316 Oslo 3, Norway [tom@ifi.uio.no]

DETLEF H. MACHE, Universität Dortmund, Fachbereich Mathematik, Postfach 50050, D-4600 Dortmund 50, Germany

ATHENA MAKROGLOU, Dept. of Mathematics, 400 Carver Hall, Iowa State University, Ames, IA 50011 [s1axm@ccvax.iastate.edu]

JEFF MARASOVICH, Dept. of Mathematics, Vanderbilt University, Nashville, TN 37235

G. MASTRIOANNI, Inst. di Matematica, Via N. Sauro 85, Univ. della Basilicata, Potenza, Italy

ALLAN W. MCINNES, Dept. of Mathematics, University of Canterbury, Christchurch, New Zealand [awm@math.canterbury.ac.nz]

VALDIR A. MENEGATTO, Dept. of Mathematics, The University of Texas at Austin, Austin, TX 78712 [menegatt@icmsc.usp.ansp.br]

F. T. METCALF, Department of Mathematics, University of California - Riverside, Riverside, CA 92521

H. N. MHASKAR, Dept. of Mathematics, California State University, Los Angeles, CA 90032 [hmhaska@csula.bitnet]

G. MICULA, Faculty of Mathematics, University of Cluj-Napoca, 3400 Cluj-Napoca, Romania

Participants xv

LAURA BACCHELLI MONTEFUSCO, *Via Dante 32/2, 40125- Bologna, Italy*
[montelau@md.unibo.it]

WILLIAM F. MOSS, *Dept. of Mathematical Sciences, Clemson University, Clemson, SC 29634-1907* [bmoss@math.clemson.edu]

THANDWA MTHEMBU, *100 Mathematics Building, Ohio State University, 231 West 18th Avenue, Columbus, OH 43210*
[mthembu@function.mps.ohio-state.edu]

BERND MULANSKY, *Institute of Numerical Mathematics, Technical University of Dresden, Mommsenstr. 13, 0-8027 Dresden, Germany*

MANFRED W. MÜLLER, *Universität Dortmund, Lehrstuhl Mathematik VIII, Postfach 500500, D-4600 Dortmund 50, Germany*

EDMOND NADLER, *Department of Mathematics, Wayne State University, Detroit, MI 48202-9861* [enadler@math.wayne.edu]

M. Z. NASHED, *Dept. of Mathematical Sciences, University of Delaware, Newark, DE 19716* [patsy@chopin.udel.edu]

M. NEAMTU, *Dept. of Applied Mathematics, University of Twente, P.O. Box 217, 7500 AE Enschede, The Netherlands* [neamtu@utwente.nl]

D. J. NEWMAN, *Department of Mathematics, Temple University, Philadelphia, PA 19122*

TOSHIHIKO NISHISHIRAHO, *Dept. of Mathematics, University of the Ryukyus, Nishihara-Chom Ijubawa 903-01, Japan*

G. NÜRNBERGER, *Universität Mannheim, Lehrstuhl für Mathematik IV - A5, D-W-6800 Mannheim 1, Germany* [fm18c45@dmarum8.bitnet]

V. OPERSTEIN, *Dept. of Mathematics, Technion, 3200 Haifa, Israel*

GERHARD OPFER, *Universität Hamburg, Institut für Angewandte Mathematik, Bundesstrasse 55, D-2000 Hamburg 13, Germany*
[am50150@dhhuni4.bitnet]

JUDITH PALAGALLO, *Dept. of Mathematical Sciences, The University of Akron, Akron, OH 44325-4002* [r1pala@akronvm.bitnet]

K. C. PAN, *Department of Mathematics, University of California - Riverside, Riverside, CA 92521*

DENNIS D. PENCE, *Department of Mathematics & Statistics, Western Michigan University, Kalamazoo, MI 49008-5152* [pence@gw.wmich.edu]

JÜRGEN PRESTIN, *FB Mathematik, Universität Rostock, Universitätsplatz 1, D-0-2500 Rostock, Germany*
[prestin@mathematik.uni-rostock.dbp.de]

THOMAS PRICE, *Dept. of Mathematical Sciences, The University of Akron, Akron, OH 44325-4002* [r1price@akronvm.bitnet]

JOAO B. PROLLA, *IMECC - UNICAMP, Caixa Postal 6065, 13081 Campinas, SP, Brazil*

LUIGIA PUCCIO, *Dipartimento di Matematica, Salita Sperone, 31, Contrada Papardo, B8166 Messina, Italy* [laura@imeuniv.bitnet]

EWALD QUAK, *Center for Approximation Theory, Texas A&M University, College Station, TX 77843-3368* [egq4460@tamvenus.bitnet]

LOUISE RAPHAEL, *Dept. of Mathematics, Howard University, Washington, DC 20059* [alama38@humain.bitnet]

SZILARD GY. REVESZ, *Math. Institute of the Hungarian Academy of Sciences, PO Box 127, Budapest, H-1364 Hungary* [h1163rev@ella.hu]

AMOS RON, *Department of Computer Science, University of Wisconsin - Madison, Madison, WI 53706* [amos@cs.wisc.edu]

ALEXANDER RUSSAKOVSKII, *Mathematisches Institut, Heinrich-Heine Universität Düsseldorf, Universitätsstrasse 1, D-4000 Düsseldorf 1, Germany* [russakov@dd0rud81.bitnet]

E. B. SAFF, *Dept. of Mathematics, Univ. of South Florida, Tampa, FL 33620* [dkgavaa@cfrvm.bitnet]

KESTUTIS SALKAUSKAS, *Dept. of Math. and Statistics, University of Calgary, Calgary, AB, Canada T2N 1N4* [salkauskas@uncamult.bitnet]

ROBERT SCHABACK, *Institut für Numerische, und Angewandte Mathematik, Georg-August-Universität Göttingen, Lotzestrasse 16-18, D-W-3400 Göttingen, Germany* [u0195@dgogwdg5]

DARRELL SCHMIDT, *Dept. of Math. Science, Oakland University, Rochester, MI 48098* [schmidt@vela.acs.oakland.edu]

JOCHEN W. SCHMIDT, *Technical University of Dresden, Dept. of Mathematics, Mommsenstr. 13, D-0-8027 Dresden, Germany* [numath@urzdfn.mathematik.tu-dresden.dbp.de]

EBERHARD SCHMITT, *Institut für Numerische und Angewandte Mathematik, Lotzestrasse 16-18, D-3400 Göttingen, Germany* [eschmitt@namu01.gwdg.de]

LARRY L. SCHUMAKER, *Dept. of Mathematics, Vanderbilt University, Nashville, TN 37240* [s@mars.cas.vanderbilt.edu]

RIVKA SENDEROVIZH, *P.O. Box 2250, Haifa, Israel*

BORIS SHEKHTMAN, *Dept. of Mathematics, University of South Florida, Tampa, FL 33620*

ZUOWEI SHEN, *Center for the Mathematical Sciences, University of Wisconsin - Madison, 610 Walnut Street, Madison, WI 53705* [shen@cms.cms6.wisc.edu]

Participants xvii

X. L. SHI, *Department of Mathematics, Texas A&M University, College Station, TX 77843-3368*

MEHRDAD SIMKANI, *Dept. of Mathematics, The University of Michigan - Flint, Flint, MI 48502*

N. SIVAKUMAR, *Center for Approximation Theory, Department of Mathematics, Texas A&M University, College Station, TX 77843-3368* [n0s9437@venus.tamu.edu]

PHILIP SMITH, *IMSL, 2500 Park W. Tower 1, 2500 City W. Blvd., Houston, TX 77042* [imsl!smith@uunet.uu.net]

MANFRED SOMMER, *Katholische Universität Eichstätt, Math.-Geographische Fakultät, Ostenstrasse 26-28, D-8078 Eichstätt, Germany* [sommer@urz.ku-eichstatt.dbp.de]

V. P. SREEDHARAN, *Department of Mathematics, Michigan State University, East Lansing, MI 48824*

SONYA STEPHENSON STANLEY, *1410 Hillmeade Drive, Nashville, TN 37221* [sstanley@athena.cas.vanderbilt.edu]

JOACHIM STÖCKLER, *Department of Mathematics, University of Duisburg, Lotharstr. 65, D-4100, Germany* [hn277st@unidu.uni-duisburg.de]

HANS STRAUSS, *Institut für Angewandte Mathematik, Universität Erlangen, Martensstr. 3, D-8520 Erlangen, Germany* [strauss@am.uni-erlangen.de]

WEIYI SU, *Dept. of Mathematics, Nanjing Univ., Nanjing, 210008, China*

XINGPING SUN, *Dept. of Mathematics, SMSU, Springfield, MO 65804* [xis280F@smsvma.bitnet]

WIM SWELDENS, *Dept. of Computer Science, K.U. Leuven, Celestunenlaan 200A, B-3001 Leuven, Belgium* [wim@cs.kuleuven.ac.be]

J. SWETITS, *Dept. of Mathematics and Statistics, Old Dominion University, Norfolk, VA 23529* [swetits@mathsun.math.odu.edu]

J. SZABADOS, *Mathematical Institute, P.O.B. 127, H-1364 Budapest, Hungary* [h1142sza@ella.uucp]

MANFRED TASCHE, *FB Mathematik, Universität Rostock, Universitätsplatz 1, D-0-2500 Rostock, Germany* [tasche@mathematik.uni-rostock.dbp.de]

GERALD D. TAYLOR, *Dept. of Mathematics, Colorado State University, Fort Collins, CO 80523* [taylor@taylor.math.colostate.edu]

JEAN-PIERRE THIRAN, *Dept. of Mathematics, Facultés ND de le Paine, Remfort de la Vièrge, 8, B-5000 Namur, Belgium* [jpthiran@bnandp51.bitnet]

V. TOTIK, Dept. of Mathematics, Univ. South Florida, Tampa, FL 33620 [totik@gauss.math.usf.edu]

LEONARDO TRAVERSONI, Universidad Autonoma Metropolitana (Iztapalapa), Division de Ciencias Basicas e Ingenieria, Ap Post 55-532, C.P. 09340 Mexico D.F. Mexico

VASANT A. UBHAYA, Dept. of Computer Sci. and Operations Research, 300 Minard Hall, North Dakota State University, Fargo, ND 58105 [nu043919@ndsuvm1.bitnet]

WALTER VAN ASSCHE, Katholieke Universiteit Leuven, Dept. of Mathematics, Celestijnenlaan 200 B, B-3001 Heverlee (Leuven) Belgium [fgaee03@cc1.kuleuven.ac.be]

PATRICK J. VAN FLEET, Dept. of Mathematics, P.O. Box 1543 Station B, Vanderbilt University, Nashville, TN 37235 [vanfleet@athena.cas.vanderbilt.edu]

P. VERTESI, Mathematical Institute of the Hungarian Academy of Sciences, P.O.B. 127, H-1364, Budapest, Hungary [h1145ver@ella.uucp]

JOSEPH D. WARD, Department of Mathematics, Texas A&M University, College Station, TX 77843-3368 [jdw6562@tamvenus.bitnet]

G. ALISTAIR WATSON, Dept. of Math. and Computer Science, University of Dundee, Dundee DD14HN Scotland [gawatson@mcs.dund.ac.uk]

NORMAN WEYRICH, FB Mathematik, Universität Rostock, Universitätsplatz 1, D-0-2500 Rostock, Germany [weyrich@mathematik.uni-rostock.dbp.de]

ADAM P. WOJCIK, Dept. of Mathematics, University of Umeå, S-90187 Umeå, Sweden

SHUNTANG WU, Dept. of Mathematics, Texas A&M University, College Station, TX 77843-3368

TAILIANG XIE, Dept. of Mathematics, University of Arizona, Tucson, AZ 85721 [tailiang@math.arizona.edu]

YUAN XU, Dept. of Math. and Statistics, Univ. of Oregon, Eugene, OR 97403 [yuan@math.uoregon.edu]

YUESHENG XU, Dept. of Mathematics, North Dakota State University, Fargo, ND 58105 [xu@plains.nodak.edu]

CHENGMIN YANG, Dept. of Mathematics, University of Rhode Island, Kingston, RI 02881 [huv101@uriacc]

XIANG MING YU, Dept. of Mathematics, Southwest Missouri State Univ., Springfield, MO 65807 [xmy944F@smsvma.bitnet]

J. ZHANG, Dept. of Mathematics, Ohio State University, 231 West 18th Avenue, Columbus, OH 43210 [johnz@mps.ohio-state.edu]

Participants

KANG ZHAO, *Dept. of Mathematics, University of Wisconsin - Madison, Madison, WI 53706* [zhao@math.wisc.edu]

JUN ZHONG, *Dept. of Mathematics, The Pennsylvania State University, University Park, PA 16802* [jun@math.psu.edu]

SONGPING ZHOU, *Dept. of Math., Stat. and Comp. Science, Dalhousie University, Halifax, Nova Scotia, Canada B3H 3J5* [zhou@cs.dal.ca]

CHANG ZHONG ZHU, *Dept. of Computer Science, Univ. of Western Ontario, London, Canada N6A 5B7* [czhu@uwovax.uwo.ca]

ZVI ZIEGLER, *Dept. of Mathematics, Technion, Haifa, Israel* [mar71aa@technion.bitnet]

DAN ZWICK, *45 Helen Avenue, So. Burlington, VT 05403* [zwick@uvm.edu]

Approximation Order without Quasi-Interpolants

Carl de Boor

Abstract. In the study of approximation order, particularly in a multivariable setting, quasi-interpolants have played a major role. This report points out some limitations of quasi-interpolants and describes some recent results on approximation order obtained without the benefit of the quasi-interpolant idea.

§1. Approximation Order

In most general terms, "approximation order" is defined as follows.

Definition 1.1. *The indexed collection* (S_h) *(with* $h \to 0$*) of linear subspaces of some normed linear space* X *has (exact) approximation order* k*, in symbols:*

$$\mathbf{ao}((S_h)_h) = k ,$$

provided

(i) *for all "smooth"* f, $\operatorname{dist}(f, S_h) = O(h^k)$ *(lower bound)*
(ii) *for some "smooth"* f, $\operatorname{dist}(f, S_h) \neq o(h^k)$ *(upper bound)*

This definition raises many questions.

• **norm?** In this report, I will usually consider $X = L_p(G)$, with G some 'suitable' subset of $\mathbb{R}^d$, e.g., either a bounded convex body, or else all of $\mathbb{R}^d$. In fact, the major results reported are for $G = \mathbb{R}^d$ and $p = 2$ or $p = \infty$. With X such a function space,

$$X_c$$

denotes the subspace of *compactly supported* $f \in X$.

• **"smooth"?** With X as chosen, a typical choice for "smooth" is that f be in the Sobolev space $W_p^k(G)$ (written $W^{k,p}(G)$ in [1]). If G is bounded and there is no specification of the expected constant in $O(h^k)$, then it is usually sufficient to define "smooth" to mean "polynomial". In that case, it is usually a polynomial of homogeneous degree k that furnishes the upper bound.

- **how does $O(h^k)$ depend on f?** The definition of approximation order permits, offhand, the possibility that the constant in the $O(h^k)$ term of 1.1(i) depends in some entirely unspecified way on f. It is more satisfactory, though, if this dependence can be made explicit, for example in the terms that specify "smoothness". Thus a desirable strengthening of 1.1(i) is that

$$\sup_{h,f} \frac{\text{dist}(f, S_h)}{h^k \|f\|_{(k)}} < \infty ,$$

with the finiteness of $\|f\|_{(k)}$ defining that f is "smooth". Theorems 7.1, 6.3 and 6.4 below give such results.

- **S_h?** In this report, I will deal only with the following choices:
  ○ Each S_h is a space of piecewise polynomial (=: pp) or, more generally, piecewise exponential (=: pe) or piecewise analytic (=: pa) functions, and h is the "meshsize" of the underlying partition Δ (consisting, typically, of convex bodies, such as simplices and the like).
  ○ (S_h) is a *scale*, i.e.,

$$S_h = \sigma_h S := \{f(\cdot/h) : f \in S\} ,$$

  with S some fixed space of functions. In this case, I will use the abbreviation

$$\mathbf{ao}(S) := \mathbf{ao}((\sigma_h S)_h) .$$

  Such an indexed collection (S_h) is called *stationary*, in order to distinguish it from the next example.
  ○ More generally, we might have $S_h = \sigma_h S^h$, a case referred to as *non-stationary* (in case the S^h do change with h). Note that, in either case, the space S_h is given as the h-dilate of some space. This is done since, in certain arguments, it is more efficient to deal with the scale-ups $\sigma_{1/h} S_h$ than with the spaces S_h themselves. In the stationary case, this amounts to considering the approximation of

$$f_h := \sigma_{1/h} f = f(\cdot h)$$

  from the *fixed* space S.
  ○ Of particular interest in this report (and in much current work in approximation theory, in part because of the current interest in wavelets) is the case when each S^h is shift-invariant, i.e., closed under *shifts* := integer translations.

§2. Shift-invariance

A collection S of functions on $\mathbb{R}^d$ is called *shift-invariant* if
$$g \in S \implies g(\cdot + \alpha) \in S \text{ for all } \alpha \in \mathbb{Z}^d$$
(where $\mathbb{Z}^d$ is the set of d-vectors whose entries are integers).

For example, the space
$$\Pi^\rho_{<k,\Delta}$$
of all pp C^ρ-functions of total degree $< k$ on some partition Δ is shift-invariant in case the partition is shift-invariant in the sense that
$$\Delta + \alpha = \Delta \text{ for all } \alpha \in \mathbb{Z}^d .$$
Examples of interest include the three- and four-direction mesh popular in the bivariate box spline literature.

The simplest (nontrivial) example of a shift-invariant space is the space
$$\mathcal{S}_0(\varphi) := \left\{ \sum_{\alpha \in \mathbb{Z}^d} \varphi(\cdot - \alpha)\, c(\alpha) : c \in \ell_0(\mathbb{Z}^d) \right\}$$
of all finite linear combinations of the shifts of one function, φ. This is the *shift-invariant space generated by* φ since it is the smallest shift-invariant space containing φ. If $\mathcal{S}_0(\varphi)$ is contained in our normed linear space X of interest, then we follow [6] and write
$$\mathcal{S}(\varphi) := \mathcal{S}_0(\varphi)^-$$
for the closure of $\mathcal{S}_0(\varphi)$ in X and call it the *principal shift-invariant*, or *PSI*, space generated by φ.

More generally, if Φ is a finite collection of functions on $\mathbb{R}^d$, then we set
$$\mathcal{S}_0(\Phi) := \sum_{\varphi \in \Phi} \mathcal{S}_0(\varphi)$$
and call
$$\mathcal{S}(\Phi) := \mathcal{S}_0(\Phi)^-$$
a *finitely generated shift-invariant*, or *FSI*, space, and call Φ its set of generators. The structure of PSI and FSI spaces in $L_2(\mathbb{R}^d)$ is detailed in [6] and [7], with particular emphasis on the construction of generating sets for a given FSI space having good properties (such as 'stability' or 'linear independence').

It is natural to consider approximations from $\mathcal{S}(\varphi)$ in the form
$$\varphi * c := \sum_{\alpha \in \mathbb{Z}^d} \varphi(\cdot - \alpha)\, c(\alpha)$$
for a suitable coefficient sequence c. However, offhand, such a sum makes sense only for finitely supported c, and one of the technical difficulties in ascertaining the approximation order of $\mathcal{S}(\varphi)$ derives from the fact that, in general, $\mathcal{S}(\varphi)$ may contain elements which cannot be represented in the form $\varphi * c$ for some sequence c, with the series $\varphi * c$ converging in norm.

§3. Quasi-interpolants

In the spline and finite-element literature, lower bounds for $\mathbf{ao}((S_h)_h)$ are usually obtained with the aid of a corresponding sequence $(Q_h)_h$ of linear maps, with $\operatorname{ran} Q_h \subseteq S_h$, which is a 'good quasi-interpolant sequence of order k' in the sense of the following definition.

Definition 3.1. $(Q_h)_h$ *is a good quasi-interpolant sequence of order k if it satisfies the following two conditions:*
(i) *uniformly local: For some h-independent finite ball B and all $x \in G$,*
$$|(Q_h f)(x)| \le \operatorname{const} \|f_{|x+hB}\|;$$
(ii) *polynomial reproduction: $Q_h f = f$ for all $f \in \Pi_{<k}$.*

Here,
$$\Pi_{<k}$$
denotes the collection of all polynomials in d arguments of total degree $< k$.

The term 'quasi-interpolant' is used in the finite element literature (see, e.g., [26]) to stress the fact that $Q_h f$ does not necessarily match function values at all the nodes of the finite elements used, but 'merely' reproduces certain polynomials. [4] contains a recent survey of the use of quasi-interpolants in spline theory.

To recall, the standard use made of such a good quasi-interpolant sequence is to observe that, for arbitrary f and arbitrary $g \in \Pi_{<k}$,
$$|f(x) - Q_h f(x)| = |(1 - Q_h)(f - g)(x)| \le \operatorname{const} \|(f - g)_{|x+hB}\|,$$
which provides a bound on $\|f - Q_h f\|$ in terms of how well f can be approximated from $\Pi_{<k}$ on a set of the form $x + hB$, giving the error bound $\operatorname{const}_B h^k \|f\|_{(k)}$ in which $\|f\|_{(k)}$ measures the 'size' of the k-th derivatives of f and which provides the desired $O(h^k)$. If our space X is L_p for some $p < \infty$, then this argument has to be fleshed out a bit (see, e.g., [20]).

Since this argument is so simple and effective, there have been various generalizations. For example, since the argument relies on how well f can be approximated locally from $\Pi_{<k}$, it has been observed (e.g., in [15], [11], [22]) that it is sufficient to have Q_h reproduce a translation-invariant space H (e.g., a space of exponentials) which is 'locally close' to $\Pi_{<k}$ (in the sense defined at this section's end).

As another example, if $S_h = \sigma_h S(\varphi)$, then it is natural to construct $Q_h f$ in the form
$$\sigma_h Q f_h$$
(recall that $f_h := \sigma_{1/h} f$) with
$$Qf := \sum_{\alpha \in \mathbb{Z}^d} \varphi(\cdot - \alpha) \lambda f(\cdot + \alpha)$$
for some suitable linear functional λ. Since, for any linear functional λ (defined at least on $\Pi_{<k}$) and any $f \in \Pi_{<k}$, $\alpha \mapsto \lambda f(\cdot + \alpha)$ is polynomial of degree $< k$

in α, this approach requires that $\varphi * c$ be at least well-defined for sequences c with some growth at infinity. In the original context of a compactly supported φ (e.g., as in [27]), this is no problem. However, with the recent interest in radial basis functions (see, e.g., [24]) and wavelets, also noncompactly supported φ have to be considered and, for these, the quasi-interpolant approach (as used, e.g., in [23], [16], [20], and [2]) requires that φ satisfy the condition $\varphi(x) = O(|x|^{-d-k-\varepsilon})$ at ∞ for some positive ε (and forces one to make do with Q which is only 'essentially local'). In particular, the higher the desired approximation order, the faster must φ decay at infinity.

There are other costs associated with the quasi-interpolant approach. For example, it works, offhand, only with integer values of k. Also, it requires that

$$\cap_h S_h \neq \{0\}\ .$$

The artificiality of this last restriction is nicely illustrated by the following simple example, from [15]:

Example 3.2. (Dyn, Ron). Let $d = 1$, $p = \infty$, and let S_h be the span of the $h\mathbb{Z}$-translates of the piecewise linear function

$$\varphi_h : x \mapsto \begin{cases} x+1, & 0 \leq x < h\ ; \\ 0, & \text{otherwise}\ . \end{cases}$$

Thus S_h consists of certain piecewise linear functions, with breakpoint sequence $h\mathbb{Z}$, but the only polynomial (hence the only analytic function) it contains is the zero polynomial. In particular, it is not possible to construct a quasi-interpolant of positive order for it. Nevertheless, the approximation

$$Q_h f := \sum_{j \in h\mathbb{Z}} \varphi_h(\cdot - j) f(j)$$

has the error

$$f - Q_h f = f - \sum_{j \in h\mathbb{Z}} \chi_h(\cdot - j) f(j) + \sum_{j \in h\mathbb{Z}} (\chi_h - \varphi_h)(\cdot - j) f(j)\ ,$$

with χ_h the characteristic function of the interval $[0 \mathinner{\ldotp\ldotp} h)$. Since $\|\chi_h - \varphi_h\|_\infty = h$,

$$\|f - Q_h f\|_\infty \leq \omega_f(h) + \|f\|_\infty h\ ,$$

where ω_f is the modulus of continuity of f. It follows that $Q_h f$ converges to f uniformly in case f is uniformly continuous and bounded. ∎

This example could still be treated by an appropriate generalization of the notion of quasi-interpolant. Specifically, one could consider a good quasi-interpolant sequence (Q_h) of positive *local* order k, meaning that (Q_h) is uniformly local and that

$$Q_h f = f + O(\|f_{|B}\| |h|^k)$$

on hB for any $f \in \Pi_{<k}$. A sufficient condition for this is that $Q_h = 1$ on a D-invariant space H of entire functions which is *locally close to* $\Pi_{<k}$ in the sense that its 'limit at the origin' (cf. [10]), $H_\downarrow$, contains $\Pi_{<k}$. Here,

$$H_\downarrow := \operatorname{span}\{f_\downarrow : f \in H\}, \qquad (3.3)$$

where $f_\downarrow$ is the *initial*, i.e., the first nonzero, term in the expansion $f = f_0 + f_1 + f_2 + \cdots$ of f into homogeneous polynomials f_j of degree j, all j. Thus, for any $f \in \Pi_{<k}$, there exists $g \in H$ with $f = g + O(|h|^k)$ on hB, hence, on hB, $Q_h f = Q_h g + O(|h|^k) = g + O(|h|^k) = f + O(|h|^k) = f + O(\|f_{|B}\| |h|^k)$ (the last equality by the fact that $\Pi_{<k}$ is finite-dimensional).

Still, the point of the example should be clear.

Finally, the quasi-interpolant approach is of no help with upper bounds.

§4. Upper Bounds

Upper bounds for $\mathbf{ao}((S_h)_h)$ have to be fashioned separately for each case (much as the details of a quasi-interpolant sequence have to be so fashioned). The general principle employed is duality, which provides the following well-known observation.

If Y is a linear subspace of the normed linear space X, and $\lambda \in X^*$ with $\lambda \perp Y$ (i.e., λ is a continuous linear functional on X which vanishes on all of Y), then, for any $x \in X$ and any $y \in Y$, $\lambda x = \lambda(x - y) \le \|\lambda\| \|x - y\|$, hence $|\lambda x| \le \|\lambda\| \operatorname{dist}(x, Y)$. In other words,

$$\lambda \perp Y \implies \operatorname{dist}(x, Y) \ge \frac{|\lambda x|}{\|\lambda\|}.$$

As a simple application, consider $\mathbf{ao}(S)$ for

$$X = L_\infty(G), \quad S = \Pi^\rho_{<k,\Delta}.$$

Assume without loss of generality that G is the d-dimensional cube,

$$G = C := [-1 \mathinner{.\,.} 1]^d,$$

let δ be any element in the partition Δ, and let g be any nontrivial homogeneous polynomial of degree k. If e is the error in the best $L_2(\delta)$-approximation to g from $\Pi_{<k}$, then the mapping

$$\lambda : L_\infty \to \mathbb{R} : f \mapsto \int_\delta ef$$

(i) is a bounded linear functional;
(ii) is orthogonal to S, since all λ sees of $f \in S$ is its restriction to δ, and on δ each $f \in S$ is just a polynomial of degree $< k$;
(iii) satisfies $\lambda g = \int_\delta ee > 0$.

Now consider $\lambda_h f := \int_\delta ef(h\cdot)$. Then
(i) λ_h is a bounded linear functional, with h-independent norm

$$\|\lambda_h\| = \int_\delta |e| = \lambda \operatorname{signum}(e) ,$$

where $\operatorname{signum}(e) : x \mapsto \operatorname{signum}(e(x))$.
(ii) $\lambda_h \perp S_h := \sigma_h S$, since $g \in S_h$ is of the form $f(\cdot/h)$ for some $f \in S$.
(iii) Using the homogeneity of g, one computes that $\lambda_h g = \int_\delta eg(h\cdot) = h^k \int_\delta eg = h^k \lambda g$ with $\lambda g > 0$.

So, altogether,

$$\operatorname{dist}(g, S_h) \geq h^k(\lambda g/\lambda \operatorname{signum}(e)) ,$$

showing that $\mathbf{ao}(\Pi^\rho_{<k,\Delta}) \leq k$.

If we try the same argument for $p < \infty$, we hit a little snag. Take, in fact, p at the other extreme, $p = 1$. There is no difficulty with (ii) or (iii), but the conclusion is weakened because (i) now reads

(i)' $\|\lambda_h\| = \sup_{f \in L_1} |\int_\delta ef(h\cdot)|/\|f\|_1 \leq \|e_{|\delta}\|_\infty \sup_{f \in L_1(\delta)} \int_\delta |f(h\cdot)|/\|f\|_1$,

and the best we can say about that last supremum is that it is at most h^{-d} since $\int_\delta f(h\cdot) = \int_{h\delta} f/h^d$. Hence, altogether, $\|\lambda_h\| \leq \operatorname{const}/h^d$.

Thus, now our bound reads

$$\operatorname{dist}_1(g, S_h) \geq h^k \operatorname{const}/(\operatorname{const}/h^d) \neq o(h^{k+d})$$

which is surely correct, but not very helpful.

What we are witnessing here is the fact that the error in a max-norm approximation is indeed localized, i.e., it occurs at a point, while, for $p < \infty$, the error 'at a point' is less relevant; the error is more global; one needs to consider the error over a good part of G. Further, in the argument below, I need some kind of uniformity of the partition Δ, of the following (very weak) sort (in which $|A|$ denotes the d-dimensional volume of the set A, and C continues to denote the cube $[-1..1]^d$):

Assumption 4.1. *There exists an open set b and a locally finite set $I \subset \mathbb{R}^d$ (meaning that I meets any bounded set only in finitely many points) so that*
(α) $b + I$ is the disjoint union of $b + i$, $i \in I$, with each $b + i$ lying in some $\delta \in \Delta$ (the possibility of several lying in the same δ is not excluded);
(β) for some const > 0 and all n, $|(b+I) \cap nC| \geq \operatorname{const}|nC|$.

For example, any uniform partition of $\mathbb{R}$ satisfies this condition. As another example, if $d = 2$ and Δ is the three-direction mesh, then Δ consists of triangles of two kinds, and taking b to be the interior of one of these triangles and $I = \mathbb{Z}^2$ guarantees (α), while (β) holds with const $= 1/2$. On the other hand, Shayne Waldron (a student at Madison) has constructed a neat example to show that the Assumption 4.1 is, in general, necessary for the conclusion that $\mathbf{ao}(\Pi^\rho_{<k,\Delta}) \leq k$. The example uses $\rho = -1$ and arbitrary k,

$d = 1$, $G = [-1..1]$, $p = 1$, and Δ obtained from $\mathbb{Z}$ by subdividing $[j..j+1]$ into $2^{|j|}$ equal pieces, $j \in \mathbb{Z}$.

With Assumption 4.1 holding, define λ as before, but with b replacing the element δ of Δ. Further, assume without loss that $C \subseteq G$, and define

$$\lambda_h f := \int_b e \sum_{i \in I_h} f(h \cdot + i),$$

where

$$I_h := \{i \in I : b + i \subseteq C/h\}.$$

This gives:

(i)$_1$
$$\|\lambda_h\| \leq \sup_{f \in L_1} \frac{\sum_{i \in I_h} \int_{b+i} |e| |f(h \cdot)|}{\sum_{i \in I_h} \int_{h(b+i)} |f|} = \|e_{|b}\|_\infty / h^d,$$

using the fact that the sum $b + I_h$ is disjoint.

Hence, we didn't worsen our situation here. We also didn't sacrifice (ii) because, by assumption, each $b + i$ lies in the interior of some $\delta \in \Delta$, and therefore $\int_b ef(h \cdot + i) = 0$ for every $f \in S_h$. But we materially improved the situation as regards (iii), for we now obtain

(iii)$_1$
$$\lambda_h g = \int_b e \sum_{i \in I_h} g(h \cdot + i) = h^k \int_b e \sum_{i \in I_h} g = h^k \text{ const } \#I_h$$

with

$$\#I_h = |b + I_h|/|b| \geq \text{const}\, |C/h|/|b| = \text{const}\, /h^d.$$

With this, our conclusion is back to what we want:

$$\text{dist}_1(g, S_h) \geq (h^k \text{ const}\, /h^d)/(\text{const}\, /h^d) \neq o(h^k).$$

Note that this lower bound on the distance only sees S as a space of pp's of degree $< k$, hence is valid even when we take the biggest such space, i.e., the space

$$\Pi_{<k,\Delta}$$

of all pp functions of degree $< k$ on the partition Δ. For this space, it is not hard to show that the approximation order is at least k, since approximations can be constructed entirely locally. Thus,

$$\text{ao}(\Pi_{<k,\Delta}) = k.$$

For this reason, this is called the *optimal* approximation order for a pp space of degree $< k$.

Approximation Order without Quasi-Interpolants 9

Such a local construction of approximations is still possible for $\Pi^0_{<k,\Delta}$ (at least in the uniform norm; it would be interesting to run down this argument for the 1-norm), hence, at least in the uniform norm,

$$\mathbf{ao}(\Pi^\rho_{<k,\Delta}) = k \quad \text{for} \quad \rho \leq 0 \, .$$

However, for $\rho > 0$, the story is largely unknown, with first results in [5] and [19].

I became sensitized to the issue that the derivation of upper bounds for the approximation order from pa spaces requires much more care for $p < \infty$ than for $p = \infty$ by the paper [22] in which $\mathbf{ao}((S_h)_h)$ is carefully studied for the case that each S_h is a piecewise exponential space. Here is their result concerning upper bounds (in which the term 'exponential' is meant to describe any function which is a linear combination, with polynomial coefficients, of functions of the form $x \mapsto \exp(\theta \cdot x)$).

Theorem 4.2. (Lei, Jia). *Let $(S_h)_h$ be an indexed collection of piecewise exponential spaces on $\mathbb{R}^d$ with the property that, for some open subset Ω of $(0..1)^d$ and every h and every $\alpha \in \mathbb{Z}^d$, $S_{h|(\Omega+\alpha)h} \subseteq H_{|(\Omega+\alpha)h}$ for some fixed D-invariant finite-dimensional space H of exponentials for which $\Pi_k \not\subseteq H_\downarrow$ (as defined in (3.3)). Then, for any p in the range $1 \leq p \leq \infty$, $\mathbf{ao}((S_h)_h) \leq k$.*

Here is my version of their proof (in which $\|f\|_p(B)$ denotes the $L_p(B)$-norm of $f_{|B}$ while $\|a\|$ is any norm of the n-vector a, and B_h is the Euclidean ball with radius h centered at the origin). The special case of pp S_h treated earlier is simpler since, in that case, H is also scale-invariant.

Let

$$V := [v_1, v_2, \ldots, v_n] : \mathbb{R}^n \to H_\downarrow : a \mapsto \sum_j v_j a(j)$$

be any homogeneous basis for $H_\downarrow$.

I claim that any $F = [f_1, f_2, \ldots, f_n] : \mathbb{R}^n \to H$ with $f_{j\downarrow} = v_j$, all j, is a basis for H. For the proof (which also proves the inequality (4.3) of use later), observe that $\|Va\|_p(B_h) = \|Va^h\|_p(B_1) \geq \|a^h\|/\|V^{-1}\|$, where

$$a^h := (h^{d/p + \deg v_j} a(j))_{j=1}^n \, , \qquad \|V^{-1}\| := \sup_c \|c\|/\|Vc\|_p(B_1) \, ,$$

and $\|V^{-1}\|$ is certainly finite. On the other hand, $(v_j - f_j)(x) = O(|x|^{\deg v_j + 1})$ since $v_j = f_{j\downarrow}$, hence

$$\|(F - V)a\|_p(B_h) \leq h \operatorname{const}_F \|a^h\| \, .$$

Therefore, altogether,

$$\|Fa\|_p(B_h) \geq \|Va\|_p(B_h) - \|(F - V)a\|_p(B_h)$$
$$\geq (1/\|V^{-1}\| - h \operatorname{const}_F)\|a^h\| =: \operatorname{const}_{h,F} \|a^h\| \, , \qquad (4.3)$$

which shows that F is one-to-one (since the last expression is positive for all sufficiently small h). Since $\dim H_\downarrow = \dim H$ by [10], this finishes the proof.

Now let q be a homogeneous polynomial not in the range of V. Then $[q, V]$ is one-to-one, and is made up of the initial terms of the columns of $[q, F]$. This permits substitution of $[q, F]$ for F in (4.3) (with $\mathrm{const}_{[q,F]} = \mathrm{const}_F$), and so gives the conclusion that, for all $a \in \mathbb{R}^n$,

$$\|q - Fa\|_p(B_h) = \|[q, F](1, -a)\|_p(B_h) \geq \mathrm{const}_{h,F} \|(h^{d/p + \deg q}, a^h)\|$$
$$\geq \mathrm{const}_{h,F} h^{d/p + \deg q},$$

hence

$$\mathrm{dist}_p(q, H)(B_h) = \min_a \|q - Fa\|_p(B_h) \geq \mathrm{const}_{h,F} h^{d/p + \deg q}, \quad (4.4)$$

with $\lim_{h \to 0} \mathrm{const}_{h,F} = 1/\|V^{-1}\| > 0$. Since we can choose $\deg q = k$ by assumption, this proves the upper bound when $p = \infty$. (I note in passing that this argument could also have been phrased explicitly in terms of annihilating linear functionals.)

As to the L_p-argument, start with the observation that it is sufficient to prove an upper bound for the L_1-approximation order on any bounded G since this implies the same upper bound for any $p > 1$ (including $p = \infty$) and for any G, bounded or not.

Thus, to establish the desired upper bound, it is sufficient to prove that

$$\mathrm{dist}_1(q, S_h)(B_\rho) \geq \mathrm{const}\, h^k$$

for some smooth q, some positive const, and any particular positive ρ.

For this, we now choose q to be any homogeneous polynomial of *minimal* degree not in $H_\downarrow$. Then, for any z, $q(\cdot + z) = q + Va_z$, with $\|a_z\| \leq \mathrm{const}\,\|z\|$, and $q(\cdot + z) - Fa = q - F(a - a_z) + (V - F)a_z$, therefore

$$\mathrm{dist}_1(q(\cdot + z), H)(B_h) \geq \mathrm{dist}_1(q, H)(B_h) - h\,\mathrm{const}_F\,\|a_z^h\|.$$

This implies with (4.4) that there exist positive constants const, h_0, R (depending on F and q) so that

$$\mathrm{dist}_1(q(\cdot + z), H)(B_h) \geq \mathrm{const}\, h^{d + \deg q} \quad (4.5)$$

for all $h < h_0$, $\|z\| < R$.

By the translation-invariance of H (which follows from the assumed D-invariance),

$$\mathrm{dist}(q, H)(\Omega h + z) = \mathrm{dist}(q(\cdot + z), H)(\Omega h)$$

while, by assumption, $S_h \subseteq H$ on each $(\Omega + \alpha)h$ with $\alpha \in \mathbb{Z}^d$. Thus, from (4.5) and using the fact that Ω contains some ball of positive radius, we find that

$$\mathrm{dist}_1(q, S_h)(B_\rho) \geq \sum_{\alpha \in N} \mathrm{dist}_1(q(\cdot + \alpha h), H)(\Omega h) \geq \mathrm{const}\, h^{\deg q}\, h^d\, \#N,$$

Approximation Order without Quasi-Interpolants

with
$$N := \{\alpha \in \mathbb{Z}^d : (\Omega + \alpha)h \subseteq B_\rho, \|\alpha h\| < R\}$$
and with $h < h_0$, where const > 0 and $R > 0$ are independent of h. Since $\#N = O(h^{-d})$ for all small h, and $\deg q \le k$, we are done. ∎

Further illustrations of the use of duality in the derivation of upper bounds on $\mathbf{ao}(S)$ (albeit only for bivariate pp S) can be found in [9] and its references.

§5. The Strang-Fix Condition

The literature on $\mathbf{ao}(\mathcal{S}(\varphi))$ for a compactly supported φ has been dominated by the Strang-Fix condition. It concerns the behavior of the Fourier transform
$$\widehat{\varphi} : \xi \mapsto \int_{\mathbb{R}^d} \varphi e_{-\xi}$$
of φ at the points of $2\pi\mathbb{Z}^d$. Here and below,
$$e_\theta : \mathbb{R}^d \to \mathbb{C} : x \mapsto \exp(i\theta \cdot x)$$
denotes the exponential function (with purely imaginary frequency $i\theta$). In one of its many versions, the Strang-Fix condition reads as follows.

Definition 5.1. We say that φ satisfies SF_k in case
(i) $\widehat{\varphi}(0) = 1$;
(ii) For all multi-indices α satisfying $|\alpha| < k$ we have $D^\alpha \widehat{\varphi} = 0$ on $2\pi\mathbb{Z}^d \backslash 0$.

Its importance derives from the following theorem, in which we use the convenient notation
$$\varphi *' f := \sum_{j \in \mathbb{Z}^d} \varphi(\cdot - j) f(j)$$
for the *semidiscrete convolution* of the two functions φ and f even if it requires further discussion of just what exactly is meant by it when the sum is not (locally) finite. Also, recall that $L_1(\mathbb{R}^d)_c$ denotes the compactly supported functions in $L_1(\mathbb{R}^d)$.

Theorem 5.2. (Schoenberg ($d=1$), Fix and Strang). *For $\varphi \in L_1(\mathbb{R}^d)_c$, the following are equivalent:*
(a) $\varphi *'$ *is degree-preserving on* $\Pi_{<k}$: *for all p in $\Pi_{<k}$, $\varphi *' p \in p + \Pi_{<\deg p}$;*
(b) φ *satisfies* SF_k.

The proof is via the Poisson summation formula. Starting with [27], the theorem is used to construct a good quasi-interpolant sequence (Q_h) of order k with ran $Q_h \subseteq \sigma_h \mathcal{S}(\varphi)$. More than that, it forms part of an argument that seems to show that $\mathbf{ao}(\mathcal{S}(\varphi)) \ge k$ if and only if $\varphi/\widehat{\varphi}(0)$ satisfies SF_k. The precise statement of this equivalence for $X = L_2(\mathbb{R}^d)$ (see [27]) involves, unfortunately, a restricted notion of approximation order called 'controlled' approximation. (According to [25], this restriction can be dropped for $X = L_\infty(\mathbb{R}^d)$ provided $\widehat{\varphi}(0) \ne 0$.)

On a related issue, [27] reports the following

Conjecture 5.3. (Babuška). *The approximation order in $L_2(\mathbb{R}^d)$ of the FSI space $S(\Phi)$ with $\Phi \subset L_2(\mathbb{R}^d)_c$ is already attained by some PSI space $S(\varphi)$ with $\varphi \in S_0(\Phi)$.*

The actual version of this conjecture reported in [27] involves controlled approximation and was eventually shown to be invalid by Jia in [18]. The following correct version, involving yet another restricted notion of approximation order called 'local' approximation, can be found in [8], with some details actually attended to only in [20].

Theorem 5.4. (de Boor, Jia). *Let Φ be a finite subset of $L_p(\mathbb{R}^d)_c$, and let $X = L_p(\mathbb{R}^d)$. Then the following are equivalent:*
(a) $(\sigma_h S(\Phi))$ *has 'local' approximation order k;*
(b) *some $\varphi \in S_0(\Phi)$ satisfies SF_k.*

This theorem verifies the version of the Babuška conjecture used in [14]. Further, [21] shows that (b) is equivalent to the statement

(b)' *Some sequence (φ_n) in $S_0(\Phi)$ satisfies SF_k "in the limit".*

Finally, [19] contains the following extension of work begun in [5]:

Theorem 5.5. (Jia). *Let S be a univariate, shift-invariant, locally finite-dimensional set of functions, closed under convergence on compact sets. Then the following are equivalent:*
(a) $\mathbf{ao}(S \cap L_p(\mathbb{R})) \geq k$;
(b) *Some $\varphi \in S_c$ satisfies SF_k.*

§6. Approximation Order in L_∞

In [13], Chui, Jetter and Ward introduce the *commutator* for $\varphi \in C(\mathbb{R}^d)_c$ as the linear map
$$C(\mathbb{R}^d) \to C(\mathbb{R}^d) : f \mapsto \varphi *' f - f *' \varphi$$
and use it for the construction of a good quasi-interpolant sequence (Q_h) of order k with $\operatorname{ran} Q_h \subseteq \sigma_h S(\varphi)$. For this, they prove the following.

Proposition 6.1. (Chui, Jetter, Ward). *If φ belongs to $C(\mathbb{R}^d)_c$ and satisfies SF_k, then*
$$\text{for all } f \in \Pi_{<k}, \quad \varphi *' f = f *' \varphi.$$

Subsequently, it was observed in [3] that actually
$$\text{for all } f \in S(\varphi), \quad \varphi *' f = f *' \varphi, \tag{6.2}$$

and this observation was exploited by A. Ron in [25] in the following simple and surprising way. He observes that, as a consequence of (6.2),
$$\text{for all } f \in S(\varphi), \quad \varphi *' e_\theta - e_\theta *' \varphi = \varphi *' (e_\theta - f) - (e_\theta - f) *' \varphi,$$

(recall that $e_\theta : x \mapsto \exp(i\theta \cdot x)$), and this leads to the conclusion that
$$\|\varphi *' e_\theta - e_\theta *' \varphi\|_\infty \le 2 \|\varphi *'\|_\infty \, \mathrm{dist}_\infty(e_\theta, \mathcal{S}(\varphi))$$
(with $\|\varphi *'\|_\infty = \|\sum_{\alpha \in \mathbb{Z}^d} |\varphi(\cdot - \alpha)|\|_\infty$). Since (as pointed out by A. Ron)
$$\frac{\varphi *' e_\theta - e_\theta *' \varphi}{e_\theta} \sim c + \sum_{\alpha \in \mathbb{Z}^d \setminus 0} \widehat{\varphi}(\theta + 2\pi\alpha) \, e_\alpha \,,$$
this throws new light on the connection between $\mathbf{ao}(\mathcal{S}(\varphi))$ in L_∞ and the behavior of $\widehat{\varphi}$ 'at' $2\pi \mathbb{Z}^d$.

[12] exploits this idea in the more general context of a $\varphi \in X := L_\infty(\mathbb{R}^d)$ with the only requirement that $\varphi *'$ be a bounded map from ℓ_∞ to X. Further, while $\mathcal{S}(\varphi)$ is still taken to be the 'closure' of $\mathcal{S}_0(\varphi)$, it is not taken as the norm-closure but, in effect, as the largest shift-invariant space containing $\mathcal{S}_0(\varphi)$ and satisfying (6.2).

Here is the main result of [12] concerning *upper bounds*.

Theorem 6.3. ([12]). *Let (φ_h) be an indexed collection of elements of $X := L_\infty(\mathbb{R}^d)$. Assume that $\varphi_h *' : \ell_\infty \to X$ is defined and bounded independently of h, and that $\theta \in \mathbb{R}^d$. If $\mathrm{dist}(e_\theta, \sigma_h \mathcal{S}(\varphi_h)) = O(h^k)$, then*
$$\sum_{\alpha \in \mathbb{Z}^d \setminus 0} |\widehat{\varphi}_h(h\theta + 2\pi\alpha)|^2 \le \mathrm{const}_\theta \, h^{2k} \,.$$

In particular, then
$$|\widehat{\varphi}_h(h\theta + 2\pi\alpha)| \le \mathrm{const}_\theta \, h^k \quad \text{for all nonzero } \alpha \text{ in } \mathbb{Z}^d \,.$$

The following points should be stressed:
- There is some latitude here for the definition of "smooth" since it need only include complex exponentials.
- Only mild decay of φ_h is needed (enough to make $\varphi *' : \ell_\infty \to L_\infty$ well-defined).
- Nothing is said here about $\widehat{\varphi}_h(0)$ (which is particularly important if $\widehat{\varphi}_h(0)$ is zero).
- It is easy to recover the rest of SF_k in the stationary case, i.e., in case $\varphi_h = \varphi$, for all h.
- Even if "smooth" is taken to mean "compactly supported, but infinitely smooth", the same condition is obtained, provided φ_h has a certain decay at ∞.

The results of [12] concerning *lower bounds* on $\mathbf{ao}(\mathcal{S}(\varphi))$ make use of the following definition of "smooth": $f \in X$ is "smooth" if its Fourier transform is a Radon measure for which
$$\|f\|_{(k)} := \|(1 + |\cdot|^k)\widehat{f}\|_1 < \infty \,,$$
with the suffix '1' intended to indicate that the total variation of the measure in question is meant.

Here is a sample result.

Theorem 6.4. ([12]). *Assume that $\varphi_h *' : \ell_\infty \to L_\infty$ is bounded for every h. Then, for any positive η,*

$$\operatorname{dist}(f, \sigma_h \mathcal{S}(\varphi_h)) \leq h^k (2\pi)^{-d} \|f\|_{(k)} A + o(h^k)$$

with

$$A := \sup_h \sum_{\alpha \in \mathbb{Z}^d \backslash 0} \left\| \frac{1}{(h^k + |\cdot|^k)} \frac{\widehat{\varphi}_h(\cdot + 2\pi\alpha)}{\widehat{\varphi}_h} \right\|_{L_\infty(B_\eta)}.$$

Since this theorem gives $\mathbf{ao}((\sigma_h \mathcal{S}(\varphi_h))_h) \geq k$ only if $A < \infty$, this focuses attention on the behavior near zero of each of the functions

$$\widehat{\varphi}_h(\cdot + 2\pi\alpha)/\widehat{\varphi}_h, \qquad \alpha \in \mathbb{Z}^d \backslash 0. \tag{6.5}$$

Specifically, in the stationary case, if this ratio is a smooth function in a neighborhood of 0, then the finiteness of A would require the ratio to have a zero of order k at 0, and conversely, provided $\widehat{\varphi}$ has some decay. From this vantage point, the Strang-Fix condition SF_k is seen to be neither necessary nor sufficient for $\mathbf{ao}(\mathcal{S}(\varphi)) \geq k$, but to come close to being necessary and sufficient for appropriately restricted φ.

Note that the finiteness of A requires the infinite sum in its definition to be finite, and such finiteness can, in general, only be deduced when $\widehat{\varphi}_h$, in addition to being "small" near $2\pi\mathbb{Z}^d \backslash 0$, decays appropriately (and this requires some smoothness of φ_h).

The fact that the finiteness of A involves only the *ratios* (6.5) makes the conclusion of the theorem *independent of localization*, i.e., independent of which difference operators were applied to the original generator for $\mathcal{S}(\varphi_h)$ in order to obtain the appropriately decaying φ_h.

The proof in [12] of results like this theorem makes use of an approximation from $\mathcal{S}(\varphi)$ of the form

$$f \approx Rf := (2\pi)^{-d} \int_{\mathbb{R}^d} \varepsilon_\theta \widehat{f}(\theta) \, d\theta \in \mathcal{S}(\varphi)$$

in which the approximation

$$e_\theta \approx \varepsilon_\theta := \varphi *' e_\theta / \widetilde{\varphi}(\theta) \in \mathcal{S}(\varphi)$$

is suggested by

$$e_\theta *' \varphi = e_\theta \sum_j \exp(-ij)\varphi(j) =: e_\theta \, \widetilde{\varphi}(\theta).$$

§7. Approximation Order in L_2

For an arbitrary $\varphi \in X := L_2(\mathbb{R}^d)$, the approximation order of $\mathcal{S}(\varphi)$ can be characterized completely, in terms of $\widehat{\varphi}$. This is due to the fact (proved in [6] but also derivable from more general results in [17]) that, if P_S is the orthogonal projector onto $\mathcal{S}(\varphi)$, then

$$\widehat{P_S f} = \frac{[\widehat{f}, \widehat{\varphi}]}{[\widehat{\varphi}, \widehat{\varphi}]} \widehat{\varphi},$$

where

$$[\widehat{f}, \widehat{\varphi}] : \mathbb{T}^d \to \mathbb{C} : x \mapsto \sum_{\alpha \in 2\pi \mathbb{Z}^d} \widehat{f}(x+\alpha)\overline{\widehat{g}}(x+\alpha)$$

is the very convenient "bracket product" of $\widehat{f}, \widehat{\varphi} \in X$, and $\mathbb{T}^d$ is the d-dimensional torus, i.e.,

$$\mathbb{T}^d := [-\pi .. \pi]^d$$

with the appropriate identification of boundary points.

The definition of f being "smooth" employed in [6] is that

$$\|f\|_{W_2^k(\mathbb{R}^d)} := \|(1+|\cdot|)^k \widehat{f}\|_2 < \infty.$$

The characterization uses the following abbreviation

$$\Lambda_\varphi := 1 - \frac{|\widehat{\varphi}|^2}{[\widehat{\varphi},\widehat{\varphi}]} = \frac{\sum_{\alpha \in \mathbb{Z}^d \setminus 0} |\widehat{\varphi}(\cdot + 2\pi\alpha)|^2}{\sum_{\alpha \in \mathbb{Z}^d} |\widehat{\varphi}(\cdot + 2\pi\alpha)|^2}.$$

Theorem 7.1. ([6]). *For any* $(\varphi_h)_h$ *in* $X = L_2(\mathbb{R}^d)$,

$$\mathbf{ao}((\sigma_h \mathcal{S}(\varphi_h))_h) \geq k \iff \sup_h \left\| \frac{\Lambda_{\varphi_h}}{(h+|\cdot|)^{2k}} \right\|_{L_\infty(\mathbb{T}^d)} < \infty.$$

This result focuses attention on the behavior of Λ_φ near 0, hence, if $\widehat{\varphi}$ is bounded away from zero near 0, it focuses, once again, attention on the ratios (6.5). Here is a typical

Corollary 7.2. ([6]). *If* $\varphi \in L_2(\mathbb{R}^d)$, *and* $1/\widehat{\varphi}$ *is essentially bounded near 0, and* $\widehat{\varphi} \in W_2^\rho(U)$ *for some* $\rho > k + d/2$ *and some nbhd* U *of* $2\pi\mathbb{Z}^d \setminus 0$, *and if* φ *satisfies* SF_k, *then* $\mathbf{ao}(\mathcal{S}(\varphi)) \geq k$.

For a general closed shift-invariant subspace of $L_2(\mathbb{R}^d)$, there is the following result.

Theorem 7.3. ([6]). *Let* S *be a closed shift-invariant subspace of* $L_2(\mathbb{R}^d)$, *and let* $f, g \in L_2(\mathbb{R}^d)$. *Then*

$$\mathrm{dist}(f, S) \leq \mathrm{dist}(f, \mathcal{S}(P_S g)) \leq \mathrm{dist}(f, S) + 2\,\mathrm{dist}(f, \mathcal{S}(g)).$$

This theorem shows that the approximation power of a general shift-invariant subspace of L_2 is already attained by some PSI subspace of it, provided we can, for given k, supply an element $g \in L_2(\mathbb{R}^d)$ for which $\mathbf{ao}(\mathcal{S}(g)) > k$. But that is easy to do:

Lemma 7.4. *There are simple functions g (e.g., the inverse Fourier transform of the characteristic function of some small neighborhood of the origin) for which, for any k,*

$$\operatorname{dist}(f, \sigma_h S(g)) = o(h^k \|f\|_{W_2^k(\mathbb{R}^d)}) .$$

§8. The Babuška Conjecture Revisited

Theorem 7.1 is used in [7] to provide a proof of the Babuška Conjecture 5.3, as follows.

Let $S = \mathcal{S}(\Phi)$, where Φ is a finite subset of $L_2(\mathbb{R}^d)_c$.

(i) Since each $\varphi \in \Phi$ is compactly supported, hence $\widehat{\varphi}$ is analytic, it can be assumed, after going to a subset of Φ if need be, that, for almost every $x \in \mathbb{T}^d$, the set of $\ell_2(\mathbb{Z}^d)$-vectors

$$\widehat{\varphi}_{\|x} := (\widehat{\varphi}(x + 2\pi\alpha))_{\alpha \in \mathbb{Z}^d} , \qquad \varphi \in \Phi ,$$

is linearly independent, hence is a basis for $\widehat{S}_{\|x}$.

(ii) For any $g \in L_2(\mathbb{R}^d)$,

$$\widehat{P_S g} = \sum_{\varphi \in \Phi} \frac{\det G_\varphi(g)}{\det G(\Phi)} \widehat{\varphi}$$

where

$$G(\Phi) := ([\widehat{\varphi}, \widehat{\psi}])_{\varphi, \psi \in \Phi}$$

and $G_\varphi(g)$ is obtained from this by replacing the row $[\widehat{\varphi}, \cdot]$ by the row $[\widehat{g}, \cdot]$.

(iii) Since

$$[\widehat{f}, \widehat{g}] = \sum_{j \in \mathbb{Z}^d} \langle f, g(\cdot + j) \rangle e_j ,$$

each entry of $G(\Phi)$ is a trigonometric polynomial, hence so is $\det G(\Phi)$, and $\det G(\Phi) \neq 0$ a.e. (by (i)).

(iv) If $g \in L_2(\mathbb{R}^d)_c$, then $\mathcal{S}(P_S g) = \mathcal{S}(g_\star)$ (it is shown in [6] that $\mathcal{S}(\psi') = \mathcal{S}(\psi)$ in case $\psi' \in \mathcal{S}(\psi)$ and $\operatorname{supp}\widehat{\psi}' \supseteq \operatorname{supp}\widehat{\psi}$), where

$$\widehat{g}_\star := \det G(\Phi) \widehat{P_S g} = \sum_{\varphi \in \Phi} \det G_\varphi(g) \widehat{\varphi} ,$$

by (ii), hence $g_\star \in \mathcal{S}_0(\Phi)$, by (iii).

(v) By Theorem 7.3 and Lemma 7.4, we can choose g so that

$$\operatorname{dist}(f, \mathcal{S}(g_\star)) \sim \operatorname{dist}(f, \mathcal{S}(\Phi)) ,$$

hence *Babuška was right*.

Acknowledgments. I am indebted to Ron DeVore, Rong-Qing Jia, and Amos Ron for helpful comments on a version of this report.

References

1. Adams, R.A., *Sobolev Spaces*, Academic Press, New York, 1975.
2. Beatson, R.K. and W.A. Light, Quasi-interpolation in the absence of polynomial reproduction, Univ. of Leicester Mathematics Tech. Report 1992/15, May 1992.
3. de Boor, C., The polynomials in the linear span of integer translates of a compactly supported function, Constr. Approx. **3** (1987), 199–208.
4. de Boor, C., Quasiinterpolants and approximation power of multivariate splines, in *Computation of Curves and Surfaces*, W. Dahmen, M. Gasca and C. A. Micchelli (eds.), Kluwer, 1990, 313–345.
5. de Boor, C. and R. DeVore, Partitions of unity and approximation, Proc. Amer. Math. Soc. **93** (1985), 705–709.
6. de Boor, C., R. DeVore, and A. Ron, Approximation from shift-invariant subspaces of $L_2(\mathbb{R}^d)$, CMS-TSR University of Wisconsin-Madison **92-2**, 1991.
7. de Boor, C., R. DeVore, and A. Ron, The structure of finitely generated shift-invariant spaces in $L_2(\mathbb{R}^d)$, CMS-TSR University of Wisconsin-Madison **92-8**, 1992.
8. de Boor, C. and Rong-Qing Jia, Controlled approximation and a characterization of the local approximation order, Proc. Amer. Math. Soc. **95** (1985), 547–553.
9. de Boor, C. and Rong-Qing Jia, A sharp upper bound on the approximation order of smooth bivariate pp functions, J. Approx. Theory, to appear.
10. de Boor, C. and A. Ron, On multivariate polynomial interpolation, Constr. Approx. **6** (1990), 287–302.
11. de Boor, C. and A. Ron, The exponentials in the space of the integer translates of a compactly supported function, London Math. J., to appear.
12. de Boor, C. and A. Ron, Fourier analysis of approximation orders from principal shift-invariant spaces, Constr. Approx. , to appear.
13. Chui, C. K., K. Jetter, and J. D. Ward, Cardinal interpolation by multivariate splines, Math. Comp. **48** (1987), 711–724.
14. Dahmen, W. and C. A. Micchelli, On the approximation order from certain multivariate spline spaces, J. Austral. Math. Society, Ser. B **26** (1984), 233–246.
15. Dyn, N. and A. Ron, Local approximation by certain spaces of multivariate exponential-polynomials, approximation order of exponential box splines and related interpolation problems, Trans. Amer. Math. Soc. **319** (1990), 381–404.

16. Halton, E. J. and W. A. Light, On local and controlled approximation order, J. Approx. Theory, to appear.
17. Helson, H., *Lectures on Invariant Subspaces*, Academic Press, New York, 1964.
18. Jia, Rong-Qing, A counterexample to a result concerning controlled approximation, Proc. Amer. Math. Soc. **97** (1986), 647–654.
19. Jia, Rong-Qing, Approximation order of translation invariant subspaces of functions, in *Approximation Theory VI*, C. C. Chui, L. L. Schumaker, and J. D. Ward (eds.), Academic Press, New York, 1989, 349–352.
20. Jia, Rong-Qing and Junjiang Lei, Approximation by multiinteger translates of functions having global support, J. Approx. Theory, to appear.
21. Jia, Rong-Qing and Junjiang Lei, A new version of the Strang-Fix conditions, J. Approx. Theory, to appear.
22. Lei, Junjiang and Rong-Qing Jia, Approximation by piecewise exponentials, SIAM J. Math. Anal. **22** (1991), 1776–1789.
23. Light, W. A. and E. W. Cheney, Quasi-interpolation with translates of a function having noncompact support, Constr. Approx. **8** (1992), 35–48.
24. Powell, M. J. D., The theory of radial basis function approximation in 1990, in *Advances in Numerical Analysis II: Wavelets, Subdivision Algorithms and Radial Functions*, W. Light (ed.), Oxford University Press, 1992, 105–210.
25. Ron, A., A characterization of the approximation order of multivariate spline spaces, Studia Math. **98** (1991), 73–90.
26. Strang, G. and J. Fix, *An Analysis of the Finite Element Method*, Prentice-Hall, Englewood Cliffs, N.J., 1973.
27. Strang, G. and G. Fix, A Fourier analysis of the finite element variational method, in *Constructive Aspects of Functional Analysis*, G. Geymonat (ed.), C.I.M.E. II Ciclo 1971, 1973, 793–840.

Carl de Boor
Center for Mathematical Sciences
University of Wisconsin - Madison
Madison, WI 53706
deboor@cs.wisc.edu

Supported in part by the United States Army under Contract DAAL03-G-90-0090 and by the National Science Foundation under grant DMS-9000053.

Wavelets and Signal Analysis

Charles K. Chui

Abstract. The notion of multiresolution analysis (MRA) is a familiar concept to the approximation theorist. In fact, the family of mth order cardinal spline spaces V_j^m, $j \in \mathbb{Z}$, with knot sequences $2^{-j}\mathbb{Z}$, is the most typical example being used to demonstrate the structure of an MRA. What has been somewhat neglected in approximation theory in the past is the study of the structure and implications of the complementary subspaces of an MRA. Any function that belongs to a set of generators of such complementary subspaces may be called a wavelet. The objective of this contribution is to give a brief review of wavelet analysis, with special emphasis on how it fits in and enhances the field of approximation theory, and why it has recently created much excitement in the mathematical and engineering communities. From the mathematical point of view, we will use cardinal splines to demonstrate the importance of the subject of wavelet analysis to the approximation theorist; and for engineering applications, we will restrict our discussion to why wavelet analysis has revolutionized the field of signal processing. Since there is already a vast literature, including at least three monographs and several edited volumes devoted to wavelet analysis, our presentation will be very brief.

§1. Introduction

In approximation theory, one frequently studies various problems that arise from the approximation of functions in some normed linear space X from a nested sequence of subspaces V_j of a possibly different space Y, by using the same norm of the space X. Typical examples of V_j are algebraic polynomials, trigonometric polynomials, and splines. For convenience, let us write

$$\cdots \subset V_{-1} \subset V_0 \subset V_1 \subset \cdots$$

(although the nested sequence may not be doubly infinite), and set

$$Y = \text{clos}_X \left(\bigcup_j V_j \right).$$

Let us also consider a corresponding sequence of (linear) projection operators

$$P_j\colon Y \to V_j .$$

A typical example is the family defined by (certain centers of) best approximations. For each $f \in Y$, consider

$$f_j = P_j f \quad \text{and} \quad g_j = f_{j+1} - f_j .$$

Then

$$W_j := \{f - P_j f\colon f \in V_{j+1}\}$$

is a subspace of V_{j+1} which is complementary to V_j, relative to the projection operator P_j. We denote this direct-sum decomposition of V_{j+1} by

$$V_{j+1} = V_j \oplus W_j .$$

Since $\{V_j\}$ is nested, under the additional assumption that

$$\bigcap_j V_j = \{0\} ,$$

we may conclude that

$$Y = \bigoplus_j W_j .$$

That is, every $f \in Y$ has a unique decomposition

$$f = \sum_j g_j , \qquad g_j \in W_j .$$

What is so special about this decomposition of f? In approximation theory, if f is "sufficiently smooth," then we may usually conclude that

$$f_j \to f \quad \text{as } j \to \infty ,$$

very rapidly. So, $g_j = f_{j+1} - f_j \to 0$ quite rapidly also. What if f is smooth outside some region B but not so smooth on B? If each V_j has a local basis, and the locality "shrinks to zero" as $j \to \infty$, then one still expects that

$$g_j = f_{j+1} - f_j \to 0 \quad \text{outside} \quad B ,$$

but since g_j is not small on B, it reveals the "details" of f on B. For instance, if f has some sharp changes such as "singularities," then g_j detects the location B of such changes! Furthermore, if each V_j has a "good" local basis, then not only the location B of sharp changes of f is identified, but the degree of both slow and sharp changes can be analyzed and localized by studying the various components g_j of f. In addition to locating and studying the singularities of functions, there are many other problem areas that can benefit from or even require a good knowledge of the location and magnitude of changes. In engineering applications, such problems include pattern recognition, edge detection, radar and sonar signal analysis, echo-removal, detection and removal of noise, data compression, and so on.

In image compression, for example, since very small changes cannot be recognized by the human eye, "thresholding" of each component g_j of the original image f usually amounts to a substantial saving of image data. Further saving can be achieved by grouping and using symbolic representation of the image data that survive the thresholds. Other special tricks can also be applied as long as the locations and magnitudes of the variations for each component g_j can be identified. Of course, a recovery scheme from the filtered compressed components is eventually needed for image reconstruction.

Hence, while in approximation theory one is interested in the study of the "approximants" f_j of f from V_j, we note that it is also very rewarding to study the differences of consecutive errors of approximation, namely:

$$g_j = (f_{j+1} - f) - (f_j - f).$$

If the spaces V_j have good local bases, then the decomposition of f into its components g_j is extremely valuable. Efficient algorithms for such decomposition and eventually the corresponding reconstruction of f from its components g_j are therefore of utmost importance. This is what "wavelet analysis" is about. The structure of both decomposition and reconstruction is dictated by the projection operators P_j, which, in turn, are governed by the problem of approximation. The "best" local bases to the approximation theorist are, perhaps, the B-splines, at least in the univariate setting. This motivates our study of spline-wavelets.

§2. Cardinal Spline-wavelets

For simplicity, we only discuss the one-dimensional problem. As mentioned in the previous section, the most attractive local basis, at least to the approximation theorist, is a B-spline basis. Although the following discussion is valid for splines with arbitrary knot sequences, we only consider those with equally-spaced knots. That is, we will be concerned with cardinal splines. Let m be any positive integer and let S_m denote the space of all cardinal splines of order m and with knot sequence $\mathbb{Z}$, the set of integers. It is well-known (cf. [58,326]) that any $f \in S_m$ can be written as

$$f(x) = \sum_k c_k N_m(x-k), \qquad (2.1)$$

where N_m is the mth order cardinal B-spline defined by $N_m = N_{m-1} * N_1$ with $N_1 = \chi_{[0,1)}$, the characteristic function of $[0,1)$.

We consider the situation
$$X = Y = L^2,$$
where $L^2 := L^2(\mathbb{R})$, and define a spline subspace
$$V_0^m = \operatorname{clos}_{L^2}(S_m). \tag{2.2}$$
Hence, a function f is in V_0^m if and only if it has a B-spline series representation as in (2.1) with coefficient sequence $\{c_k\} \in \ell^2$. The other spaces V_j^m are defined by
$$f(x) \in V_j^m \Leftrightarrow f(2x) \in V_{j+1}^m, \quad j \in \mathbb{Z}, \tag{2.3}$$
using V_0^m in (2.2) as a reference subspace. In other words, V_j^m is the closed subspace of mth order cardinal splines with knot sequence $2^{-j}\mathbb{Z}$, so that $\{V_j^m\}$ possesses the following properties:
$$\cdots \subset V_{-1}^m \subset V_0^m \subset V_1^m \subset \cdots, \tag{2.4}$$
$$L^2 = \operatorname{clos}_{L^2}\left(\bigcup_{j \in \mathbb{Z}} V_j^m\right), \tag{2.5}$$
$$\bigcap_{j \in \mathbb{Z}} V_j^m = \{0\}, \tag{2.6}$$
$$f(x) \in V_j^m \Leftrightarrow f(x + 2^{-j}) \in V_j^m, \tag{2.7}$$
and $\{N_m(\cdot - k): k \in \mathbb{Z}\}$ is a Riesz (or unconditional) basis of V_0^m. This means that there exist positive constants A and B such that
$$A\|c\|_{\ell^2}^2 \leq \left\|\sum_{k \in \mathbb{Z}} c_k N_m(\cdot - k)\right\|_2^2 \leq B\|c\|_{\ell^2}^2, \tag{2.8}$$
for all $\{c_k\} \in \ell^2$, where $0 < A \leq B < \infty$. Here, the smallest value of B is 1, and the largest value of A can be expressed in terms of the roots of the Euler-Frobenius polynomial
$$E_{2m-1}(z) := (2m-1)! z^{m-1} \sum_{k=-m+1}^{m-1} N_{2m}(m+k) z^k \tag{2.9}$$
(cf. [74, Th. 4.5]). The properties (2.3)–(2.8) of the B-spline N_m qualify it as a "scaling function" that generates the "multiresolution analysis" $\{V_j^m\}$ of L^2. (See Mallat [277] and Meyer [294], where this notion was introduced.)

It is easy to see that (2.8) is equivalent to
$$A \leq \sum_{k \in \mathbb{Z}} |\widehat{N}_m(\omega + 2\pi k)|^2 \leq B, \quad \omega \in \mathbb{R}. \tag{2.10}$$

On the other hand, it follows from the Poisson Summation formula that

$$\sum_{k \in \mathbb{Z}} |\widehat{N}_m(\omega + 2\pi k)|^2 = \frac{z^{-m+1}}{(2m-1)!} E_{2m-1}(z), \quad z = e^{-i\omega}, \qquad (2.11)$$

where the notion of Euler-Frobenius polynomials as defined in (2.9) is used. We also recall, from the definition of N_m, that

$$\langle N_m, N_m(\cdot - k) \rangle = N_{2m}(m - k). \qquad (2.12)$$

As for the nested property of $\{V_j^m\}$ in (2.4), it can be described explicitly by the so-called two-scale relation:

$$N_m(x) = \sum_{k=0}^{m} p_k N_m(2x - k), \qquad (2.13)$$

where

$$p_k = p_k^{(m)} := 2^{-m+1} \binom{m}{k}. \qquad (2.14)$$

Recall that (2.13) is used in constructing subdivision schemes for cardinal spline functions.

Let us now consider the orthogonal projection operators P_j from L^2 onto the spaces V_j^m of mth order cardinal splines, and study the orthogonal complementary subspaces W_j^m of V_{j+1}^m relative to V_j^m. (Recall that if the projection P_j is not orthogonal, then the direct-sum decomposition $V_{j+1} = V_j \oplus W_j$ is not an orthogonal decomposition.) Our first goal is to characterize the bases of W_j^m.

In view of the equivalence between (2.8) and (2.10), we see that any Riesz basis of V_0^m is of the form

$$\left\{ \sum_k \alpha_k N_m(x - k - \ell) \right\}_{\ell \in \mathbb{Z}}, \qquad (2.15)$$

where $\{\alpha_k\} \in \ell^2$ satisfies

$$0 < C \le \left| \sum_k \alpha_k e^{-ik\omega} \right|^2 \le D < \infty, \quad \omega \in \mathbb{R}. \qquad (2.16)$$

So, if we consider

$$f(x) := \sum_k \alpha_k N_m(2x - k) \in V_1^m \qquad (2.17)$$

and

$$g_0 := f - P_0 f, \qquad (2.18)$$

then $\{g_0(x-k): k \in \mathbb{Z}\}$ is a Riesz basis of W_0^m if and only if $\{f(x-\ell): \ell \in \mathbb{Z}\}$ is a Riesz basis of V_1^m, or equivalently, $\{\alpha_k\} \in \ell^2$ satisfies (2.16) for some positive constants C and D. To determine g, observe that the relation $W_0^m \perp V_0^m$ yields

$$\langle f - P_0 f, N_m(\cdot - \ell)\rangle = 0, \qquad \ell \in \mathbb{Z}. \tag{2.19}$$

Hence, writing

$$P_0 f := \sum_k c_k N_m(\cdot - k), \tag{2.20}$$

and applying (2.12)–(2.13), we can reformulate the "normal equations" (2.19) as

$$\sum_k c_k N_{2m}(m - \ell + k) = \langle f, N_m(\cdot - \ell)\rangle \tag{2.21}$$

$$= \frac{1}{2} \sum_k \alpha_k \sum_{j=0}^m p_j N_{2m}(m - j - 2\ell + k), \qquad \ell \in \mathbb{Z}.$$

Since the Euler-Frobenius polynomial E_{2m-1}, as defined in (2.9), is zero-free on the unit circle, the Toeplitz matrix $[N_{2m}(m-\ell+k)]$, $\ell, k \in \mathbb{Z}$, is invertible; and this allows unique solution of the coefficient sequence $\{c_k\}$ in the B-spline series representation of $P_0 f$ by using (2.21). Hence, for any $\{\alpha_k\}$ satisfying (2.16), the function g_0 as defined in (2.18) generates a Riesz basis of W_0^m.

By applying very simple algebraic manipulation, we can even obtain a compact formula for the symbol (or z-transform) of $\{c_k\}$ (cf. [94]). From this formula, by using the basic property of the zero structure of the Euler-Frobenius polynomials, it is easy to identify those $\{\alpha_k\}$ that correspond to compactly supported g_0. In particular, the g_0 with minimum (compact) support is given by

$$\psi_m(x) := \sum_{k=0}^{3m-2} q_k N_m(2x - k), \tag{2.22}$$

with

$$q_k = q_k^{(m)} := \frac{(-1)^k}{2^{m-1}} \sum_{\ell=0}^m \binom{m}{\ell} N_{2m}(k - \ell + 1), \tag{2.23}$$

where ψ_m is unique up to a multiplicative constant and integral translation (cf. [97]). Graphs of ψ_m for $m = 1, 2, 3, 4$ are shown in Figure 1.

In view of the definition of W_j^m, namely

$$V_{j+1}^m = V_j^m + W_j^m \quad \text{and} \quad V_j^m \perp W_j^m, \qquad j \in \mathbb{Z},$$

where V_j^m satisfies (2.3), it is clear that, for each $j \in \mathbb{Z}$, the family

$$\psi_{m;j,k}(x) := 2^{j/2} \psi_m(2^j x - k), \qquad k \in \mathbb{Z}, \tag{2.24}$$

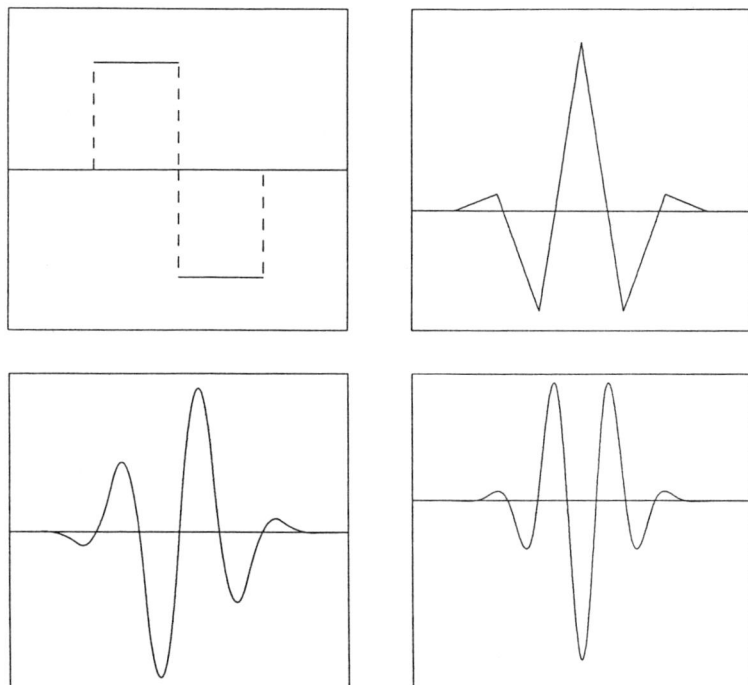

Figure 1. B-wavelets ψ_m ($m = 1, \ldots, 4$).

is a Riesz basis of W_j^m. Hence, it follows from (2.4)–(2.6), that the family

$$\{\psi_{m;j,k}: \ j, k \in \mathbb{Z}\} \tag{2.25}$$

is a Riesz basis of L^2. That is, ψ_m generates a Riesz basis of all of L^2, simply by binary dilations and dyadic translations.

In [97], where it was first constructed, ψ_m is called the mth order B-wavelet. We remark that since $\{\psi_m(\cdot - k)\}$ is not an orthogonal sequence, ψ_m is also called a prewavelet (cf. [36,301]), while since $\psi_{m;j,k} \perp \psi_{m;j',k'}$ for all $j, j', k, k' \in \mathbb{Z}$ with $j \neq j'$, it is also called a semi-orthogonal wavelet (cf. [78]). For more details, the reader is referred to the monograph [74].

It is important to point out the role of ψ_m as compared to that of the cardinal B-spline N_m. They are generators of W_0^m and V_0^m, respectively, and they both have minimum supports. In fact,

$$\begin{cases} \operatorname{supp} \psi_m = [0, 2m - 1] \ ; \\ \operatorname{supp} N_m = [0, m] \ . \end{cases} \tag{2.26}$$

While N_m is used for "approximation" and "smoothing," ψ_m is used for "decomposition" and "detecting irregularities." What is so special about the local basis (of V_0^m) generated by N_m is the so-called "total positivity" property. It "diminishes" sign changes, and hence, possesses a smoothing capability. Hence, it is a very important project to extract the essence of the mth order B-wavelet ψ_m also. In [95], the notion of "complete oscillation" is introduced. Instead of diminishing the number of sign changes, it actually increases this count. A detailed analysis of the linear B-wavelet ψ_2 is given in [95]. With this precise count, it is possible to derive a characterization of a B-wavelet series by its "zero-crossings" (which has important applications to signal analysis and data compression). See [74,95].

§3. Dual Wavelets and Integral Wavelet Transforms

If $\{\psi_{j,k}: j,k \in \mathbb{Z}\}$ is a Riesz basis of L^2, then it has a unique dual basis

$$\{\psi^{j,k}: j,k \in \mathbb{Z}\} \tag{3.1}$$

defined by

$$\langle \psi_{j,k}, \psi^{j',k'} \rangle = \delta_{j,j'}\delta_{k,k'}, \qquad j,k,j',k' \in \mathbb{Z}. \tag{3.2}$$

Suppose that $\{\psi_{j,k}\}$ is generated by a single function ψ, say, in the sense that

$$\psi_{j,k}(x) = 2^{j/2}\psi(2^j x - k), \qquad j,k \in \mathbb{Z}. \tag{3.3}$$

Then it seems reasonable to expect that its dual basis is generated by a single function also. Somewhat surprisingly, this is not true in general, and a simple class of counterexamples is given in [85]. If ψ is a semi-orthogonal wavelet, however, then indeed the dual basis $\{\psi^{j,k}\}$ is generated by some $\tilde{\psi} \in L^2$, in the sense that

$$\psi^{j,k}(x) = \tilde{\psi}_{j,k}(x) := 2^{j/2}\tilde{\psi}(2^j x - k). \tag{3.4}$$

To see this, we first note that semi-orthogonality already implies (3.2) for $j \neq j'$ by using (3.4) for $\psi^{j,k}$. Hence, it is sufficient to verify

$$\langle \psi_{0,k}, \tilde{\psi}_{0,k'} \rangle = \delta_{k,k'} \qquad k,k' \in \mathbb{Z}. \tag{3.5}$$

In terms of Fourier transforms, (3.5) can be written as

$$\sum_{k \in \mathbb{Z}} \hat{\psi}(\omega + 2\pi k)\overline{\hat{\tilde{\psi}}(\omega + 2\pi k)} = 1 \quad \text{a.e.,} \tag{3.6}$$

and this, in turn, shows that

$$\hat{\tilde{\psi}}(\omega) = \frac{\hat{\psi}(\omega)}{\sum_{k \in \mathbb{Z}} |\hat{\psi}(\omega + 2\pi k)|^2}. \tag{3.7}$$

For the mth order B-wavelets ψ_m, $m \geq 1$, which are semi-orthogonal, we can give explicit formulations of their duals (cf. [97]). Let L_{2m} denote the $(2m)^{\text{th}}$-order fundamental cardinal spline defined by

$$\begin{cases} L_{2m} \in V_0^{2m} \; ; \\ L_{2m}(k) = \delta_{k,0} \, , \qquad k \in \mathbb{Z} \, . \end{cases} \qquad (3.8)$$

Write
$$L_{2m}(x) = \sum_{k \in \mathbb{Z}} c_k^{(2m)} N_{2m}(x + m - k) \, . \qquad (3.9)$$

It was shown in [93] that the function

$$\psi_{I,m}(x) := L_{2m}^{(m)}(2x - 1) \qquad (3.10)$$

is in W_0^m and that
$$\{2^{j/2} \psi_{I,m}(2^j x - k) \colon j, k \in \mathbb{Z}\}$$
is a Riesz basis of L^2. Hence, $\psi_{I,m}$ is also a cardinal spline-wavelet, although it has infinite support. Now, the dual $\widetilde{\psi}_m$ of the B-wavelet ψ_m can be written in terms of $\psi_{I,m}$ using the coefficient sequence $\{c_k^{(2m)}\}$ in (3.9) as follows:

$$\widetilde{\psi}_m(x) = \frac{(-1)^{m+1}}{2^{m-1}} \sum_{k \in \mathbb{Z}} c_k^{(2m)} \psi_{I,m}(x + m + 1 - k) \, . \qquad (3.11)$$

The graphs of $\psi_{I,m}$ and $\widetilde{\psi}_m$ for $m = 1, 2, 3, 4$ are shown in Figures 2 and 3, respectively.

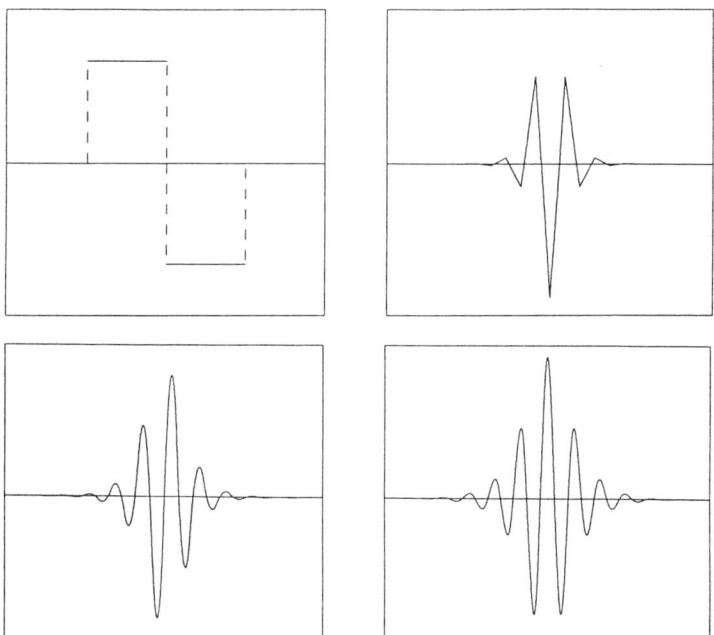

Figure 2. Interpolatory spline-wavelets $\psi_{I,m}$ ($m = 1, \ldots, 4$).

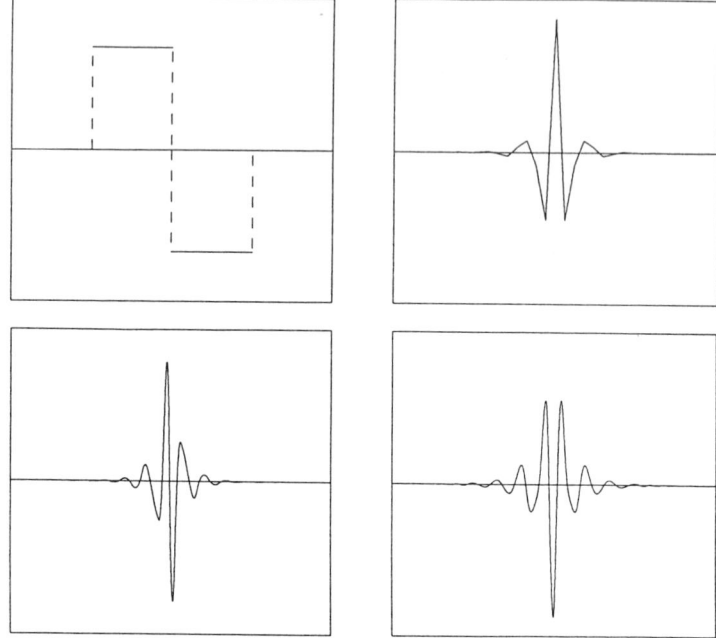

Figure 3. Dual B-wavelets $\widetilde{\psi}_m$ ($m = 1, \ldots, 4$).

Observe that if $\widetilde{\psi}$ is the dual wavelet of a wavelet ψ as defined in (3.3) and (3.4) using the duality condition (3.2), then ψ is also the dual wavelet of $\widetilde{\psi}$. That is, ψ and $\widetilde{\psi}$ are dual to each other. Now, since each of $\{\psi_{j,k}\}$ and $\{\widetilde{\psi}_{j,k}\}$ is a Riesz basis of L^2, we have

$$f(x) = \sum_{j,k \in \mathbb{Z}} \langle f, \widetilde{\psi}_{j,k} \rangle \psi_{j,k}(x) \tag{3.12}$$
$$= \sum_{j,k \in \mathbb{Z}} \langle f, \psi_{j,k} \rangle \widetilde{\psi}_{j,k}(x) \, ,$$

for every $f \in L^2$. Each of the series in (3.12) is a wavelet series representation of $f \in L^2$. In this respect, a wavelet series representation may be considered as an $L^2(\mathbb{R})$ analog of the Fourier series representation for periodic functions. What is really important in a wavelet series representation, however, is that its coefficient sequence, say

$$c_{j,k} := \langle f, \psi_{j,k} \rangle \tag{3.13}$$

has very significant physical meaning. To discuss this aspect, we introduce the notion of integral wavelet transform IWT (cf. [203]) with respect to a "basic wavelet" (or "mother wavelet") ψ, namely:

$$(W_\psi f)(b, a) := \frac{1}{\sqrt{a}} \int_{-\infty}^{\infty} f(x) \overline{\psi\left(\frac{x-b}{a}\right)} dx \, . \tag{3.14}$$

Wavelets and Signal Analysis

The coefficient sequence in (3.13) is a sequence of values of the IWT of f, given by

$$c_{j,k} = \langle f, \psi_{j,k} \rangle = (W_\psi f)\left(\frac{k}{2^j}, \frac{1}{2^j}\right) . \tag{3.15}$$

Observe that if ψ is somewhat localized in the sense that it is negligibly small in magnitude outside a compact interval (a more precise measure of the locality of ψ can be formulated by using the notion of standard deviation), then $(W_\psi f)(b,a)$ localizes f around $b + at^*$, where t^* is the "center" of ψ. The localization improves as $a \to 0^+$. Hence, if f is considered as an analog signal, then $(W_\psi f)(b,a)$ gives local "information" on f in the so-called "time- or space-domain." This information becomes "sharper" as the value of $a > 0$ is decreased. We may call this the "zoom-in" effect. But what information of f do we actually receive? To answer this question, let us view the IWT, $(W_\psi f)(b,a)$, as the inner product of f with the "window function"

$$\psi_{b;a}(x) := |a|^{-\frac{1}{2}} \psi\left(\frac{x-b}{a}\right) , \tag{3.16}$$

where a and b are fixed in this consideration. Note that the Fourier transform of $\widehat{\psi}_{b;a}$ is given by

$$\widehat{\psi}_{b;a}(\omega) = |a|^{-\frac{1}{2}} a e^{-ib\omega} \widehat{\psi}(a\omega) . \tag{3.17}$$

So, if the center ω^* of $\widehat{\psi}_{b;a}$ is assumed to be positive, and if

$$\eta(\omega) := \widehat{\psi}(\omega + \omega^*) , \tag{3.18}$$

then by the Parseval identity for L^2, we have

$$\begin{aligned}(W_\psi f)(b,a) &= \langle f, \psi_{b;a}\rangle \\ &= \frac{1}{2\pi}\langle \hat{f}, \widehat{\psi}_{b;a}\rangle \\ &= \frac{a|a|^{-\frac{1}{2}}}{2\pi}\int_{-\infty}^{\infty} \hat{f}(\omega)\overline{\eta\left(a\left(\omega - \frac{\omega^*}{a}\right)\right)} d\omega .\end{aligned} \tag{3.19}$$

That is, the local information $(W_\psi f)(b,a)$ of f in a neighborhood of $b + at^*$ in the time-domain is the localized "value" of the Fourier transform $\hat{f}$ of f in a neighborhood of ω^*/a in the "frequency- or phase-domain." So, for small positive values of a (when we zoom-in to investigate f), we obtain the local content of $\hat{f}$ at high frequency (when ω^*/a is large). This agrees with the actual need in time-frequency and phase-space analysis, namely: to investigate fine details (of contrast), we zoom-in (small $a > 0$), and to see a more global picture (with less contrast), we zoom-out (large $a > 0$).

Returning to the wavelet series representation of f in (3.12), we see that the wavelet series partitions of f into components, each of which gives the local time-frequency (or phase-space) information of f in a neighborhood (or time-frequency window) of the point

$$\left(\frac{k}{2^j} + \frac{t^*}{2^j},\ 2^j\omega^*\right) \qquad (3.20)$$

in the time-frequency domain, where $j, k \in \mathbb{Z}$. While the area of this time-frequency window is a constant as governed by the Uncertainty Principle, the time-window narrows when high-frequency (i.e., large j) phenomena are detected, and it widens to study the environment of low-frequency (i.e., $j \to -\infty$).

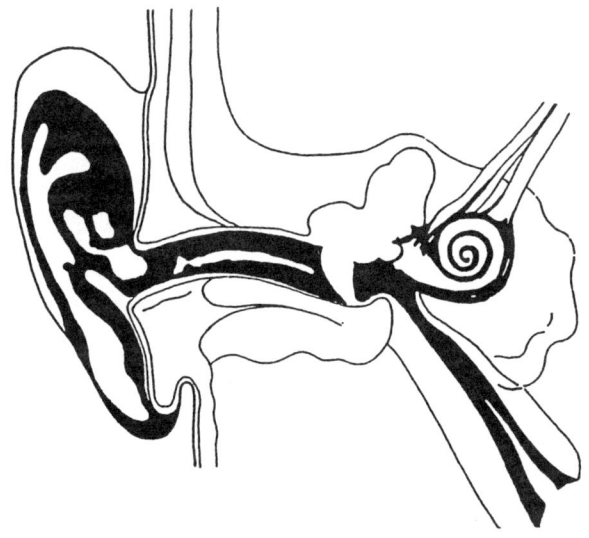

Figure 4. The Human Ear.

Let us demonstrate the need for these adaptive time-frequency windows by modeling the cochlea of the human ear (see Figure 4). The simplest approach is to consider the cochlea as a linear filter with transfer function $H_x(\omega)$, where x corresponds to the location in the cochlea. Hence, the input-output relationship is described by the equation

$$\hat{g}(\omega) = H_x(\omega)\hat{f}(\omega)\ .$$

The corti inside the cochlea consists of many receptor cells which, in turn, are partitioned by sensory hair cells (see Figure 5). So, if the spiral geometric shape of the cochlea is stretched and then mapped onto the positive half of

Figure 5. The Corti inside the Cochlea.

the real x-axis, $\mathbb{R}_+$, with x representing the location of the images on $\mathbb{R}_+$ of the receptor cells, then we may write

$$H_x(\omega) = \overline{\widehat{\Theta}(x - \ln \omega)}\,, \qquad (3.21)$$

where Θ is some real-valued function with Fourier transform $\widehat{\Theta}$. Since the Fourier transform of $f(t-b)$ is given by

$$\hat{f}(\cdot - b)|_\omega = \hat{f}(\omega)e^{-ib\omega}\,,$$

the output $g(\cdot, x)$ at the position x, corresponding to the input signal f in a neighborhood of $t = b$ is the inverse Fourier transform of the function

$$H_x(\omega)\{\hat{f}(\omega)e^{-ib\omega}\} = \hat{f}(\omega)e^{-ib\omega}\overline{\widehat{\Theta}(x - \ln \omega)}\,. \qquad (3.22)$$

Because both f and Θ are real-valued, the inverse Fourier transform of (3.12) is given by

$$g(t-b, x) = \frac{1}{\pi} \mathrm{Re} \int_0^\infty \{\hat{f}(\omega)e^{-ib\omega}\overline{\widehat{\Theta}(x - \ln \omega)}\}e^{it\omega}d\omega\,,$$

or equivalently,
$$g(b, x) = \frac{1}{\pi} Re \int_0^\infty \hat{f}(\omega)\overline{\hat{\Theta}(x - \ln \omega)} e^{-ib\omega} d\omega \ .$$

Now, by setting $x = -\ln a$, $0 < a \leq 1$, we arrive at
$$G(b, a) := g\left(b, \ln \frac{1}{a}\right) = \frac{1}{\pi} Re \int_0^\infty \hat{f}(\omega)\overline{\hat{\Theta}\left(\ln \frac{1}{a\omega}\right)} e^{-ib\omega} d\omega \ ;$$

and by defining
$$\hat{\psi}(\omega) := \hat{\Theta}\left(\ln \frac{1}{\omega}\right) \ , \tag{3.23}$$

we finally obtain
$$\begin{aligned} G(b, a) &= \frac{1}{\pi} Re \int_0^\infty \hat{f}(\omega)\overline{\hat{\psi}(a\omega)} e^{ib\omega} d\omega \\ &= \frac{1}{2\pi} \left\{ \int_0^\infty \hat{f}(\omega)\overline{\hat{\psi}(a\omega)} e^{ib\omega} d\omega \right. \\ &\quad \left. + \int_0^\infty \overline{\hat{f}(\omega)}\hat{\psi}(a\omega) e^{ib\omega} d\omega \right\} \\ &= \frac{1}{a} \int_{-\infty}^\infty f(t)\overline{\psi\left(\frac{t-b}{a}\right)} dt \ . \end{aligned} \tag{3.24}$$

Hence, in view of (3.14), we have
$$\sqrt{a}\, g\left(b, \ln \frac{1}{a}\right) = (W_\psi f)(b, a) \ . \tag{3.25}$$

Observe that according to this model, high-frequency signals are received at the receptor cells with small $a > 0$, or large x.

§4. Affine Frames

In the wavelet series representation (3.12), we consider only binary dilation and dyadic translation, namely:
$$\psi_{j,k}(x) = 2^{j/2}\psi(2^j x - k) \quad \text{and} \quad \tilde{\psi}_{j,k}(x) = 2^{j/2}\tilde{\psi}(2^j x - k) \ ,$$

where $j, k \in \mathbb{Z}$. In general, for any $a > 1$ and $b > 0$, we can extend this notion to
$$\psi_{a,b;j,k}(x) = a^{j/2}\psi(a^j x - kb) \ , \qquad j, k \in \mathbb{Z} \ . \tag{4.1}$$

However, if $\{\psi_{j,k}\}$ is a Riesz basis of L^2, one cannot expect $\{\psi_{a,b;j,k}\}$ to be a basis of L^2 in general. After all, $\{\psi_{2,1/2;j,k}\}$ properly contains $\{\psi_{j,k}\}$ and must be linearly dependent. On the other hand, for $a = 2$ and $b = n^{-1}$, where n is a positive integer, the redundancy of the family $\{\psi_{2,1/n;j,k}\}$ gives more room for error in the data coefficient sequence $c_{j,k} = (W_\psi f)(k2^{-j}, 2^{-j})$ or $(W_{\widetilde\psi} f)(k2^{-j}, 2^{-j})$. This is quite important in the application of wavelets. For instance, in data reduction, the values of $c_{j,k}$ are filtered by thresholding, say, to reduce the storage and transmittance volume. In order to be able to reconstruct f from the data

$$(W_\psi f)\left(\frac{kb}{a^j}, a^{-j}\right), \qquad j, k \in \mathbb{Z},$$

however, we require "stability" of the family $\{\psi_{a,b;j,k}\}$. In other words, we require this family to form an (affine) frame of L^2, namely:

$$A\|f\|_2^2 \leq \sum_{j,k} |\langle f, \psi_{a,b;j,k}\rangle|^2 \leq B\|f\|_2^2, \qquad f \in L^2. \tag{4.2}$$

Here, A and B, satisfying $0 < A \leq B < \infty$, are called *frame bounds* of the frame. It is easy to see that any Riesz basis is a frame, but the converse is false in general because of linear dependency. An (affine) frame as in (4.2) is called a *tight* (affine) frame if $A = B$; that is,

$$\|f\|_2^2 = \frac{1}{A} \sum_{j,k} |\langle f, \psi_{a,b;j,k}\rangle|^2, \qquad f \in L^2. \tag{4.3}$$

By a standard argument (applying the identity relating $\|\ \|_2^2$ and $\langle\ ,\ \rangle$), we see that (4.3) is equivalent to

$$\langle f, g\rangle = \frac{1}{A} \sum_{j,k} \langle f, \psi_{a,b;j,k}\rangle \langle \psi_{a,b;j,k}, g\rangle, \qquad f, g \in L^2. \tag{4.4}$$

Hence, by replacing g with an approximation of the identity (or delta function), and passing to the limit, (4.4) yields

$$f(x) = \frac{1}{A} \sum_{j,k} \langle f, \psi_{a,b;j,k}\rangle \psi_{a,b;j,k}(x), \qquad f \in L^2. \tag{4.5}$$

This representation formula may be called a frame series. From this formula, we see that a tight frame behaves somewhat like an orthonormal basis, and that the larger the value of the constant $A > 1$, the more room is available to allow for error in the data sequence $\{\langle f, \psi_{a,b;j,k}\rangle\}$.

If ψ generates a frame $\{\psi_{a,b;j,k}\}$ which, however, is not necessarily tight, and if ψ does have some dual $\widetilde\psi$ in the sense that the family

$$\widetilde\psi_{a,b;j,k}(x) := a^{j/2}\widetilde\psi(a^j x - kb) \tag{4.6}$$

generated by $\tilde{\psi}$ satisfies

$$\langle f, g \rangle = \sum_{j,k} \langle f, \psi_{a,b;j,k} \rangle \langle \tilde{\psi}_{a,b;j,k}, g \rangle, \qquad f, g \in L^2, \qquad (4.7)$$

then we still have the frame series representation:

$$f(x) = \sum_{j,k} \langle f, \psi_{a,b;j,k} \rangle \tilde{\psi}_{a,b;j,k}(x) \qquad (4.8)$$
$$= \sum_{j,k} \langle f, \tilde{\psi}_{a,b;j,k} \rangle \psi_{a,b;j,k}(x), \qquad f \in L_2.$$

Note that every ψ that generates a tight frame, with frame bound $A = B$, has a dual $\tilde{\psi}$ given by $\tilde{\psi} = \frac{1}{A}\psi$.

Recently [89], we have shown that if a is an integer at least 2 and $\{\psi_{a,b;j,k}\}$ is a frame, then $\{\psi_{a,b/n;j,k}\}$ is a frame of L^2 for any positive integer n, provided that ψ satisfies certain mild decay and smoothness conditions. Hence, "oversampling" does not destroy a frame, and from any Riesz basis $\{\psi_{2,1;j,k}\}$, we can create infinitely many frames

$$\{\psi_{2,1/n;j,k}\}, \qquad n \in \mathbb{N}.$$

But what happens to the frame bounds? This seems to be a very difficult question to answer. Indeed, if H is the Haar function

$$H(x) = \begin{cases} 1 & \text{for} & 0 < x \leq \frac{1}{2} \\ -1 & \text{for} & \frac{1}{2} < x \leq 1 \\ 0 & \text{otherwise,} \end{cases} \qquad (4.9)$$

then the functions $H_{j,k} := H_{2,1;j,k}$, $j, k \in \mathbb{Z}$, constitute an orthonormal basis of L^2, so that $\{H_{j,k}\}$ is a tight frame with bound $A = 1$. However, the functions

$$\begin{cases} f_1(x) = H\left(x + \frac{1}{2}\right) & \text{and} \\ f_2(x) = |f_1(x)| \end{cases}$$

satisfy the property:

$$\begin{cases} \|f_1\|_2 = \|f_2\|_2 = 1 \quad \text{and} \\ \sum_{j,k} |\langle f_1, H_{2,\frac{1}{2};j,k}\rangle|^2 = 3 \sum_{j,k} |\langle f_2, H_{2,\frac{1}{2};j,k}\rangle|^2 = 9. \end{cases}$$

This shows that the frame $\{H_{2,\frac{1}{2};j,k}\}$ is no longer tight! So, oversampling may change the frame bounds nonlinearly (cf. [88]).

If the sampling rate n of a frame $\{\psi_{a,b;j,k}\}$, where $a \geq 2$ is an integer, is changed according to the rule

$$(n, a) = 1 \quad \text{(or } n \text{ and } a \text{ are relatively prime)}, \tag{4.10}$$

then the frame bounds are shown in [88,89] to change linearly. More precisely, the result in [89] states that if $a \geq 2$ is an integer, then

$$nA\|f\|_2^2 \leq \sum_{j,k} |\langle f, \psi_{a,b/n;j,k}\rangle|^2 \leq nB\|f\|_2^2, \qquad f \in L^2, \tag{4.11}$$

holds for all $n \geq 2$ satisfying (4.10), provided that it holds for $n = 1$. In particular, if $\{\psi_{a,b;j,k}\}$ is a tight frame (with $A = B$ and $n = 1$ in (4.11)), and $n \in \mathbb{N}$ satisfies (4.10), then we have

$$\langle f, g \rangle = \frac{1}{nA} \sum_{j,k} \langle f, \psi_{a,b/n;j,k}\rangle \langle \psi_{a,b/n;j,k}, g\rangle, \qquad f, g \in L^2. \tag{4.12}$$

This result is also generalized in [89] to frames which are not necessarily tight, as follows. If ψ and $\widetilde{\psi}$ generate dual frames as defined by (4.1), (4.6) and (4.7), and $a \geq 2$ is an integer, then for any $n \in \mathbb{N}$ satisfying (4.10), we have

$$\langle f, g \rangle = \frac{1}{n} \sum_{j,k} \langle f, \psi_{a,b/n;j,k}\rangle \langle \widetilde{\psi}_{a,b/n;j,k}, g\rangle, \qquad f, g \in L^2. \tag{4.13}$$

Applying (4.13) to the cardinal spline-wavelets as discussed in Sections 2 and 3, and specializing to $a = 2$ and $b = 1$, we have the frame series representation formulas

$$f(x) = \frac{1}{n} \sum_{j,k} \langle f, \widetilde{\psi}_{m;j,k/n}\rangle \psi_{m;j,k/n}(x) \tag{4.14}$$

$$= \frac{1}{n} \sum_{j,k} \langle f, \psi_{m;j,k/n}\rangle \widetilde{\psi}_{m;j,k/n}(x), \qquad f \in L^2,$$

for any positive odd integer n. Here, m is any positive integer, and ψ_m is the mth order B-wavelet given in (2.22)–(2.23) and $\widetilde{\psi}_m$ is its dual as in (3.11).

For more details on frames, such as necessary conditions, sufficient conditions, and applications, the reader is referred to [74,89,132,134,152,369].

§5. Wavelet Packets

When a wavelet is used for time-frequency localization, the area of its time-frequency window is unchanged, regardless of its location on the time-frequency domain. Since the time-window narrows in observing high-frequency signals (small $a > 0$), the frequency bandwidth (or width of the frequency window) has to widen. Hence, in some applications, there is a need to partition the high frequency bands. *Wavelet packets* can be used for this purpose. Orthonormal wavelet packets were introduced in [117], and have been applied very successfully to acoustic signal processing [117,118,119,364,366]. For certain applications which require (phase) distortion-free filtering, non-orthogonal wavelet-packets should be used. The difficulty in dropping orthogonality, however, is the possible loss of stability, as observed in [108]. In a recent paper [81], it is shown that stability can still be achieved if the wavelet packet decompositions are controlled properly.

To introduce the notion of wavelet packets, we return to the notion of MRA and wavelets as described in Section 2. Let ϕ be a scaling function that generates an MRA of L^2. Then since

$$\phi \in V_0 \subset V_1 ,$$

there is a sequence $\{p_k^0\} \in \ell^2$ such that

$$\phi(x) = \sum_k p_k^0 \phi(2x - k) . \tag{5.1}$$

This is called the two-scale relation. Recall that such a relation for the mth order cardinal B-spline was discussed in (2.13)–(2.14). Now let W_0 be a complementary subspace of V_1 relative to V_0; that is,

$$V_1 = V_0 + W_0 \quad \text{and} \quad V_0 \cap W_0 = \{0\} ,$$

where we do not require orthogonality. Suppose that $\psi \in W_0$ generates W_0 in the sense that $\{\psi(\cdot - k): k \in \mathbb{Z}\}$ is a Riesz basis of W_0. Then for each $j \in \mathbb{Z}$, we may define W_j as the L^2-closure of $\{\psi(2^j \cdot -k): k \in \mathbb{Z}\}$, so that

$$V_{j+1} = V_j + W_j \quad \text{and} \quad V_j \cap W_j = \{0\} , \qquad j \in \mathbb{Z} .$$

Now, since $\psi \in W_0 \subset V_1$, there is another sequence $\{p_k^1\} \in \ell^2$ such that

$$\psi(x) = \sum_k p_k^1 \phi(2x - k) . \tag{5.2}$$

This is the other two-scale relation for ψ and ϕ. The two-scale relation for the mth order cardinal B-wavelet ψ_m and B-spline N_m was given by (2.22)–(2.23).

Let us introduce the notation

$$\mu_0 := \phi \quad \text{and} \quad \mu_1 := \psi . \tag{5.3}$$

Then the pair of two-scale relations (5.1) and (5.2) can be written in a compact form

$$\mu_\lambda(x) = \sum_k p_k^\lambda \mu_0(2x - k) , \qquad \lambda = 0, 1 . \tag{5.4}$$

We now extend this relation to introduce a family of functions μ_n, $n = 0, 1, \ldots$, as follows:
$$\mu_{2n+\lambda}(x) = \sum_k p_k^\lambda \mu_n(2x - k), \qquad \lambda = 0, 1. \tag{5.5}$$
This already gives a flavor of "wavelet packets." To be more precise, we need certain assumptions. First, we assume that the functions
$$\psi_{j,k}(x) := 2^{j/2} \psi(2^j x - k), \qquad j, k \in \mathbb{Z}, \tag{5.6}$$
constitute a Riesz basis of L^2; and secondly, we assume the existence of "duals" $\tilde{\phi}$ and $\tilde{\psi}$, of ϕ and ψ respectively, in the sense that the following bi-orthogonality conditions are satisfied:
$$\langle \phi(\cdot - k), \tilde{\phi}(\cdot - k') \rangle = \delta_{k,k'}, \qquad k, k' \in \mathbb{Z}$$
and
$$\langle \psi_{j,k}, \tilde{\psi}_{j',k'} \rangle = \delta_{j,j'} \delta_{k,k'}, \qquad j, j', k, k' \in \mathbb{Z}.$$
Then, as in (5.3) and (5.5), we may also introduce the family of functions $\tilde{\mu}_n$, $n = 0, 1, \ldots$. Let us now define the spaces
$$\begin{cases} \tilde{V}_j := \operatorname{clos}_{L^2} \langle \tilde{\phi}(2^j \cdot - k) \colon k \in \mathbb{Z} \rangle; \\ \tilde{W}_j := \operatorname{clos}_{L^2} \langle \tilde{\psi}(2^j \cdot - k) \colon k \in \mathbb{Z} \rangle; \\ U_n := \operatorname{clos}_{L^2} \langle \mu_n(\cdot - k) \colon k \in \mathbb{Z} \rangle; \\ \tilde{U}_n := \operatorname{clos}_{L^2} \langle \tilde{\mu}_n(\cdot - k) \colon k \in \mathbb{Z} \rangle. \end{cases} \tag{5.7}$$
Then, as shown in [81], the following direct-sum decompositions are achieved:
$$V_j = \bigoplus_{n=0}^{2^j - 1} U_n; \quad W_j = \bigoplus_{n=2^j}^{2^{j+1} - 1} U_n, \tag{5.8}$$
for each $j = 0, 1, \ldots$, and consequently,
$$L^2 = \bigoplus_{n=0}^\infty U_n. \tag{5.9}$$
It also follows from the bi-orthogonality properties that
$$\begin{cases} U_m \perp \tilde{U}_n, & m \neq n, \quad m, n \in \mathbb{Z}_+; \\ V_j \perp \tilde{W}_j \text{ and } \tilde{V}_j \perp W_j, & j \in \mathbb{Z}. \end{cases} \tag{5.10}$$
The main result in [81] is the following.

Theorem. Let $\{\psi_{j,k}\}$ be a Riesz basis of L^2. Then for each $\ell \in \mathbb{Z}_+$, the family
$$\{2^{j/2} \mu_n(2^j x - k) \colon j, k \in \mathbb{Z}, 2^\ell \leq n \leq 2^{\ell+1} - 1\} \tag{5.11}$$
is also a Riesz basis of L^2, with dual basis given by
$$\{2^{j/2} \tilde{\mu}_n(2^j x - k) \colon j, k \in \mathbb{Z}, 2^\ell \leq n \leq 2^{\ell+1} - 1\}. \tag{5.12}$$

The functions μ_n, $n = 2, \ldots$, are called *wavelet packets* generated by $\mu_0 = \phi$ and $\mu_1 = \psi$. In Figure 6, we show the graph of the cubic B-wavelet $\mu_1 = \psi_4$ and its two wavelet packets μ_2 and μ_3; and in Figure 7, we illustrate the effect of frequency decomposition achieved by this splitting, by showing the magnitude spectra $|\hat{\psi}_4|$, $|\hat{\mu}_2(\omega)|$, and $|\hat{\mu}_3(\omega)|$.

Cubic B—wavelet

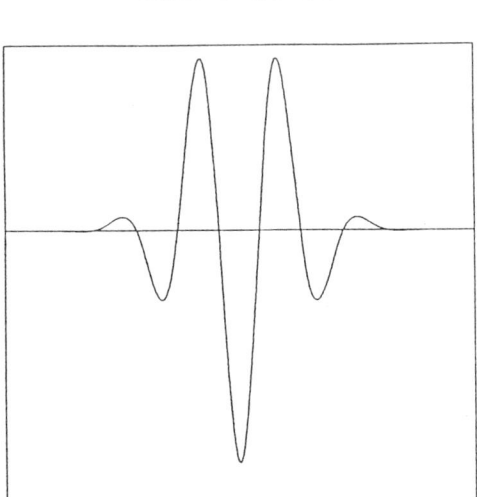

Low—pass wavelet packet   High—pass wavelet packet

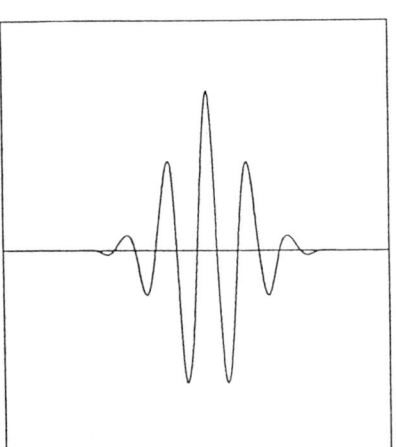 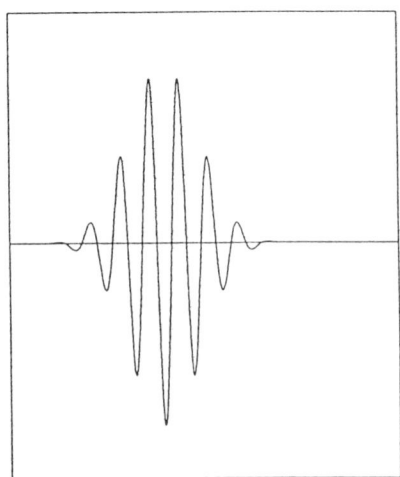

Figure 6. Cubic B-wavelet ψ_4 and its two wavelet packets.

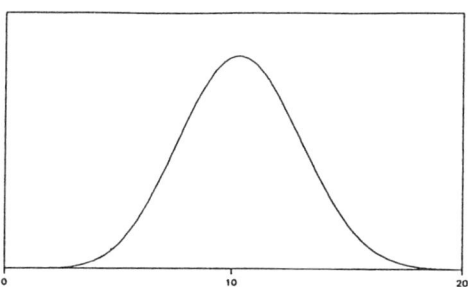
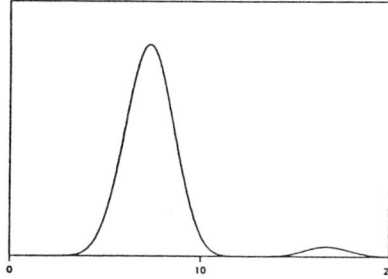
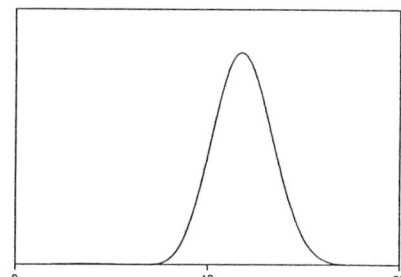

Figure 7. Magnitude spectra of ψ_4 and its two wavelet packets.

§6. What Else?

A very brief account, mainly based on the author's recent research activities, of the one-dimensional theory of wavelets was outlined in the previous sections. The exposition was somewhat biased, expressing only the point of view of an approximation theorist, with special interest in signal analysis. Other approaches and more detailed discussions can be found in the vast literature listed in the bibliography. The interested reader is also referred to the monographs [74,134,290,300], edited volumes [76,124,324], and the references therein for further study.

Acknowledgments. This research was supported by NSF Grant DMS-89-01345, ARO Grant DAAL 03-90-G-0091, and Texas Higher Education Grants TATP 32134 and TARP 32135.

References

1. Adelson, E. H., E. Simoncelli, and R. Hingorani, Orthogonal pyramid transforms for image coding, in *Proc. SPIE Conf. Visual Communication and Image Processing*, Cambridge, MA, 1987, 50–58.
2. Aldroubi, A. and M. Unser, Families of wavelet transforms in connection with Shannon's sampling theory and the Gabor transform, in *Wavelets: A Tutorial In Theory and Applications*, C. K. Chui (ed.), Academic Press, Cambridge, MA, 1992, 509–528.
3. Allen, J. B. and L. R. Rabiner, A unified theory of short-time spectrum analysis and synthesis, Proc. IEEE **65** (11) (1977), 1558–1564.
4. Alpert, B., Construction of simple multiscale bases for fast matrix operations, in *Wavelets and Their Applications*, M. B. Ruskai, G. Beylkin, R. Coifman, I. Daubechies, S. Mallat, Y. Meyer, and L. Raphael (eds.), Jones and Bartlett, Boston, 1992, 211–226.
5. Alpert, B., *Sparse Representation of Smooth Linear Operators*, Ph. D. Thesis, Yale University Department of Computer Science Report #814, August 1990.
6. Alpert, B., Wavelets and other bases for fast numerical linear algebra, in *Wavelets: A Tutorial In Theory and Applications*, C. K. Chui (ed.), Academic Press, Cambridge, MA, 1992, 181–216.
7. Alpert, B., G. Beylkin, R. Coifman, and V. Rokhlin, Wavelets for the fast solution of second-kind integral equations, SIAM J. Sci. Statist. Comput., to appear.
8. Antoine, J. P., P. Carrette, R. Murenzi, and B. Piette, Image analysis with two-dimensional continuous wavelet transform, 1991, submitted to IEEE Trans. Inform. Theory.
9. Antonini, M., M. Barlaud, and P. Mathieu, Image coding using lattice vector quantization of wavelet coefficients, Proc. IEEE Int. Conf. ASSP 1991, 2273–2276.
10. Antonini, M., M. Barlaud, P. Mathieu, and I. Daubechies, Image coding using vector quantization in the wavelet domain, in *Proc. IEEE Int. Conf. ASSP*, Albuquerque, NM, 1990, 2297–2300.
11. Antonini, M., M. Barlaud, P. Mathieu, and I. Daubechies, Image coding using wavelet transforms, IEEE Trans. ASSP (1991), to appear.
12. Arneodo, A., F. Argoul, J. Elezgaray and G. Grasseau, Wavelet transform analysis of fractals: application to nonequilibrium phase transitions, in *Nonlinear Dynamics*, G. Turchetti (ed.), World Scientific, Singapore, 1988, 130.
13. Arneodo, A., F. Argoul, E. Freysz, J. F. Muzy, and B. Pouligny, The optical wavelet transform, in *Wavelets and Their Applications*, M. B. Ruskai, G. Beylkin, R. Coifman, I. Daubechies, S. Mallat, Y. Meyer, and L. Raphael (eds.), Jones and Bartlett, Boston, 1992, 241–273.
14. Arneodo, A., F. Argoul, E. Bacry, J. Elezgaray, E. Freysz, G. Grasseau, J. F. Muzy, and B. Pouligny, Wavelet transform of fractals: I, From the transition to chaos to fully developed turbulence, preprint.

15. Arneodo, A., F. Argoul, E. Bacry, J. Elezgaray, E. Freysz, G. Grasseau, J. F. Muzy, and B. Pouligny, Wavelet transform of fractals: II, Optical wavelet transform of fractal growth phenomena, preprint.
16. Argoul, F., A. Arneodo, J. Elezgaray, G. Grasseau, and R. Murenzi, Wavelet transform of two-dimensional fractal aggregates, Phys. Lett. A. **135** (1989), 327–336.
17. Argoul, F., A. Arneodo, G. Grasseau, Y. Gagne, E. J. Hopfinger, and U. Frisch, Wavelet analysis of turbulence reveals the multifractal nature of the Richardson cascade, Nature **338** (1989), 51–53.
18. Auscher, P., Ondelettes à support compact et conditions aux limites, preprint.
19. Auscher, P., Ondelettes fractales et applications, Thèse de Doctorat, Univ. Paris-Dauphine, 1989.
20. Auscher, P., Symmetry properties for Wilson bases and new examples with compact support, Université de Rennes, 1990, preprint.
21. Auscher, P., Wavelet bases for $L^2(\mathbb{R})$, with rational dilation factor, in *Wavelets and Their Applications*, M. B. Ruskai, G. Beylkin, R. Coifman, I. Daubechies, S. Mallat, Y. Meyer, and L. Raphael (eds.), Jones and Bartlett, Boston, 1992, 439–451.
22. Auscher, P., Wavelets with boundary conditions on the interval, in *Wavelets: A Tutorial In Theory and Applications*, C. K. Chui (ed.), Academic Press, Cambridge, MA, 1992, 217–236.
23. Auscher, P. and Ph. Tchamitchian, Bases d'ondelettes sur les courbes corde-arc, noyau de Cauchy et espaces de Hardy associés, Rev. Mat. Iberoamericana **5** cenk (1989), 139–170.
24. Auscher, P. and Ph. Tchamitchian, Ondelettes, pseudoacerétivité, noyan de Cauchy et espaces de Hardy, preprint.
25. Auscher, P., G. Weiss, and M. V. Wickerhauser, Local sine and cosine bases of Coifman and Meyer and the construction of smooth wavelets, in *Wavelets: A Tutorial In Theory and Applications*, C. K. Chui (ed.), Academic Press, Cambridge, MA, 1992, 237–256.
26. Auslander, L. and I. Gertner, Wide-band ambiguity function and $ax + b$ group, in *Signal Processing Part I: Signal Processing Theory*, L. Auslander, F. A. Grünbaum, J. W. Helton, T. Kailath, P. Khargonekar, S. Mitter (eds.), Springer-Verlag, New York, 1990, 1–12.
27. Auslander, L. and R. Tolimieri, Radar ambiguity functions and group theory, SIAM J. Math. Anal. **16** (1985), 577–601.
28. Bacry, H., A. Grossmann, and J. Zak, Proof of the completeness of lattice states in the kq-representation, Phys. Rev. B **12** (1975), 1118–1120.
29. Bacry, E., S. Mallat, G. Papanicolaou, Time and space wavelet adaptative scheme for partial differential equations, Communication at "Wavelets and Turbulence", Princeton, June 1991.
30. Bacry, E., J. F. Muzy, and A. Arneodo, Fractal dimensions from wavelet analysis: exact results, in preparation.
31. Bandt, C., Self-similar sets 5-Integer matrices and fractal tilings of $\mathbb{R}^n$, Proc. Amer. Math. Soc. **112** (1991), 549–562.

32. Basdevant, C., M. Holschneider, and V. Perrier, Méthode des ondelettes mobiles, C. R. Acad. Sci. Paris, Série I (1990), 647–652.
33. Bastiaans, M. J., A sampling theorem for the complex spectrogram, and Gabor's expansion of a signal in Gaussian elementary signals, Opt. Eng. **20** (1981), 594–598.
34. Bastiaans, M. J., Gabor's expansion of a signal into Gaussian elementary signals, Proc. IEEE **68** (1980), 538–539.
35. Bastiaans, M. J., On the sliding-window representation in digital signal processing, IEEE Trans. ASSP **33** (1985), 868–873.
36. Battle, G., A block spin construction of ondelettes. Part I: Lemarié functions, Comm. Math. Phys. **110** (1987), 601–615.
37. Battle, G., Heisenberg proof of the Balian-Low theorem, Lett. Math. Phys. **15** (1988), 175–177.
38. Battle, G., Phase space localization theorem for ondeletts, J. Math. Phys. **30** (1989), 2195–2196.
39. Battle, G., Wavelets: a renormalization group point of view, in *Wavelets and Their Applications*, M. B. Ruskai, G. Beylkin, R. Coifman, I. Daubechies, S. Mallat, Y. Meyer, and L. Raphael (eds.), Jones and Bartlett, Boston, 1992, 323–349.
40. Battle, G., Cardinal spline interpolation and the block spin construction of wavelets, in *Wavelets: A Tutorial In Theory and Applications*, C. K. Chui (ed.), Academic Press, Cambridge, MA, 1992, 73–90.
41. Benassi A., S. Jaffard, and D. Roux, Bases d'ondelettes orthonormées pour un opérateur différentiel, 1991, preprint.
42. Benedetto, J., Gabor representations and wavelets, in *Commutative Harmonic Analysis*, D. Colella (ed.), Contemp. Math. **19** Am. Math. Soc., Providence, 1989, 9–27.
43. Benedetto, J., Stationary frames and spectral estimation, in *Probabilistic and Stochastic Advances in Analysis, with Applications*, J. S. Byrnes and J. L. Byrnes (eds.), NATO-ASI Series C, Kluwer Academic Publishers, 1991, to appear.
44. Benedetto, J., Uncertainty principle inequalities and spectrum estimation, in *Recent Advances in Fourier Analysis and its Applications*, J. S. Byrnes and J. L. Byrnes (eds.), NATO-ASI Series C, Kluwer Academic Publishers, **315**, 1990, 143–182.
45. Benedetto, J., Irregular sampling frames, in *Wavelets: A Tutorial In Theory and Applications*, C. K. Chui (ed.), Academic Press, Cambridge, MA, 1992, 445–507.
46. Benedetto, J., C. Heil, and D. Walnut, Remarks on the proof of the Balian-Low theorem, Canad. J. Math., to appear.
47. Benedetto, J. and W. Heller, Irregular sampling and the theory of frames, I, Note Math., 1991.
48. Benedetto, J. and W. Heller, Irregular sampling and the theory of frames, II, preprint.
49. Berger, M. A., Wavelets as attractors of random dynamical systems, in *Stochastic Analysis: Liber Amicorum for Moshe Zakai*, E. Mayer-Wolf,

M. Zakai, E. Merzbah, and A. Shwartz (eds.), Academic Press, NY, 1991.
50. Berger, M. A. and Y. Wang, Multi-scale dilation equations and iterated function systems, Report Math: 011491-001, Georgia Institute of Technology, preprint.
51. Berger, M. A. and Y. Wang, Multidimensional two-scale dilation equations, in *Wavelets: A Tutorial In Theory and Applications*, C. K. Chui (ed.), Academic Press, Cambridge, MA, 1992, 295-323.
52. Bertrand, J. and P. Bertrand, Time-frequency representations of broadband signals, in *Wavelets*, J. M. Combes, A. Grossmann and Ph. Tchamitchian (eds.), 1989, 164-171.
53. Beylkin, G., On the representation of operators in compactly supported wavelet bases, SIAM J. on Numerical Analysis, to appear.
54. Beylkin, G., R. Coifman, and V. Rokhlin, Fast wavelet transforms and numerical algorithms I, Comm. Pure and Appl. Math. **44** (1991), 141-183.
55. Beylkin, G., R. Coifman, and V. Rokhlin, Fast wavelet transforms and numerical algorithms II, in progress.
56. Beylkin, G., R. R. Coifman, and V. Rokhlin, Wavelets in numerical analysis, in *Wavelets and Their Applications*, M. B. Ruskai, G. Beylkin, R. Coifman, I. Daubechies, S. Mallat, Y. Meyer, and L. Raphael (eds.), Jones and Bartlett, Boston, 1992, 181-210.
57. Boashash, B., Time-frequency signal analysis, in *Advances in Spectrum Analysis and Array Processing*, S. Haykin (ed.), Prentice Hall, Englewood Cliffs, NJ, 1990, 418-517.
58. de Boor, C., *A Practical Guide to Splines*, Springer-Verlag, NY, 1978.
59. de Boor, C, R. DeVore, and A. Ron, On the construction of multivariate (pre)wavelets, 1992, preprint.
60. Bourgain J., A remark on the uncertainty principle of Hilbertian basis, J. of Funct. Anal. **79** (1988), 136-143.
61. Bovik, A., T. Emmoth, and A. Restrepo, Localized measurement of emergent image frequencies by Gabor wavelets, IEEE Trans. Inform. Theory **38** (2) (1992), to appear.
62. Buhmann, M. and C. A. Micchelli, Spline prewavelets for non-uniform knots, IBM Research Report, 1990.
63. Burt, P. J., Fast algorithms for estimating local image properties, Comput. Graphics and Image Processing **21** (1983), 368-382.
64. Burt, P. J. and E. H. Adelson, The Laplacian pyramid as a compact image code, IEEE Trans. Comm. **31** (1983), 482-540.
65. Calderón, A. P., Intermediate spaces and interpolation, the complex method, Studia Math. **24** (1964), 113-190.
66. Calderón, A. P. and A. Zygmund, Singular integral operators and differential equations, Amer. J. Math. **79** (1957), 901-921.
67. Cavaretta, A. S., W. Dahmen, and C. A. Micchelli, Stationary subdivision, Mem. Amer. Math. Soc., to appear.
68. Cenker, C., Master thesis, Kohärente reihendarstellungen von distributionen, Vienna, 1989.

69. Cenker, C., H. G. Feichtinger, and K. Gröchenig, Non-orthogonal expansions of signals and some of their applications, in *Image Acquisition and Real-Time Visualization*, 14, OEAGM Treffen, ÖCG, Bd. **56**, G. Bernroider and A. Pinz (eds.), May 1990, 129–138.
70. Chan, A. K. and C. K. Chui, Real-time signal analysis with quasi-interpolatory splines and wavelets, in *Curves and Surfaces*, P. J. Laurent, A. Le Méhauté, and L. L. Schumaker (eds.), Academic Press, 1991, 75–82.
71. Chan, A. K., C. K. Chui, J. Z. Wang, Q. Liu, and J. Zha, Introduction to B-wavelets and applications to signal processing, CAT Report #245, Texas A&M University, 1991.
72. Chui, C. K., Curve design and analysis using splines and wavelets, Trans. Eighth Army Conf. on Appl. Math. and Computing, 1990, 471–481.
73. Chui, C. K., An overview of wavelets, in *Approximation Theory and Functional Analysis*, C. K. Chui (ed.), Academic Press, Boston, MA, 1991, 47–72.
74. Chui, C. K., *An Introduction to Wavelets*, Academic Press, Boston, MA, 1992.
75. Chui, C. K., On cardinal spline-wavelets, in *Wavelets and Their Applications*, M. B. Ruskai, G. Beylkin, R. Coifman, I. Daubechies, S. Mallat, Y. Meyer, and L. Raphael (eds.), Jones and Bartlett, Boston, MA, 1992, 419–438.
76. Chui, C. K. (ed.), *Wavelets: A Tutorial in Theory and Applications*, Academic Press, Boston, 1992.
77. Chui, C. K., On wavelet analysis, in *U.S.-U.S.S.R. Workshop Vol.*, A. Gonchar and E. Saff (eds.), Springer-Verlag, 1992, to appear.
78. Chui, C. K., Wavelets–with emphasis on spline-wavelets and applications to signal analysis, in NATO Advanced Institute Studies, *Approximation Theory, Splines, and Applications*, S. Singh (ed.), to appear.
79. Chui, C. K., Wavelets: with Emphasis on Time-Frequency Analysis, in preparation.
80. Chui, C. K., Wavelets and spline interpolation, in *Advances in Numerical Analysis II: Wavelets, Subdivisions, and Radial Basis Functions*, W. Light (ed.), Oxford University Press, Oxford, 1992, 1–35.
81. Chui, C. K. and C. Li, Non-orthogonal wavelet packets, CAT Report #261, Texas A&M University, 1991.
82. Chui, C. K. and C. Li, A general framework of multivariate wavelets with duals, CAT Report #268, Texas A&M University, 1992.
83. Chui, C. K. and H. N. Mhaskar, On trigonometric wavelets, Constructive Approx., to appear.
84. Chui, C. K. and E. Quak, Wavelets on a bounded interval, in *Numerical Methods of Approximation Theory*, D. Braess and L. L. Schumaker (eds.), Birkhäuser Verlag, Basel, 1992, to appear.
85. Chui, C. K. and X. L. Shi, Inequalities of Littlewood-Paley type for frames and wavelets, SIAM J. Math. Analysis, to appear.
86. Chui, C. K. and X. L. Shi, On a Littlewood-Paley identity and characterization of wavelets, J. Math. Analysis and Applic., to appear.

87. Chui, C. K. and X. L. Shi, Characterization of scaling functions and wavelets, CAT Report #259, Texas A&M University, 1991.
88. Chui, C. K. and X. L. Shi, $N\times$ oversampling preserves any tight affine-frame for odd N, CAT Report #264, Texas A&M University, 1992.
89. Chui, C. K. and X. L. Shi, Bessel sequences and affine frames, CAT Report #267, Texas A&M University, 1992.
90. Chui, C. K. and X. L. Shi, Wavelets and multiscale interpolation, in *CAGD and Signal Analysis*, T. Lyche and L. L. Schumaker (eds.), Academic Press, 1992, 111-133.
91. Chui, C. K., J. Stöckler, and J. D. Ward, Compactly supported box-spline wavelets, CAT Report #230, Texas A&M University, 1990.
92. Chui, C. K., J. Stöckler, and J. D. Ward, Polynomial expansions for cardinal interpolants and orthonormal wavelets, in *Curves and Surfaces*, P. J. Laurent, A. Le Méhauté, and L. L. Schumaker (eds.), Academic Press, 1991, 83–90.
93. Chui, C. K. and J. Z. Wang, A cardinal spline approach to wavelets, Proc. Amer. Math. Soc. **113** (1991), 785–793.
94. Chui, C. K. and J. Z. Wang, A general framework of compactly supported splines and wavelets, J. Approx. Theory, to appear.
95. Chui, C. K. and J. Z. Wang, An analysis of cardinal-spline wavelets, J. Approx. Theory, to appear.
96. Chui, C. K. and J. Z. Wang, Computational and algorithmic aspects of cardinal-spline wavelets, CAT Report #235, Texas A&M University, 1990.
97. Chui, C. K. and J. Z. Wang, On compactly supported spline wavelets and a duality principle, Trans. Amer. Math. Soc., 1992, to appear.
98. Cohen, A., Construction de bases d'ondelettes a-Hölderiennes, Revista Matematica Iberoamericana, (1991).
99. Cohen, A., Ondelettes, analyses multirésolutions et filtres miroir en quadrature, Annales de l'IHP, analyse non linéaire **7** (5) (1990), 439–459.
100. Cohen, A., Ondelettes, Analyses multirésolutions et traitement numérique du signal, Ph. D. Thesis, Univ. Paris-Dauphine, 1990.
101. Cohen, A., Wavelets and digital signal processing, in *Wavelets and Their Applications*, M. B. Ruskai, G. Beylkin, R. Coifman, I. Daubechies, S. Mallat, Y. Meyer, and L. Raphael (eds.), Jones and Bartlett, Boston, 1992, 105–121.
102. Cohen, A., Biorthogonal wavelets, in *Wavelets: A Tutorial In Theory and Applications*, C. K. Chui (ed.), Academic Press, Cambridge, MA, 1992, 123–152.
103. Cohen, A., Wavelet bases, approximation theory, and subdivision schemes, *Approximation Theory VII*, E. W. Cheney, C. K. Chui, and L. L. Schumaker (eds.), Academic Press, New York, 1992,
104. Cohen, A. and J. P. Conze, Régularité des bases d'ondelettes et mesures ergodiques, in *Comptes rendus de l'Académie des Sciences*, Paris, (1991), to appear.

105. Cohen, A. and I. Daubechies, Orthonormal bases of compactly supported wavelets III: Better frequency localization, AT&T Bell Laboratories, preprint, submitted to SIAM J. Math. Anal.
106. Cohen, A. and I. Daubechies, A stability criterion for biorthogonal wavelet bases and their related subband coding schemes, AT&T Bell Laboratories, 1991, preprint.
107. Cohen, A. and I. Daubechies, Nonseparable bidimensional wavelet bases, AT&T Bell Laboratories, 1991, preprint.
108. Cohen, A. and I. Daubechies, Private communication.
109. Cohen, A., I. Daubechies, and J. C. Feauveau, Biorthogonal bases of compactly supported wavelets, Comm. Pure and Appl. Math., 1991, to appear.
110. Cohen, A., I. Daubechies, and P. Vial, Wavelets and fast wavelet transform on the interval, AT&T Bell Laboratories, 1991, preprint.
111. Cohen, A. and J. Johnston, Joint optimization of wavelet and impulse response constraints for biorthogonal filter pairs with exact reconstruction, AT&T Bell Laboratories, 1991, preprint.
112. Cohen, A. and J. M. Schlenker, Compactly supported wavelets with haxagonal symmetry, AT&T Bell Laboratories, 1991, preprint.
113. Cohen, L., Time-frequency distributions - a review, Proc. IEEE **77** (7) (1989), 941–981.
114. Coifman, R. R., Adaptive multiresolution analysis, computation, signal processing, and operator theory, ICM 90, Kyoto.
115. Coifman, R. R., Wavelet analysis and signal processing in *Signal Processing I: Signal Processing Theory*, L. Auslander, T. Kailath, and S. Mitter (eds.), Springer-Verlag, 1990.
116. Coifman, R. R. and Y. Meyer, Remarques sur l'analyse de Fourier à fenêtre, C. R. Acad. Sci. Paris Série I **312** (1991), 259–261.
117. Coifman, R., Y. Meyer, S. Quake, and M. V. Wickerhauser, Signal processing and compression with wave packets, in *Proc. Conf. on Wavelets*, Marseilles, Spring 1989.
118. Coifman, R., Y. Meyer, and M. V. Wickerhauser, Size properties of wavelet packets, in *Wavelets and Their Applications*, M. B. Ruskai, G. Beylkin, R. Coifman, I. Daubechies, S. Mallat, Y. Meyer, and L. Raphael (eds.), Jones and Bartlett, Boston, 1992, 453–470.
119. Coifman, R. R., Y. Meyer, and M. V. Wickerhauser, Wavelet analysis and signal processing, in *Wavelets and Their Applications*, M. B. Ruskai, G. Beylkin, R. Coifman, I. Daubechies, S. Mallat, Y. Meyer, and L. Raphael (eds.), Jones and Bartlett, Boston, 1992, 153–178.
120. Coifman, R. R. and R. Rochberg, Representation theorems for holomorphic and harmonic functions in L^p, Astérisque **77** (1980), 11–66.
121. Coifman, R. R. and M. V. Wickerhauser, Entropy-based algorithms for best basis selection, IEEE Trans. Inform. Theory **38** (2) (1992), to appear.
122. Colella, D. and C. Heil, Characterizations of scaling functions, I: continuous solutions, MITRE Corporations, McLean, Virginia, 1991.

123. Colella, D. and C. Heil, The characterization of continuous, four-coefficient scaling functions and wavelets, IEEE Trans. Inform. Theory **38** (2) (1992), to appear.
124. Combes, J. M., A. Grossmann, and Ph. Tchamitchian (eds.), *Wavelets, Time-Frequency Methods and Phase Space*, Lecture Notes on IPTI, Springer-Verlag, Berlin, New York, 1989.
125. Conze, J. P. and A. Raugi, Fonctions harmoniques pour un opérateur de transition et applications, Dept. de Math., Université de Rennes, France, 1990, preprint.
126. Dahmen, W., and A. Kunoth, Multilevel preconditioning, preprint.
127. Dahmen, W. and C. A. Micchelli, On stationary subdivision and the construction of compactly supported orthonormal wavelets, in *Multivariate Approximation and Interpolation*, W. Haußmann and K. Jetter (eds.), Birkhäuser Verlag, Basel, 1990, 69–89.
128. Dahmen, W. and C. A. Micchelli, Banded matrices with banded inverses II: locally finite decomposition of spline spaces, preprint.
129. Dahmen, W. and C. A. Micchelli, Using the refinement equations for evaluating integrals of wavelets, preprint.
130. Daubechies, I., Orthonormal bases of compactly supported wavelets, Comm. Pure and Appl. Math. **41** (1988), 909–996.
131. Daubechies, I., Orthonormal bases of compactly supported wavelets, II: variations on a theme, AT&T Bell Laboratories, 1989, preprint, SIAM J. Math. Anal., to appear.
132. Daubechies, I., The wavelet transform, time-frequency localization and signal analysis, IEEE Trans. Inform. Theory **36** (1990), 961–1005.
133. Daubechies, I., Time-frequency localization operators: a geometric phase space approach, IEEE Trans. Inform. Theory **34** (1988), 605–612.
134. Daubechies, I., *Ten Lectures on Wavelets*, CBMS/NSF Series in Applied Mathematics #61, SIAM Publ., 1992.
135. Daubechies, I. and A. Grossmann, Frames in the Bargmann space of entire functions, Comm. Pure and Appl. Math. **41** (1988), 151–164.
136. Daubechies, I., A. Grossmann, and Y. Meyer, Painless nonorthogonal expansions, J. Math. Phys. **27** (1986), 1271–1283.
137. Daubechies, I., S. Jaffard, and J. L. Journé, A simple Wilson orthonormal basis with exponential decay, SIAM J. Math. Anal. **22** (2) (1991), 554–572.
138. Daubechies, I. and J. C. Lagarias, Sets of matrices all infinite products of which converge, Linear Algebra Appl., to appear.
139. Daubechies, I. and J. C. Lagarias, Two-scale difference operations I: existence and global regularity of solutions, SIAM J. Math. Anal. **22** (1991), 1388–1410.
140. Daubechies, I. and J. C. Lagarias, Two-scale difference operations II: infinite matrix products, local regularity and fractals, SIAM J. Math. Anal., to appear.
141. Daubechies, I., and J. C. Lagarias, Two-scale difference equations, Part I and II, SIAM J. Math. Anal., (1990), to appear.

142. Daubechies, I. and T. Paul, Time-frequency localization operators - a geometric phase space approach II, Inverse Problems **4** (1988), 661–680.
143. Daubechies, I. and T. Paul, Wavelets–some applications, in *Proc. Inter. Conf. Mathematical Physics*, M. Mebkkout and R. Seneor (eds.), World Scientific, 1986, 675–686.
144. Daugman, J. G., Complete discrete 2-D Gabor transforms by neural networks for image analysis and compression, IEEE ASSP **36** (1988), 1169–1179.
145. DeVore, R., B. Jawerth, and B. Lucier, Surface compression, preprint.
146. DeVore, R., B. Jawerth, and B. L. Lucier, Image compression through wavelet transform coding, IEEE Trans. Inform. Theory **38** (2) (1992), to appear.
147. DeVore, R., B. Jawerth, and V. Popov, Compression of wavelet decompositions, preprint.
148. Delprat, N., et al., Asymptotic wavelet and Gabor analysis: extraction of instantaneous frequencies, IEEE Trans. Inform. Theory **38** (2) (1992), to appear.
149. Doganata, Z., P. P. Vaidyanathan, and T. Q. Nguyen, General synthesis procedures for FIR lossless transfer matrices, for perfect reconstruction multirate filter bank applications, IEEE Trans. ASSP **36** (1988), 1561–1574.
150. Dubuc, S., Interpolation through an iterative scheme, J. Math. Anal. and Appl. **114** (1986), 185–204.
151. Dufaux, F. and M. Kunt, Massively parallel implementation for real-time Gabor decompositions, in *SPIE Conf. in Visual Comm. and Image Processing*, Boston, November 1991,
152. Duffin, R. J. and A. C. Schaeffer, A class of nonharmonic Fourier series, Trans. Amer. Math. Soc. **72** (1952), 341–366.
153. Dutilleux, P., An implementation of the 'algorithme 'a trous' to compute the wavelet transform, in *Wavelets, Time-Frequency Methods and Phase Space*, J. M. Combes, A. Grossmann, and Ph. Tchamitchian (eds.), Springer-Verlag, Berlin, New York, 1989.
154. Duval-Destin, M., M. A. Muschietti, and B. Torrésani, From continuous wavelets to wavelet packets, in preparation.
155. Eirola, T., Sobolev characterization of solutions of dilation equations, Helsinki University of Technology, submitted to SIAM J. Math. Anal. , 1991.
156. Escudié, B., Wavelet analysis of asymptotic signals: A tentative model for bat sonar receiver, in *Wavelets and Applications*, Y. Meyer (ed.), Mason/Springer, to appear.
157. Escudié, B. and B. Torrésani, Wavelet analysis of asymptotic signals, Centre de Physique Théorique, CNRS-Luminy, 1989, preprint.
158. Escudié, B. and B. Torrésani, Wavelet representation and time-scaled matched receiver for asymptotic signals, *Proc. Conf. EUSIPCO V*, Barcelona, North Holland, 1990, 305–308.

159. Esteban, D. and C. Galand, Applications of quadrature mirror filters to split band voice coding schemes, in *Proc. ICASSP*, Hartford, Connecticut, 1977, 191–195.
160. Farge, M., The continuous wavelet transform of two-dimensional turbulent flows, in *Wavelets and Their Applications*, M. B. Ruskai, G. Beylkin, R. Coifman, I. Daubechies, S. Mallat, Y. Meyer, and L. Raphael (eds.), Jones and Bartlett, Boston, 1992, 275–302.
161. Farge, M., Transformée en ondelettes et application á la turbulence, J. Annuelle, Societé Mathématique de France, Mai 1990.
162. Farge, M. and D. Rabreau, Transformée en ondelettes pour détecter et analyser les structures cohérentes dans les écoulements turbulents bidimensionnels, C. R. Acad. Sci. **307** Série 2 (1988), 1479–1486.
163. Farge, M., Y. Guezennec, C. M. Ho, and C. Nineveau, Continuous wavelet analysis of coherent structures, Proceedings, Summer Program, Center for Turbulence Research, Stanford University and NASA-Ames, 1990.
164. Farge, M., M. Holschneider, and J. F. Colonna, Wavelet analysis of coherent structures in two-dimensional turbulent flows, in *Topological Fluid Mechanics*, Moffatt (ed.), Cambridge Univ. Press, 1990.
165. Feauveau, J. C., Analyse multirésolution par ondelettes non orthogonales et bancs de filtres numériques, Ph. D. Thesis, Univ. Paris-Sud, 1990.
166. Feauveau, J. C., Wavelets for Quincunx pyramid, in *Wavelets and Their Applications*, M. B. Ruskai, G. Beylkin, R. Coifman, I. Daubechies, S. Mallat, Y. Meyer, and L. Raphael (eds.), Jones and Bartlett, Boston, 1992, 53–66.
167. Feauveau, J. C., Nonorthogonal multiresolution analysis using wavelets, in *Wavelets: A Tutorial In Theory and Applications*, C. K. Chui (ed.), Academic Press, Cambridge, MA, 1992, 153–178.
168. Feauveau, J. C., P. Mathieu, M. Barlaud, and M. Antonini, Recursive biorthogonal wavelet transform for image coding, in *Proc. Int. Conf. Acoust., Speech and Signal Processing*, Toronto, Canada, 1991, 2649–2652.
169. Federbush, P., Navier and Stokes meet the wavelet, 1991, preprint.
170. Feichtinger, H., Atomic characterizations of modulation spaces through Gabor-type representations, *Proc. Conf. in Constructive Function Theory*, Edmonton, July 1986, Rocky Mount. J. Math. **19** (1989), 113–126.
171. Feichtinger, H., Coherent frames and irregular sampling, *Proc. Conf. in Recent Advances in Fourier Analysis and Its Applications*, NATO conference, PISA, July 1989, J. S. Byrnes and J. L. Byrnes (eds.), Kluwer Acad. Publ., NATO ASI Series C **315** (1990), 427–440.
172. Feichtinger, H. and K. Gröchenig, A unified approach to atomic decompositions via integrable group representations. *Proc. Conf. in Function Spaces and Applications*, Lund, 1986, Lecture Notes in Math. **1302** (1988), 52–73.
173. Feichtinger, H. and K. Gröchenig, Banach spaces related to integrable group representations and their atomic decompositions I, J. Funct. Anal. **86** (1989), 307–340.

174. Feichtinger, H. and K. Gröchenig, Banach spaces related to integrable group representations and their atomic decompositions II, Monatsh. f. Math. **108** (1989), 129–148.
175. Feichtinger, H. and K. Gröchenig, Error analysis in regular and irregular sampling theory, Applicable analysis, to appear.
176. Feichtinger, H. and K. Gröchenig, Iterative reconstruction of multivariate band-limited functions from irregular sampling values, SIAM J. Math. Anal., to appear.
177. Feichtinger, H. and K. Gröchenig, Non-orthogonal wavelet and Gabor expansions, and group representations, in *Wavelets and Their Applications*, M. B. Ruskai, G. Beylkin, R. Coifman, I. Daubechies, S. Mallat, Y. Meyer, and L. Raphael (eds.), Jones and Bartlett, Boston, 1992, 353–375.
178. Feichtinger, H., K. Gröchenig, and D. Walnut, Wilson bases and modulation spaces, Math. Nachrichten, 1991, to appear.
179. Feichtinger, H. G. and K. Gröchenig, Gabor wavelets and the Heisenberg group: Gabor expansions and short time Fourier transform from the group theoretical point of view, in *Wavelets: A Tutorial In Theory and Applications*, C. K. Chui (ed.), Academic Press, Cambridge, MA, 1992, 359–397.
180. Flandrin, P., Some aspects of non-stationary signal processing with emphasis on time-frequency and time-scale methods, in *Wavelets: Time-frequency Methods and Phase Space*, J. M. Combes, A. Grossmann, and Ph. Tchamitchian (eds.), Springer-Verlag, New York, 1989, 68–98.
181. Flandrin, P., Wavelet analysis and synthesis of fractional Brownian motion, IEEE Trans. Inform. Theory **38** (2) (1992), to appear.
182. Folland, G. B., *Harmonic Analysis in Phase Space*, Princeton University Press, Princeton, NJ, 1989.
183. Frazier, M. and B. Jawerth, Decomposition of Besov spaces, Indiana Univ. Math. J. **34** (1985), 777–799.
184. Frazier, M. and B. Jawerth, A discrete transform and decomposition of distribution spaces, J. Func. Anal. **93** (1990), 34–170.
185. Frazier, M. and B. Jawerth, Applications of the ϕ and Wavelet Transforms to the theory of function spaces, in *Wavelets and Their Applications*, M. B. Ruskai, G. Beylkin, R. Coifman, I. Daubechies, S. Mallat, Y. Meyer, and L. Raphael (eds.), Jones and Bartlett, Boston, 1992, 377–417.
186. Frazier, M. and B. Jawerth, The ϕ-transform and applications to distribution spaces, in *Function Spaces and Application*, M. Cwikel et al. (eds.), Lecture Notes in Mathematics 1302, Springer, Berlin, 1988, 233–246.
187. Frazier, M., B. Jawerth, and G. Weiss, *Littlewood-Paley Theory and the Study of Function Spaces*, CBMS Regional Conference Series 79, AMS, Providence, RI, 1991.
188. Frisch, M. and H. Messer, The use of wavelet transform in the detection of an unknown transient signal, IEEE Trans. Inform. Theory **38** (2) (1992), to appear.
189. Froment, J. and S. Mallat, Second generation compact image coding with wavelets, in *Wavelets: A Tutorial In Theory and Applications*, C. K. Chui

(ed.), Academic Press, Cambridge, MA, 1992, 655–678.
190. Gabor, D., Theory of communication, J. IEE (London) **93** (1946), 429–457.
191. Glowinski R., W. Lawton, M. Ravachol, and E. Tenenbaum, Wavelet solution of linear and nonlinear elliptic, parabolic, and hyperbolic problems in one space dimension, in *Proc. Ninth Inter. Conf. on Computing Methods in Appl. Sciences and Engineering*, R. Glowinski and A. Lichnewsky (eds.), SIAM Publ., Philadelphia, PA, 1990.
192. Gopinath, R. A., Moment theorem for the wavelet-Galerkin method, Technical Report, Aware, Inc., Cambridge, MA, 1990.
193. Gopinath, R. A., Wavelet transforms and time-scale analysis of signals, Master Thesis, Rice University, 1990.
194. Gopinath, R. A. and C. S. Burrus, Wavelet transforms and filter banks, in *Wavelets: A Tutorial In Theory and Applications*, C. K. Chui (ed.), Academic Press, Cambridge, MA, 1992, 603–654.
195. Gopinath, R. A., W. M. Lawton, and C. S. Burrus, Wavelet-Galerkin approximation of linear translation invariant operators, in Proc. ICASSP, Toronto, Canada **3** (1991) 2021–2023.
196. Goodman, T. N. T., S. L. Lee, and W. S. Tang, Wavelets in wandering subspaces, CAT Report #256, Texas A&M University, 1991.
197. Gröchenig, K., Analyse multiéchelles et bases d'ondelettes, *CRAS Paris, Série 1* **305** (1987), 13–15.
198. Gröchenig, K., Describing functions: atomic decompositions versus frames, Monatsh. Math. **112** (1991), 1–42.
199. Gröchenig, K. and W. R. Madych, Multiresolution analyses, Haar bases, and self-similar tilings of R^n, IEEE Trans. Inform. Theory **38** (2) (1992), to appear.
200. Gröchenig, K. and D. Walnut, A Riesz basis for Bargmann-Fock space related to sampling and interpolation, Ark. f. Mat., to appear.
201. Grossmann, A., M. Holschneider, R. Kronland-Martinet, and J. Morlet, Detection of abrupt changes in sound signals with the help of wavelet transforms, in *Inverse problems: An interdisciplinary study; Advances in electronics and electron physics*, Supplement nr. 19, Academic Press, 1987.
202. Grossmann, A., R. Kronland-Martinet, and J. Morlet, Reading and understanding continuous wavelet transforms, in *Wavelets*, J. M. Combes, A. Grossmann, and Ph. Tchamitchian (eds.), Springer-Verlag, NY, 1989, 2–20.
203. Grossmann, A. and J. Morlet, Decomposition of Hardy functions into square integrable wavelets of constant shape, SIAM J. Math. Anal. **15** (1984), 723–736.
204. Grossmann, A. and J. Morlet, Decomposition of functions into wavelets of constant shape, and related transforms, in *Mathematics and Physics, Lectures on Recent Results*, L. Streit (ed.), World Scientific Publ., Singapore, 1985.

205. Grossmann, A., J. Morlet, and T. Paul, Transforms associated to square integrable group representations, I, J. Math. Phys. **26** (1985), 2473–2479
206. Grossmann, A., J. Morlet, and T. Paul, Transforms associated to square integrable group representations, II, Annales l'IHP, Phys. Theor. **45** (1986), 293–309.
207. Grossmann, A., G. Saracco, and Ph. Tchamitchian, Study of propagation of transient acoustic signals across a plane interface with the help of the wavelet transform, preprint.
208. Haar, A., Zur Theorie der orthogonalen Funktionen-systeme, Math. Ann. **69** (1910), 331–371.
209. Healy, D. M. and J. B. Weaver, Two applications of wavelet transforms in magnetic resonance imaging, IEEE Trans. Inform. Theory **38** (2) (1992), to appear.
210. Heil, C. and D. Walnut, Continuous and discrete wavelet transforms, SIAM Review **31** (1989), 628–666.
211. Heller, W., Frames of exponentials and applications, Ph. D. Thesis, University of Maryland, College Park, MD, 1991.
212. Heller, P. N., Higher multiplier Daubechies scaling coefficients, Technical Report, Aware, Inc., Cambridge, MA.
213. Heller, P., H. L. Resnikoff, and R. O. Wells, Jr., Wavelet matrices and the representation of discrete functions, in *Wavelets: A Tutorial In Theory and Applications*, C. K. Chui (ed.), Academic Press, Cambridge, MA, 1992, 15–50.
214. Herley, C. and M. Vetterli, Linear phase wavelets: theory and design, in *Proc. IEEE Int. Conf. ASSP*, Toronto, Canada, 1991, 2017–2020.
215. Herley, C. and M. Vetterli, Wavelets and recursive filter banks, Technical Report, Columbia University, 1992, submitted to IEEE Trans. Signal Processing.
216. Hervé, L., Ph. D. Thesis, University of Rennes I, France, 1991.
217. Holschneider, M. R. Kronland-Marttinet, J. Morlet, and Ph. Tchamitchian, A real-time algorithm for signal analysis with the help of the wavelet transform, in *Wavelets, Time-Frequency Methods and Phase Space*, J. M. Combes, A. Grossmann, and Ph. Tchamitchian, eds., Lecture Notes on IPTI, Springer-Verlag, Berlin, New York, 1989, 286–297.
218. Holschneider, M. R. and Ph. Tchamitchian, On the wavelet analysis of the Riemann's function, Centre de Physique Théorique, CNRS-Luminy, 1988, preprint.
219. Hummel, R. and R. Moniot, Reconstruction from zero-crossing in space-space, in IEEE Trans. ASSP **37** (1989).
220. Jaffard, S., Analyse par ondelettes d'un problème elliptique singulier, J. Math. Pures et Appl., to appear.
221. Jaffard, S., Construction et propriétés des bases d'ondelettes, Remarques sur la controlabilité exacte, Thèse de l'école Polytechnique, 1989.
222. Jaffard, S., Construction of wavelets on open sets, in *Wavelets*, J. M. Combes, A. Grossmann, and Ph. Tchamitchian (eds.), Springer-Verlag, Berlin, 1989, 247–252.

223. Jaffard, S., Exposants de Hölder en des points donnés et coefficients d'ondelettes, C. R. Acad. Sci. **308** Série 1 (1989), 79–81.
224. Jaffard, S., Local order of approximation by wavelets, 1991, preprint.
225. Jaffard, S., Pointwise smoothness, two-microlocalization and wavelet coefficients, Publicacions Matematiques **35** (1991), 155–168.
226. Jaffard, S., Propriétés des matrices "bien localisées" près de leur diagonale et quelques applications, Annales de l'IHP, série Problèmes non linéaires **7** (5) (1990), 461–476.
227. Jaffard, S., Sur la dimension de Hausdorff des points singuliers d'une fonction, 1991, preprint.
228. Jaffard, S., Wavelet methods for fast resolution of elliptic problems, SIAM J. Numer. Anal. , to appear.
229. Jaffard, S. and Ph. Laurençot, Orthonormal wavelets, analysis of operators, and applications to numerical analysis, in *Wavelets: A Tutorial In Theory and Applications*, C. K. Chui (ed.), Academic Press, Cambridge, MA, 1992, 543–601.
230. Jaffard,, S., P. G. Lemarié, S. Mallat, and Y. Meyer, Dyadic multiscale analysis of $L^2(R^n)$, manuscript.
231. Jaffard, J. and Y. Meyer, Bases d'ondelettes dans des ouverts de $\mathbb{R}^n$, J. Math. Pures et Appl. **68** (1989), 95–108.
232. Janssen, A. J. E. M., Gabor representation of generalized functions, J. Math. Anal. Appl. **83** (1981), 377–394.
233. Janssen, A. J. E. M., The Smith-Barnwell condition and non-negative scaling functions, IEEE Trans. Inform. Theory **38** (2) (1992), to appear.
234. Jensen, H. E., T. Hoholdt, and J. Justesen, Double series representation of bounded signals, IEEE Trans. Inform. Theory **34** (1988), 613–624.
235. Jia, R.-Q. and C. A. Micchelli, Using the refinement equations for the construction of pre-wavelets II: powers of two, in *Curves and Surfaces*, P.-J. Laurent, A. Le Méhauté, and L. L. Schumaker (eds.), Academic Press, San Diego, 1991, 209–246.
236. Jia, R.-Q. and C. A. Micchelli, Using the refinement equations for the construction of pre-wavelets V: extensibility of trigonometric polynomials, 1991, preprint.
237. Kadambe, S. and G. F. Boudreaux-Bartels, Application of the wavelet transform for pitch detection of speech signals, IEEE Trans. Inform. Theory **38** (2) (1992), to appear.
238. Kaiser, G., An algebraic theory of wavelets, Part I: Operational calculus and complex structure, SIAM J. Math. Anal. **23** (1) (1992), to appear.
239. Kaiser, G., Generalized wavelet transforms, Part I: The windowed X–Ray transform, Technical Reports Series #18, University of Lowell, 1990.
240. Kaiser, G., Generalized wavelet transforms, Part II: The multivariate analytic–signal transform, Technical Reports Series #19, University of Lowell, 1990.
241. Kaiser, G., Phase–Space approach to relativistic quantum mechanics, Ph. D. Thesis, Mathematics Department, University of Toronto, 1977.

242. Kaiser, G. and R. R. Streater, Windowed radon transforms, analytic signals, and the wave equation, in *Wavelets: A Tutorial In Theory and Applications*, C. K. Chui (ed.), Academic Press, Cambridge, MA, 1992, 399–441.
243. Klauder, J. and K. Skagerstam (eds.), *Coherent States – Applications in Physics and Mathematical Physics*, Singapore, World Scientific Publ., 1985.
244. Kovaǎević, J. and M. Vetterli, Nonseparable multidimensional perfect reconstruction filter banks and wavelet bases for $\mathbb{R}^n$, IEEE Trans. Inform. Theory **38** (2) (1992), to appear.
245. Kronland-Martinet, R., J. Morlet, and A. Grossmann, Analysis of sound patterns through wavelet transforms, Int. J. of Pattern Recognition and Artificial Intelligence **1** (1987), 273–301.
246. Laeng, E., Une base orthonormale de $L^2(\mathbb{R})$, dont les éléments sont bien localisés dans l'espace de phase et leurs supports adaptés à toute partition symétrique de l'espace des fréquences, C. R. Acad. Sci. Paris **311** (1990), 677–680.
247. Landau, H., An overview of time and frequency limiting, in *Fourier Techniques and Applications*, J. Price (ed.), Plenum Press, NY, 1985, 201-220.
248. Latto, A., and E. Tenenbaum, Compactly supported wavelets and the numerical solution of the Burgers' equation, C. R. Acad. Sci. Paris **311** Série I (1990), 903-909.
249. Laurençot, Ph., Résolution par ondelettes 1-D de l'équation de Burgers avec viscosité et de l'équation de Korteweg-De Vries avec conditions limites périodiques, Rapport de DEA, Université Paris XI, 1990.
250. Lawton, W. M., Lectures on wavelets, *CBMS Conference Series*, Lowell University, 1990.
251. Lawton, W. M., Necessary and sufficient conditions for constructing orthonormal wavelets bases, J. Math. Phys. **32** (1) (1991), 57–61.
252. Lawton, W. M., Tight frames of compactly supported affine wavelets, J. Math. Phys. **31** (8) (1990), 1898–1901.
253. Lawton, W. M. and H. L. Resnikoff, Multidimensional wavelet bases, Aware, Inc. Technical Report, submitted to SIAM J. Math. Anal.
254. Lemarié, P. G., Bases d'ondelettes sur les groupes de Lie stratifiés, Bull. Soc. Math. France, to appear.
255. Lemarié, P. G., Constructions d'ondelettes-splines, 1987, unpublished.
256. Lemarié, P. G., Ondelettes à localisation exponentielles, J. Math. Pures et Appl. **67** (1988), 227–236.
257. Lemarié, P. G., *Les ondelettes en 1989*, Lecture notes in Mathematics nr. 1438, Springer, Berlin, 1990.
258. Lemarié, P. G., Existence de fonction-père pour des ondelettes à support compact, C. R. Acad. Sci. Paris, to appear.
259. Lemarié, P. G. and G. Malgouyres, Support des fonctions de base dans une analyse multirésolution, C. R. Acad. Sci. Paris, to appear.
260. Lemarié, P. and Y. Meyer, Ondelettes et bases Hilbertiennes, Rev. Mat. Iberoamericana **2** (1986), 1–18.

261. Liandrat, J., V. Perrier, and Ph. Tchamitchian, Numerical resolution of the regularized Burgers equation using the wavelet transform, 1990, preprint.
262. Liandrat, J., V. Perrier, and Ph. Tchamitchian, Numerical resolution of nonlinear partial differential equations using the wavelet approach, in *Wavelets and Their Applications*, M. B. Ruskai, G. Beylkin, R. Coifman, I. Daubechies, S. Mallat, Y. Meyer, and L. Raphael (eds.), Jones and Bartlett, Boston, 1992, 227-238.
263. Liandrat, J. and Ph. Tchamitchian, Resolution of the 1D regularized Burgers equation using a spatial wavelet approximation, ICASE Report #90-83, NASA, 1990.
264. Light, W. (ed.), *Wavelets, Subdivisions, and Radial Functions*, Oxford University Press, Oxford, 1991.
265. Littlewood, J. E. and R. E. A. C. Paley, Theorems on Fourier series and power series, I, J. London Math. Soc. **6** (1931), 230-233.
266. Littlewood, J. E. and R. E. A. C. Paley, Theorems on Fourier series and power series, II, Proc. London Math. Soc. **42** (1936), 52-89.
267. Lorentz, R. A. H. and W. R. Madych, Spline wavelets for ordinary differential equations, GMD Technical Report.
268. Lorentz, R. A. H. and W. R. Madych, Wavelets and generalized box splines, GMD Technical Report, Applicable Analysis, to appear.
269. Lyubarskii, Y., Frames in the Bargmann Space of Entire Functions, *Proc. of Seminar on Complex Analysis*, Lecture Notes in Math. #15, Springer-Verlag, NY, to appear.
270. Madych, W. R., Multiresolution analyses, tiles, and scaling functions, in *Proc. of NATO ASI II Ciocco*, 1991, to appear.
271. Madych, W. R., Multiresolution analysis, wavelets, and homogeneous functions, BRC, preprint.
272. Madych, W. R., Polyharmonic splines, multiscale analysis, and entire functions, in *Multivariate Approximation and Interpolation*, W. Haussmann and K. Jetter (eds.), Birkhäuser Verlag, Basel, 1990, 205-216.
273. Madych, W. R., Translation invariant multiscale analysis, in *Recent Advances in Fourier Analysis and its Applications*, J. S. Byrnes and J. L. Byrnes (eds.), NATO ASI Series C **315**, Kluwer, Dordrecht, 1990, 455-462.
274. Madych, W. R., Some elementary properties of multiresolution analyses of $L^2(\mathbb{R}^n)$, in *Wavelets: A Tutorial In Theory and Applications*, C. K. Chui (ed.), Academic Press, Cambridge, MA, 1992, 259-294.
275. Mallat, S., A theory of multiresolution signal decomposition: the wavelet representation, IEEE Trans. PAMI **11** (1989), 674-693.
276. Mallat, S., Multifrequency channel decompositions of images and wavelet models, IEEE Trans. ASSP **37** (1989), 2091-2110.
277. Mallat, S., Multiresolution approximations and wavelet orthonormal bases of $L^2(\mathbf{R})$, Trans. Amer. Math. Soc. **315** (1989), 69-87.
278. Mallat, S., Multiresolution representation and wavelets, Ph. D. Thesis, Univ. of Pennsylvania, Philadelphia, 1988.

279. Mallat, S., Review of multifrequency channel decomposition of images and wavelet models, Technical Report 412, Robotics Report 178, NYU, 1988.
280. Mallat, S., Zero-crossings of a wavelet transform, IEEE Trans. Inform. Theory **37** (1991), 1019–1033.
281. Mallat, S. and W. L. Hwang, Singularity detection and processing with wavelets, IEEE Trans. Inform. Theory **38** (2) (1992), to appear.
282. Mallat, S. and S. Zhong, Complete signal representation with multiscale edges, Computer Science Technical Report #483, New York University, 1989, submitted to IEEE Trans. PAMI.
283. Mallat, S. and S. Zhong, Characterization of signals from multiscale edges, IEEE Trans. PAMI (1991), to appear.
284. Mallat, S. and S. Zhong, Wavelet transform maxima and multiscale edges, in *Wavelets and Their Applications*, M. B. Ruskai, G. Beylkin, R. Coifman, I. Daubechies, S. Mallat, Y. Meyer, and L. Raphael (eds.), Jones and Bartlett, Boston, 1992, 67–104.
285. Malvar, H., Lapped transforms for efficient transform/subband coding, IEEE Trans. ASSP **38** (1990), 969–978.
286. Marks, R., *Introduction to Shannon Sampling and Interpolation Theory*, Springer-Verlag, NY, 1991.
287. Marr, D. and E. Hildreth, Theory of edge detection, Proc. Roy. Soc. London B **207** (1980), 187–217.
288. Meyer, Y., Ondelettes et calcul scientifique performant, INRIA, 1989.
289. Meyer, Y., Ondelettes et fonctions splines, Technical Report, *Séminaire EDP*, École Polytechnique, Paris, 1986.
290. Meyer, Y., *Ondelettes et Operateurs*, in two volumes, Hermann, Paris, 1990.
291. Meyer, Y., Ondelettes, fonctions splines et analyses graduées, Rapport CEREMADE **8703**, 1987.
292. Meyer, Y., Ondelettes sur l'intervalle, Rev. Mat. Iberoamericana, to appear.
293. Meyer, Y. (ed.), *Proc. Inter. Conf. on Wavelets*, May 1989, Marseille (France); to be published by Masson (Paris), (1991).
294. Meyer, Y., Principe d'incertitude, bases Hilbertiennes et algèbres d'opérateurs, Séminaire Bourbaki **662** (1985-1986).
295. Meyer, Y., Séminaire Bourbaki **38** (1985-86), 662.
296. Meyer, Y., The Franklin Wavelets, 1988, preprint.
297. Meyer, Y., (ed.), Wavelets and applications, Proceedings of the second wavelet conference, Marseille (1989), Masson/Springer, to appear.
298. Meyer, Y., Wavelets and applications, in *Proc. of the Int. Congress of Math.*, Kyoto, 1990.
299. Meyer, Y., Wavelets and operators, in *Analysis at Urbana*, Vol. 1, E. Berkson, N. T Peck, and J. Uhl (eds.), London Math Society, Lecture Notes Series 137, Cambridge Univ.Press, 1989, 256–364.
300. Meyer, Y. and R. R. Coifman, *Ondelettes et Opérateurs III–Opérateurs Multilinéaures*, Hermann, Paris, 1991.

301. Micchelli, C. A., Using the refinement equation for the construction of pre- wavelets, Numerical Algorithms **1** (1991), 75–116.
302. Micchelli, C. A. and H. Prautzsch, Uniform refinement of curves, Lin. Alg. & Appl. **114/115** (1989), 841–870.
303. Munch, J., Noise reduction in tight Weyl-Heisenberg frames, IEEE Trans. Inform. Theory **38** (2) (1992), to appear.
304. Murenzi, R., Ondelettes multidimensionelles et application á l'analyse d'images, Ph. D. Thesis, Université Catholique de Louvain, 1990.
305. Nawab, S. H. and T. Quatieri, Short-time fourier transform, in *Advanced Topics in Signal Processing*, J. S. Lim and A. V. Oppenheim (eds.), Prentice Hall Signal Processing Series, 1988.
306. Olsen, P. A. and K. Seip, A note on irregular discrete wavelet transforms, IEEE Trans. Inform. Theory **38** (2) (1992), to appear.
307. Paul, T., Ondelettes et mécanique quantique, Ph. D. Thesis, Université de Marseille, 1985.
308. Paul, T. and K. Seip, Wavelets and quantum mechanics, in *Wavelets and Their Applications*, M. B. Ruskai, G. Beylkin, R. Coifman, I. Daubechies, S. Mallat, Y. Meyer, and L. Raphael (eds.), Jones and Bartlett, Boston, 1992, 303–321.
309. Perrier, V., Ondelettes et simulation numériques, Thèse de l'Université Paris VI, February 1991.
310. Pollen, D., Parametrization of compactly supported wavelets, Aware Technical Report, Aware, Inc., Cambridge, MA.
311. Pollen, D., $SU_I(F[z, 1/z])$ for F a subfield of $\mathbb{C}$, J. Amer. Math. Soc. **3** (3) (1990), 611–624.
312. Pollen, D., Daubechies' scaling function on [0,3], in *Wavelets: A Tutorial In Theory and Applications*, C. K. Chui (ed.), Academic Press, Cambridge, MA, 1992, 3–13.
313. Porat, M. and Y. Y. Zeevi, The generalized Gabor scheme of image representation using elementary functions matched to human vision, in *Theory Foundations of Computer Graphics*, Springer Verlag, NY, 1197–1241.
314. Portnoff, M. R., Time frequency distributions of digital systems and signals based on short-time fourier analysis, IEEE Trans. ASSP **28** (1980), 55–69.
315. Quake, S., R. Coifman, Y. Meyer, and M. V. Wickerhauser, Signal processing and compression with wave packets, preprint.
316. Resnikoff, H. L., Weierstrass functions and compactly supported wavelets. Technical Report AD900810, Aware, Inc., Cambridge, MA, 1990.
317. Resnikoff, H. L. and R. O. Wells, Jr., *Wavelet Analysis*, in preparation.
318. de Rham, G., Sur une courbe plane, J. Math. Pures et Appl. **39** (1956), 25–42.
319. Riemenschneider, S. D. and Z. W. Shen, Wavelets and pre-wavelets in low dimensions, J. Approx. Theory, to appear.
320. Riemenschneider, S. D. and Z. W. Shen, Box splines, cardinal series, and wavelets, in *Approximation Theory and Functional Analysis*, C. K. Chui (ed.), Academic Press, NY, 1991, 133–149.

321. Rioul, O., A discrete-time multiresolution theory unifying octave band filter banks, Pyramid and wavelet transforms, IEEE Trans. Signal Proccessing, to appear.
322. Rioul, O., A unifying multiresolution theory for discrete wavelet transform, regular filter banks and pyramid transforms, IEEE Trans. ASSP, (1990), submitted.
323. Rioul, O. and P. Duhamel, Fast algorithms for discrete and continuous wavelet transforms, IEEE Trans. Infor. Theory 38 (2) (1992), to appear.
324. Ruskai, M. B., G. Beylkin, R. Coifman, I. Daubechies, S. Mallat, Y. Meyer, and L. Raphael (eds.), *Wavelets and Their Applications*, Jones and Bartlett, Boston, 1992.
325. Schoenberg, I. J., *Cardinal Spline Interpolation* CBMS/NSF Series in Applied Mathematics #12. SIAM Publ., 1973.
326. Schumaker, L. L., *Spline Functions: Basic Theory*, Wiley-Interscience, NY, 1981.
327. Schweinler, H. C. and E. P. Wigner, Orthogonalization methods, J. Math. Phys. 11 (1970), 1693–1694.
328. Seip, K., Mean value theorems and concentration operators in Bargmann and Bergman spaces, in *Wavelets*, J. M. Combes, A. Grossmann, and Ph. Tchamitchian (eds.), Springer-Verlag, NY, 1989, 209–215.
329. Seip, K., Reproducing formulas and double orthogonality in Bargmann and Bergman spaces, SIAM J. Math. Anal. 22 (1991), 856–876.
330. Seip, K., Wavelets in $H^2(\mathbb{R})$: Sampling, interpolation, and phase space density, in *Wavelets: A Tutorial In Theory and Applications*, C. K. Chui (ed.), Academic Press, Cambridge, MA, 1992, 529–540.
331. Smith, M. J. T. and T. P. Barnwell, III, Exact reconstruction techniques for tree-structured subband coders, IEEE Trans. ASSP 34 (1986), 434–441.
332. Stöckler, J., The construction of box-spline wavelets in arbitrary dimension, preprint.
333. Stöckler, J., Multivariate wavelets, in *Wavelets: A Tutorial In Theory and Applications*, C. K. Chui (ed.), Academic Press, Cambridge, MA, 1992, 325–355.
334. Strömberg, J.-O., A modified Franklin system and higher order spline systems on $\mathbf{R}^n$ as unconditional bases for Hardy spaces, in *Proc. Conf. in Honor of Antoni Zygmund*, Vol. II, W. Beckner, A. P. Calderón, R. Fefferman, and P. W. Jones (eds.), Wadsworth, NY, 1981, 475–493.
335. Strang, G., Wavelets and dilation equations: a brief introduction, SIAM Review 31 (1989), 614–627.
336. Strichartz, R. S., Wavelets and self-affine tilings, preprint.
337. Super, B. J. and A. C. Bovik, Shape from texture using Gabor wavelets, *SPIE Conf. in Visual Comm. and Image Processing*, Boston, November 1991.
338. Tchamitchian, Ph., Bases d'ondelettes et intégrales singulières: Analyse des fonctions et calcul sur les opérateurs, Habilitation, Université d'Aix-Marseille 2, 1989.

339. Tchamitchian, Ph., Biorthogonalité et théorie des opérateurss, Rev. Math. Iberoamericana, to appear.
340. Tchamitchian, Ph., Ondelettes et intégrale de Cauchy sur les Courbes Lipschitziennes, Annals of Math. **303** (1986), 215–218.
341. Tchamitchian, Ph., Calcul symbolique sur les opérateurs de Caldéron-Zygmund et bases inconditionnelles de $L^2(\mathbb{R}^n)$, C. R. Acad. Sci. Paris **129** (1989), 641–649.
342. Tchamitchian, Ph. and B. Torrésani, Ridge and skeleton extraction from the wavelet transform, in *Wavelets and Their Applications*, M. B. Ruskai, G. Beylkin, R. Coifman, I. Daubechies, S. Mallat, Y. Meyer, and L. Raphael (eds.), Jones and Bartlett, Boston, 1992, 123–151.
343. Tewfik, A. and M. Kim, Correlation structure of the discrete wavelet coefficients of fractional Brownian motion, IEEE Trans. Inform. Theory **38** (2) (1992), to appear.
344. Tewfik, A., D. Sinha, and P. Jorgensen, On the optimal choice of a wavelet for signal representation, IEEE Trans. Inform. Theory **38** (2) (1992), to appear.
345. Torrésani, B., Wavelet analysis of asymptotic signals: Ridge and Skeleton of the transform, *Wavelets and Applications*, Y. Meyer (ed.), Masson/Springer, to appear.
346. Torrésani, B., Wavelets associated with representations of the affine Weyl-Heisenberg group, J. Math Phys., to appear.
347. Torrésani, B., Time-frequency representations: wavelet packets and optimal decompositions, Ann. Inst. H. Poincaré, to appear.
348. Unser, M., A. Aldroubi, and M. Eden, On the asymptotic convergence of B-spline wavelets to Gabor functions, IEEE Trans. Inform. Theory **38** (2) (1992), to appear.
349. Unser, M. and A. Aldroubi, Polynomial splines and wavelets–a signal processing perspective, in *Wavelets: A Tutorial In Theory and Applications*, C. K. Chui (ed.), Academic Press, Cambridge, MA, 1992, 91–122.
350. Vetterli, M., Wavelets and filter banks for discrete-time signal processing, in *Wavelets and Their Applications*, M. B. Ruskai, G. Beylkin, R. Coifman, I. Daubechies, S. Mallat, Y. Meyer, and L. Raphael (eds.), Jones and Bartlett, Boston, 1992, 17–52.
351. Vetterli, M. and C. Herley, Wavelets and filter banks: theory and design, IEEE ASSP (1992), to appear.
352. Vetterli, M. and K. Herley, Linear phase wavelets, IEEE Trans. ASSP, to appear.
353. Vetterli, M., J. Kovaăević and Le Gall, Perfect reconstruction filter banks for HDTV representation and coding, Image Communication **2** (1990), 349–364.
354. Villemoes, L., Energy moments in time and frequency for two scale difference equation solutions and wavelets, Math. Institute, Technical University of Denmark, DK2800 Lyngby, Denmark, 1991, preprint.
355. Volkmer, H., Distributional and square summable solutions of dilation equations, preprint.

356. Volkner, H., On the regularity of wavelets, IEEE Trans. Inform. Theory **38** (2) (1992), to appear.
357. Walnut, D. F., Continuity properties of the Gabor frame operator, JMAA, 1991, to appear.
358. Walnut, D. F., Lattice size estimates for Gabor decompositions, preprint.
359. Walter, G., A sampling theorem for wavelet subspaces, 1990, preprint.
360. Walter, G., Discrete wavelets, 1989, preprint.
361. Walter, G., Wavelets and generalized functions, in *Wavelets: A Tutorial In Theory and Applications*, C. K. Chui (ed.), Academic Press, Cambridge, MA, 1992, 51–70.
362. Wells, Jr., R. O., Parametrizing smooth compactly supported wavelets, Trans. Amer. Math. Soc. (1991), to appear.
363. Wexler, J. and S. Raz, Discrete Gabor expansions, Signal Processing **21** (1990), 207–220.
364. Wickerhauser, M. V., Acoustic signal processing with wavelet packets, Yale University, 1989, preprint.
365. Wickerhauser, M. V., Picture compression by best basis subband coding, Yale University, 1990, preprint.
366. Wickerhauser, M. V., Acoustic signal compression with wavelet packets, in *Wavelets: A Tutorial In Theory and Applications*, C. K. Chui (ed.), Academic Press, Cambridge, MA, 1992, 679–700.
367. Wilson, R., A. Calway, and E. Pearson, A generalized wavelet transform for Fourier analysis: the multiresolution Fourier transform and its application to image and audio signal analysis, IEEE Trans. Inform. Theory **38** (2) (1992), to appear.
368. Wornell, G. and A. Oppenheim, Wavelet-based representations for a class of self-similar signals with application to fractal modulation, IEEE Trans. Inform. Theory **38** (2) (1992), to appear.
369. Young, R. M., *An Introduction to Nonharmonic Fourier Series*, Academic Press, New York, 1980.
370. Zhong, S., *Edges representation from wavelet transform maxima*, Ph. D. Thesis, New York University, 1990.

Charles K. Chui
Center for Approximation Theory
Texas A&M University
College Station, TX 77843-3368
cchui@tamu.edu

Wavelet Bases, Approximation Theory and Subdivision Schemes

Albert Cohen

Abstract. Almost all wavelet bases are constructed from multiresolution analysis, a concept which is well-known in approximation theory. We review the relations between wavelet bases, subband coding schemes and subdivision (or cascade) algorithms. We present several strategies that can be used to evaluate the regularity of the wavelet as well as the convergence of the associated subdivision scheme.

§1. Introduction

Our purpose is to present wavelet bases in relation to several concepts that are well-known in approximation theory, such as shift invariant spaces and subdivision schemes. Let us first give a brief introduction to these bases.

In the most general sense they are discrete families of functions obtained by dilations and translations of a finite number of well chosen "mother" functions. We shall consider here the classic dyadic case

$$\psi_k^j(x) = 2^{-j/2} \psi(2^{-j}x - k) \qquad j \in \mathbb{Z}, \, k \in \mathbb{Z}, \tag{1.1}$$

although many generalizations are possible (see [10,15]).

Wavelet bases are usually based on the concept of multiresolution analysis (introduced in [14]), i.e., a ladder of closed subspaces $\{V_j\}_{j \in \mathbb{Z}}$ which approximate $L^2(\mathbb{R})$,

$$\cap V_j = \{0\} \subset \cdots \subset V_1 \subset V_0 \subset V_{-1} \subset \cdots \subset L^2(\mathbb{R}) = \overline{\cup V_j}, \tag{1.2}$$

and have the following properties

$$f(x) \in V_j \iff f(2x) \in V_{j-1} \iff f(2^j x) \in V_0 . \tag{1.3}$$

There exists a function $\varphi(x)$ in V_0 such that the set $\{\varphi(x-k)\}_{k \in \mathbb{Z}}$ (1.4) is an orthonormal basis (or more generally a Riesz basis) for V_0.

Consequently, these spaces are shift-invariant and the function φ is refinable. More precisely, it satisfies a so-called *two-scale equation* or *refinement equation*

$$\varphi(x) = 2 \sum_{n \in \mathbb{Z}} c_n \, \varphi(2x - n) \,. \tag{1.5}$$

Shift-invariant spaces and refinable functions were considered long before the construction of wavelet bases, for example in the theory of spline functions.

The purpose of wavelets is to characterize in a sharp way the additional details needed to go from one resolution to the next finer level. If $\{\varphi(x - k)\}_{k \in \mathbb{Z}}$ is an orthonormal basis of V_0, one simply defines

$$\psi(x) = 2 \sum_{n \in \mathbb{Z}} (-1)^n \, \bar{c}_{1-n} \, \varphi(2x - n) \,, \tag{1.6}$$

and $\{\psi(x-k)\}_{k \in \mathbb{Z}}$ is then an orthonormal basis for W_0, the orthogonal complement of V_0 in V_1. By scaling arguments and by using the embedded structure of the V_j in (1.2) one can see that the whole set $\{\psi_k^j\}_{j \in \mathbb{Z}, \, k \in \mathbb{Z}}$ is an orthonormal basis for $L^2(\mathbb{R})$. If $\{\varphi(x-k)\}_{k \in \mathbb{Z}}$ is only a Riesz basis, one can either orthonormalize it or construct biorthogonal wavelet bases with a technique that we shall describe in the next section.

In all cases, the desire to have a simple way of characterizing the details between two adjacent levels of approximation has some effect on the choice of coefficients c_n that appear in the refinement equation (1.5). As we shall see in the next section, they are related to the subband coding schemes that have been studied in electrical engineering for data compression and coding problems. We will study the subdivision schemes that are associated with these coefficients, and present different strategies to estimate the regularity of the limit function $\varphi(x)$.

In all that follows we shall suppose that the coefficients c_n are finitely non-zero. This hypothesis will help in the proof of several results and it is crucial for the practical applications in which we need to use finite impulse response filters ("FIR"-filters).

§2. Wavelet Bases and Subband Coding Schemes

2a. The Orthonormal Case. Let $\{V_j\}_{j \in \mathbb{Z}}$ be a multiresolution analysis of $L^2(\mathbb{R})$. We can use the discrete Fourier transform of the finite sequence $\{c_n\}_{n=N_1}^{N_2}$, i.e., the transfer function

$$m_0(\omega) = \sum_{n \in \mathbb{Z}} c_n e^{-in\omega} = \sum_{n=N_1}^{N_2} c_n e^{-in\omega} \,, \tag{2.1}$$

to rewrite the two-scale difference equation that characterizes $\varphi(x)$. We suppose that the c_n are real. Taking the Fourier transform, we obtain

$$\widehat{\varphi}(2\omega) = m_0(\omega) \, \widehat{\varphi}(\omega) \tag{2.2}$$

$$\widehat{\psi}(2\omega) = e^{-i\omega} \, \overline{m_0(\omega + \pi)} \, \widehat{\varphi}(\omega) = m_1(\omega) \, \widehat{\varphi}(\omega) \,. \tag{2.3}$$

Wavelet Bases, Approximation Theory and Subdivision Schemes

Two fundamental properties of $m_0(\omega)$ can be derived from the multiresolution analysis properties :

I. Since $\{\varphi(x-k)\}_{k\in\mathbb{Z}}$ is an orthonormal basis of V_0, the Fourier transform $\hat{\varphi}(\omega)$ satisfies a Poisson identity

$$\sum_{n\in\mathbb{Z}}|\hat{\varphi}(\omega+2n\pi)|^2 = 1. \qquad (2.4)$$

Combined with (2.2), this leads to

$$|m_0(\omega)|^2 + |m_0(\omega+\pi)|^2 = 1, \qquad (2.5)$$

which may also be written as

$$2\sum_{n\in\mathbb{Z}} c_n c_{n+2k} = \delta_{k,0} \qquad (=1 \text{ if } k=0,\ 0 \text{ otherwise}). \qquad (2.6)$$

II. The denseness of $\cup V_j$ in $L^2(\mathbb{R})$ is equivalent to $\hat{\varphi}(0) = \int \varphi(x)\,dx = 1$ (see [15,14,2]). Consequently, we have

$$m_0(0) = 1 \quad \text{and} \quad m_0(\pi) = 0, \qquad (2.7)$$

which may also be written as

$$\sum_{n=N_1}^{N_2} c_n = 1 \quad \text{and} \quad \sum_{n=N_1}^{N_2} (-1)^n c_n = 0. \qquad (2.8)$$

The subband coding scheme associated with our multiresolution analysis appears clearly in the Fast Wavelet Transform Algorithm of S. Mallat [14]. Let us recall its main principles. The initial data are considered as the approximation of a continuous function at the scale $j = 0$,

$$S_k^0 = \langle f, \varphi(\cdot - k)\rangle, \qquad k \in \mathbb{Z}. \qquad (2.9)$$

This allows the computation of the approximations and the details at coarser scale, i.e.,

$$S_k^j = 2^{-j/2}\langle f, \varphi_k^j\rangle \quad \text{and} \quad D_k^j = 2^{-j/2}\langle f, \psi_k^j\rangle, \qquad j > 0. \qquad (2.10)$$

(The coefficients are normalized in such way that if $f \equiv 1$ locally, then $S_k^j = 1$ in that area). The sequence $\{S_k^j\}_{k\in\mathbb{Z}}$ (resp. $\{D_k^j\}_{k\in\mathbb{Z}}$) is then derived from $\{S_k^{j-1}\}_{k\in\mathbb{Z}}$ by a convolution with the filter $\overline{m_0(\omega)}$ (resp. $\overline{m_1(\omega)}$) followed by a "decimation" of one sample out of two to keep the same total amount of information, i.e.,

$$S_k^j = \sum_{n\in\mathbb{Z}} c_{n-2k}\, S_n^{j-1}, \qquad D_k^j = \sum_{n\in\mathbb{Z}} (-1)^{n-1} c_{2k+1-n}\, S_n^{j-1}.$$

The algorithm then iterates on $\{S_k^j\}_{k \in \mathbb{Z}}$. Conversely, the sequence $\{S_k^{j-1}\}_{k \in \mathbb{Z}}$ can be recovered by applying the same filters $m_0(\omega)$ and $m_1(\omega)$ on $\{S_k^j\}_{k \in \mathbb{Z}}$ and $\{D_j^k\}_{k \in \mathbb{Z}}$ after inserting a zero between every pair of consecutive samples, and summing the two components (multiplied by two for normalization purposes), i.e.,

$$S_n^{j-1} = 2 \sum_{k \in \mathbb{Z}} c_{n-2k} S_k^j + 2 \sum_{k \in \mathbb{Z}} (-1)^{n-1} c_{2k+1-n} D_k^j .$$

All these operations, decomposition - decimation - interpolation - reconstruction, constitute a complete subband coding scheme as shown in Figure 1. The property of exact reconstruction can now be derived in two ways. It is a natural consequence of the multiresolution approach, since $V_j = V_{j+1} \overset{\perp}{\oplus} W_{j+1}$, but it can also be viewed as a consequence of formula (2.7) for the filter m_0. This type of filter pair (m_0, m_1) is known as a pair of "conjugate quadrature filters" (CQF); they were first discovered by Smith and Barnwell in 1983 [17]. The design of FIR pairs, with real coefficients and perfect reconstruction, has been generalized in [9].

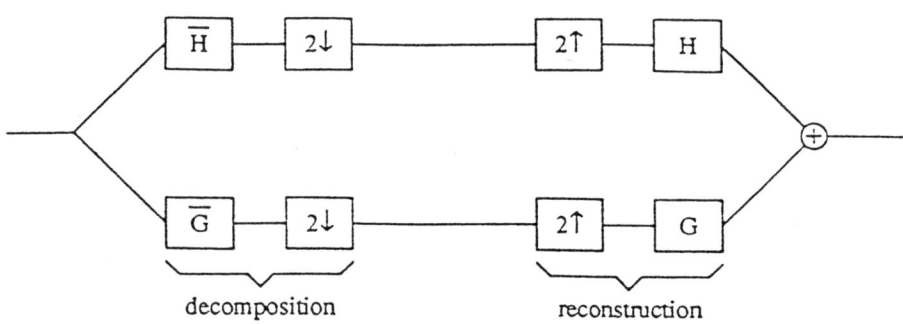

Figure 1. Subband coding scheme corresponding to the FWT algorithm The sign 2 ↓ stands for "decimation of one sample out of two" and 2 ↑ for the insertion of zeros at the intermediate values.

Since m_0 is regular (it is a trigonometric polynomial) and since $m_0(0) = 1$, we can iterate (2.2) to obtain

$$\widehat{\varphi}(\omega) = \prod_{k=1}^{+\infty} m_0(2^{-k}\omega) . \qquad (2.11)$$

Given a conjugate quadrature filter $m_0(\omega)$ — i.e., a trigonometric polynomial satisfying (2.5) and (2.7) — it is thus possible to define the scaling function, either as a solution of the two-scale difference equation (1.5), or explicitly with the above infinite product. However, this does not always lead to a multiresolution analysis. The function $\varphi(x) = \frac{1}{3}\chi_{[0,3]}$ generated by the CQF $m_0(\omega) = (1+e^{3i\omega})/2$, for example, does not have orthonormal translates. Orthonormality of the functions $\varphi(x-k)$ turns out to be equivalent to the L^2 convergence of the truncated products $\widehat{\varphi}_n(\omega) = \prod_{k=1}^{n} m_0(2^{-k}\omega)\chi_{[-2^n\pi, 2^n\pi]}(\omega)$ to $\widehat{\varphi}(\omega)$, because $\{\varphi_n(x-k)\}_{k\in\mathbb{Z}}$ is orthonormal when (2.5) is satisfied.

The following result characterizes the subclass of CQF filters that lead to a multiresolution analysis and orthonormal basis of wavelets.

Theorem 2.1. *Let $m_0(\omega)$ be a conjugate quadrature filter. The infinite product (2.11) leads to a multiresolution analysis if and only if there exist a compact neighborhood K of the origin in $\mathbb{R}$ such that*

i) $|K| = 2\pi$, *and for each $\omega \in [-\pi, \pi]$, there exists an $n \in \mathbb{Z}$ such that $\omega + 2n\pi \in K$,*

ii) *for all $n > 0$, $m_0(2^{-n}\omega)$ does not vanish on K.*

The set K is said to be "congruent to $[-\pi, \pi]$ modulo 2π" (Figure 2). The proof of this result can be found in [2]. It exploits the continuity of m_0, the compactness of K, and the equation $m_0(0) = 1$ to show that (ii) is equivalent to $\widehat{\varphi}(\omega) \geq c > 0$ on K. This is then sufficient to derive the L^2 convergence of the φ_n by Lebesgue's theorem.

Figure 2. Example of compact set congruent to $[-\pi, \pi]$ modulo 2π.

2b. The Biorthogonal Case.

The conjugate quadrature filters are a particular case of subband coding schemes with perfect reconstruction, because identical filters (up to a complex conjugation) are used for both the decomposition and the reconstruction stages. If we don't impose this restriction, then the scheme uses four different filters: $\widetilde{m}_0(\omega)$ and $\widetilde{m}_1(\omega)$ for the decomposition, $m_0(\omega)$ and $m_1(\omega)$ for the reconstruction. Perfect reconstruction for any discrete signal is then ensured if

$$\left. \begin{array}{l} \overline{m_0(\omega)}\,\widetilde{m}_0(\omega) + \overline{m_1(\omega)}\,\widetilde{m}_1(\omega) = 1 \\ \widetilde{m}_0(\omega)\,\overline{m_0(\omega+\pi)} + \widetilde{m}_1(\omega)\,\overline{m_1(\omega+\pi)} = 0 \end{array} \right\} \quad (2.12)$$

Thus $\widetilde{m}_0(\omega)$ and $\widetilde{m}_1(\omega)$ may be regarded as the solutions of a linear system. However, to avoid infinite impulse response solutions, we shall force the determinant of this system to be $\alpha e^{ik\omega}$, $\alpha \neq 0$, $k \in \mathbb{Z}$. For the sake of convenience,

we take $\alpha = -1$ and $k = 1$ (a change of these values would only mean a shift and a scalar multiplication on the impulse response of our filters). This leads to
$$\overline{m_0(\omega)}\,\tilde{m}_0(\omega) + \overline{m_0(\omega+\pi)}\,\tilde{m}_0(\omega+\pi) = 1, \tag{2.13}$$
and
$$m_1(\omega) = e^{-i\omega}\,\overline{\tilde{m}_0(\omega+\pi)}\quad,\quad \tilde{m}_1(\omega) = e^{-i\omega}\,\overline{m_0(\omega+\pi)}. \tag{2.14}$$

The formulas (2.13) and (2.14) define the most general setting for finite impulse response subband coders with exact reconstruction (in the two channel case). The functions $m_0(\omega)$ and $\tilde{m}_0(\omega)$ are called "dual filters," and were introduced in [19]. It is clear that the special case $m_0(\omega) = \tilde{m}_0(\omega)$ corresponds to the conjugate quadrature filters of 2.1.a. However, dual filters are easier to design than CQF's. For example, if m_0 is fixed, $\tilde{m}_0$ can be found as the solution of a Bezout problem that is equivalent to a linear system. The coefficients of these filters can be very simple numerically (in particular they can have a finite binary expansion, which is very useful for practical implementation); furthermore they can be chosen to be symmetric ("linear phase filter"), a property which is impossible to achieve in the CQF case.

We can mimic, in this more general framework, the construction of orthonormal wavelets from CQF. Assuming that $m_0(0) = \tilde{m}_0(0) = 1$ and $m_0(\pi) = \tilde{m}_0(\pi) = 0$, we define

$$\hat{\varphi}(\omega) = \prod_{k=1}^{\infty} m_0(2^{-k}\omega) \tag{2.15}$$

$$\hat{\psi}(2\omega) = m_1(\omega)\,\hat{\varphi}(\omega) \tag{2.16}$$

$$\hat{\tilde{\varphi}}(\omega) = \prod_{k=1}^{\infty} \tilde{m}_0(2^{-k}\omega) \tag{2.17}$$

$$\hat{\tilde{\psi}}(2\omega) = \tilde{m}_1(\omega)\,\hat{\tilde{\varphi}}(\omega). \tag{2.18}$$

In [6], the following theorem was proved,

Theorem 2.2. (i) If the truncated products

$$\hat{\varphi}_n(\omega) = \prod_{k=1}^{n} m_0(2^{-k}\omega)\,\chi_{[-2^n\pi,\,2^n\pi]}(\omega)$$

and

$$\hat{\tilde{\varphi}}_n(\omega) = \prod_{k=1}^{n} \tilde{m}_0(2^{-k}\omega)\,\chi_{[-2^n\pi,\,2^n\pi]}(\omega)$$

converge in $L^2(\mathbb{R})$ respectively to $\hat{\varphi}(\omega)$ and $\hat{\tilde{\varphi}}(\omega)$, then the following duality relations are satisfied:

$$\langle \varphi(\cdot - k), \tilde{\varphi}(\cdot - k') \rangle = \delta_{k,k'} \tag{2.19}$$

$$\langle \psi_k^j, \tilde{\psi}_{k'}^{j'} \rangle = \delta_{j,j'}\,\delta_{k,k'}. \tag{2.20}$$

Furthermore, for all f in $L^2(\mathbb{R})$ one has the unique decomposition (in the L^2 sense)

$$f = \lim_{J \to +\infty} \sum_{j=-J}^{J} \sum_{k \in \mathbb{Z}} \langle f, \psi_k^j \rangle \tilde{\psi}_k^j \qquad (2.21)$$

(ii) *If φ and $\tilde{\varphi}$ satisfy $|\hat{\varphi}(\omega)| + |\hat{\tilde{\varphi}}(\omega)| \leq C(1+|\omega|)^{-1/2-\varepsilon}$ for some $\varepsilon > 0$, then the families $\{\psi_k^j\}_{j,k \in \mathbb{Z}}$ and $\{\tilde{\psi}_k^j\}_{j,k \in \mathbb{Z}}$ are frames of $L^2(\mathbb{R})$.*

(iii) *When the hypotheses in (i) and (ii) hold, $\{\psi_k^j, \tilde{\psi}_k^j\}_{j,k \in \mathbb{Z}}$ are biorthogonal (or dual) Riesz bases of $L^2(\mathbb{R})$.*

Many examples of these systems can be found in [6] and we shall mention in Section 5 a sharp criterion that dual filters have to satisfy in order to generate stable biorthogonal wavelet bases. We now recall a practical way of constructing φ and ψ numerically from a given subband coding scheme.

§3. The Cascade Algorithm

In the last section we saw that the scaling function $\varphi(x)$ could be approximated, at least in $L^2(\mathbb{R})$, by a sequence of band-limited functions $\{\varphi_n\}_{n>0}$ defined by

$$\hat{\varphi}_n = \prod_{j=1}^{n} m_0(2^{-j}\omega) \chi_{[-2^n \pi, 2^n \pi]}(\omega) . \qquad (3.1)$$

These functions are characterized by their sampled values at the points $2^{-n}k$ ($k \in \mathbb{Z}$), i.e.,

$$s_k^n = \varphi_n(2^{-n}k) . \qquad (3.2)$$

This sequence can also be considered as the impulse response of the transfer function

$$S_n(\omega) = 2^n \prod_{j=1}^{n-1} m_0(2^j \omega) , \qquad (3.3)$$

which can be obtained recursively by the formula

$$S_{n+1}(\omega) = 2 m_0(\omega) S_n(2\omega) . \qquad (3.4)$$

In the time domain, (3.4) becomes an interpolation scheme; the sequence s_k^n is dilated by insertion of zeros ($S_n(\omega) \to S_n(2\omega)$) before being filtered (multiplication by $2m_0(\omega)$). We thus have,

$$s_p^{n+1} = 2 \sum_{k \in \mathbb{Z}} c_{p-2k} s_k^n . \qquad (3.5)$$

This iterative refinement process, which computes the $\{s_k^n\}_{k \in \mathbb{Z}}$ sequences from an initial Dirac sequence $\delta_{0,k}$, is called in [9] the "cascade algorithm." We

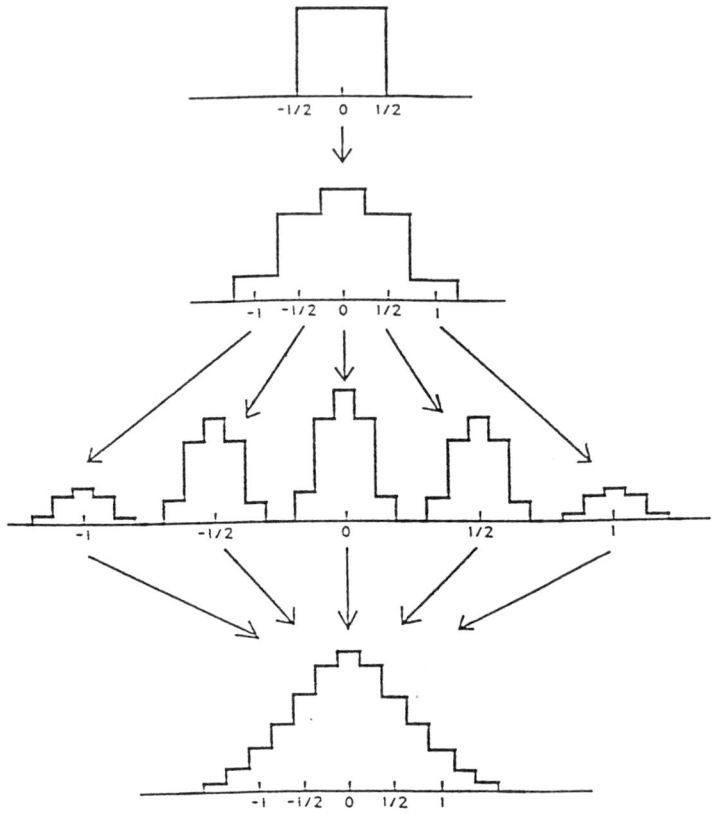

Figure 3. The cascade algorithm, from [9].

illustrate it in Figure 3 (our sequences are represented by piecewise constant functions).

Note that it identifies exactly with the reconstruction stage in the FWT algorithm described in 2a. The scaling function is thus approached by the reconstructed signal from a single approximation coefficient at a coarse scale. Similarly, the wavelet will be obtained by starting the reconstruction from a detail coefficient at a coarse scale (and thus applying $m_1(\omega)$ at the first step of the cascade).

This explains why subband coding schemes associated with regular wavelets are particularly interesting: the smoothness of the wavelet determines the appearance of the coarse scale components of the reconstructed signal. A smooth appearance is important in many applications such as data compression, where a large portion of the finer scale information is thrown away.

In the biorthogonal case, the analysis and the synthesis wavelets (ψ and $\tilde{\psi}$) need not have the same regularity. As just discussed, smoothness is important for the reconstructing function; the analyzing function needs only sufficient regularity to ensure that the wavelet bases are unconditional and that the FWT algorithm is stable. Note that an important property of the analyzing wavelet is cancellation, i.e., vanishing moments, ensuring small high scale coefficients for smooth regions in the signal being analyzed.

This type of "refinement method" is known in approximation theory as "stationary subdivision" (e.g., [1,12]). Most of the papers on this topic are motivated by interpolation problems, where smooth curves or surfaces need to be constructed, connecting (or close to) given sparse data points. Consequently, they are mainly concerned with what we call the reconstruction stage and they do not study the existence of an associated subband coding scheme. This also means that they do not address an easy way of encoding or representing the extra "detail information" ($\leftrightarrow W_j$) that can be added in going from one refinement level to the next ($V_j \to V_{j-1}$). On the other hand, the subband coding literature seldom mentions the importance of the smoothness appearing in the cascade of the reconstruction from the low scales. Orthonormal and biorthogonal wavelet bases lead to an elegant combination of these two approaches.

We now present several different methods for estimating the regularity of the wavelets associated with a given subband coding scheme. We shall concentrate on the regularity of the scaling function, which determines the regularity of the wavelet itself because $\psi(x)$ is a finite linear combination of translates of $\varphi(2x)$. Whatever method is used, if a global regularity of order r is achieved, then the cascade algorithm also converges uniformly up to this order (see [3,9,11]).

§4. Regularity: the Spectral Approach

4a. A Fourier Estimation of the Hölder Exponent. Let us denote by C^α the Hölder space defined as follows. For $\alpha = n + \beta$, $\beta \in [0,1)$, $f \in C^\alpha$ if and only if it is n times continuously differentiable and for all $x \neq y$,

$$\frac{|f^{(n)}(x) - f^{(n)}(y)|}{|x-y|^\beta} \leq C(f).$$

Set
$$\mathcal{F}_p^\alpha = \{f : (1+|\omega|)^\alpha \, \widehat{f}(\omega) \in L^p\} \qquad (\alpha \geq 0, p \geq 1). \tag{4.1}$$

It is well known (and easy to verify) that $\mathcal{F}_\infty^{\alpha+1+\varepsilon} \subset \mathcal{F}_1^\alpha \subset C^\alpha$, for $\varepsilon > 0$. For compactly supported functions f, we also have

$$f \in C^\alpha \implies f \in \mathcal{F}_\infty^\alpha. \tag{4.2}$$

Thus, the decay of the Fourier transform can be used to evaluate the global regularity. To estimate this decay in the case of the scaling function, it is possible to use the factorization of $m_0(\omega)$; due to its cancellation at $\omega = \pi$, we have

$$m_0(\omega) = \left(\frac{1+e^{i\omega}}{2}\right)^N p(\omega). \tag{4.3}$$

The infinite product (2.11) is thus divided in two parts. The first part, which comes from the factor $2^{-N}(1+e^{i\omega})^N$, i.e., the Fix-Strang rules of order $N-1$ for the scaling function φ, gives decay, since

$$\left|\prod_{k=1}^{+\infty}\left(\frac{1+e^{i2^{-k}\omega}}{2}\right)^N\right| = \left|\prod_{k=2}^{+\infty}\cos(2^{-k}\omega)\right|^N = \left|\frac{2}{\omega}\sin(\omega/2)\right|^N. \quad (4.4)$$

The second part, which involves the factor $p(\omega)$, can be controlled by a polynomial expression. Indeed, since $p(0) = 1$ and p is a regular function, the infinite product generated by the second factor satisfies

$$\left|\prod_{k=1}^{+\infty} p(2^{-k}\omega)\right| \leq C \prod_{1 \leq k < \log(1+|\omega|)/\log 2} |p(2^{-k}\omega)|. \quad (4.5)$$

Defining, for $j > 0$,

$$B_j = \sup_{\omega \in \mathbb{R}} \left|\prod_{k=0}^{j-1} p(2^k \omega)\right| \quad (4.6)$$

and

$$b_j = \frac{\log B_j}{j \log 2}, \quad (4.7)$$

we obtain

$$\left|\prod_{k=1}^{+\infty} p(2^{-k}\omega)\right| \leq C(B_j)^{\log(1+|\omega|)/\log 2} \leq C(1+|\omega|)^{b_j} \quad (4.8)$$

and finally,

$$|\widehat{\varphi}(\omega)| \leq C(1+|\omega|)^{b_j - N}. \quad (4.9)$$

Consequently, φ is in $\mathcal{F}_1^\alpha$ and in $\mathcal{C}^\alpha$ if $\alpha < N - b_j - 1$ for some $j > 0$. We see here that N must be large to allow high regularity since b_j is always positive. In fact, one can prove that if the wavelet is r times continuously differentiable, then it has at least $r+1$ vanishing moments (see [9,15], i.e., $(d/d\omega)^n(\widehat{\psi})(0) = (d/d\omega)^n(m_0)(\pi) = 0$, for $n = 0, \ldots, r+1$ and thus we have $N \geq r+1$. These cancellations are equivalent (see [18]) to the property that the polynomials of order $N-1$ can be expressed as linear combinations of $\{\varphi(x-k)\}_{k\in\mathbb{Z}}$. However, these conditions are necessary but not sufficient to ensure the regularity of the scaling function, since the effect of N may be spoiled by a large value of b_j. Fortunately, this can be avoided by a careful choice of the filter $m_0(\omega)$ (and, in the biorthogonal case, additionally $\widetilde{m}_0(\omega)$).

In the CQF-orthonormal case, a particular family of FIR filters indexed by N has been constructed in [9]. This construction uses the polynomial

$$P_N(y) = \sum_{j=0}^{N-1} \binom{N-1+j}{j} y^j \quad (4.10)$$

(with the shorthand notation $y = \sin^2(\omega/2)$), which is the lowest degree solution of the Bezout problem

$$P_N(y)(1-y)^N + y^N P_N(1-y) = 1. \quad (4.11)$$

The corresponding filters are defined by

$$m_0^N(\omega) = \left(\frac{1+e^{i\omega}}{2}\right)^N p_N(\omega) \quad (4.12)$$

and

$$|p_N(\omega)|^2 = P_N(y) = P_N\left(\frac{1-\cos\omega}{2}\right). \quad (4.13)$$

The Fejer-Riesz lemma implies that there exists an FIR filter $p_N(\omega)$ which satisfies (4.13). It is clear that the CQF condition (2.5) is equivalent to (4.11), and the conditions in Theorem 2.1 are trivially satisfied with $K = [-\pi, \pi]$. For large values of N, the regularity $\alpha(N)$ of the associated scaling function is approximately $N/5$ and the exact asymptotic ratio between $\alpha(N)$ and N can be determined. Intuitively, this means that the contribution of $p_N(\omega)$ removes "eighty percent of the regularity" brought by the factor $2^{-N}(1+e^{i\omega})^N$. For this estimation, we need to optimize the inequality (4.9); i.e., find the best possible exponent for the decay of $\hat{\varphi}(\omega)$.

4b. Optimal and Asymptotic Fourier Estimation: The Role of Fixed Points. We start by defining "the critical exponent" of $m_0(\omega)$:

$$b = \inf_{j>0} b_j = \inf_{j>0} \max_{\omega \in \mathbb{R}} \left(\frac{1}{j \log 2} \log \left|\prod_{k=0}^{j-1} p(2^k \omega)\right|\right). \quad (4.14)$$

It was proved in [3] that under the hypothesis $|p(\pi)| > 1$ (satisfied in the present case (4.13)), $\hat{\varphi}(\omega)$ cannot have a better decay at infinity than $|\omega|^{b-N}$. If the infimum b is attained for some finite j, $b = b_j$, then this estimate is optimal.

How can we estimate the critical exponent? A first method consists of evaluating b_j for large values of j. Indeed, b is the limit of the sequence b_j because the boundedness of p implies $b_J \leq b_j + O(j/J)$. This may, however, require lengthy computations.

In several cases, it is possible to use a more powerful method based on the transformation $\tau : \omega \mapsto 2\omega$ modulo 2π and the fixed points of its powers τ^n, $n > 0$. Indeed, let ω_0 be a fixed point of τ^n for $n > 0$ and define its orbit $\omega_j = \tau^j \omega_0$, for $j = 0, \ldots, n-1$. Since $p(\omega)$ has period 2π, we have

$$p(2^{nk}\omega_j) = p(\omega_j), \quad \text{for all} \quad k > 0, \quad (4.15)$$

and consequently

$$b_{nk} \geq \frac{1}{n \log 2} \log \left|\prod_{j=0}^{n-1} p(\omega_j)\right|. \quad (4.16)$$

Letting k go to $+\infty$, we are led to

$$b \geq \frac{1}{n \log 2} \log \left|\prod_{j=0}^{n-1} p(\omega_j)\right|. \tag{4.17}$$

Fixed points of τ lead therefore to lower bounds for b and upper bounds for the regularity index. In fact they can do much better and provide optimal estimates for certain types of filters. Let us consider the smallest orbit of τ different from $\{0\}$, namely the pair $\{-\frac{2\pi}{3}, \frac{2\pi}{3}\}$. Note that, because our filters have real coefficients, $|m_0(\omega)|$ and $|p(\omega)|$ are even functions so that $|p(\frac{2\pi}{3})| = |p(-\frac{2\pi}{3})|$. The following result associates the value $|p(\frac{2\pi}{3})|$ with the critical exponent b.

Theorem 4.1. *Suppose that $p(\omega)$ satisfies*

$$|p(\omega)| \leq \left|p\left(\frac{2\pi}{3}\right)\right| \quad \text{if } |\omega| \leq \frac{2\pi}{3} \tag{4.18a}$$

$$|p(\omega)p(2\omega)| \leq \left|p\left(\frac{2\pi}{3}\right)\right|^2 \quad \text{if } \frac{2\pi}{3} \leq |\omega| \leq \pi. \tag{4.18b}$$

Then

$$b = \frac{1}{\log 2} \log \left|p\left(\frac{2\pi}{3}\right)\right|. \tag{4.19}$$

Proof: We already know from (4.17) that $b \geq (1/\log 2) \log |p(2\pi/3)|$. We now use the bounds on p to find an upper bound on b_j, $j > 0$. We can regroup the factors in (4.6) by packets of one or two elements in order to apply either (4.18a) or (4.18b) on each block. Since only the last factor can miss one of these two inequalities, we obtain

$$\left|\prod_{k=0}^{j-1} p(2^k \omega)\right| \leq \left|p\left(\frac{2\pi}{3}\right)\right|^{j-1} \sup |p|, \tag{4.20}$$

and thus,

$$b_j \leq \frac{1}{\log 2} \left[\frac{j-1}{j} \log \left|p\left(\frac{2\pi}{3}\right)\right| + \frac{\sup [\log |p|]}{j}\right], \tag{4.21}$$

which leads to

$$b \leq \frac{1}{\log 2} \log \left|\left(\frac{2\pi}{3}\right)\right| \tag{4.22}$$

and to (4.19). ■

The equality (4.19) implies that the worst decay of $\widehat{\varphi}(\omega)$ occurs for the sequence $\omega_k = 2^n \pi/3$, $n > 0$. This is interesting, because (4.18a) and (4.18b) turn out to be satisfied in many cases and in particular for the whole family of

Wavelet Bases, Approximation Theory and Subdivision Schemes

CQF defined by (4.12), (4.13). This is easy to verify directly for small values of N, since the inequalities can be rewritten as

$$P_N(y) \leq P_N\left(\frac{3}{4}\right), \quad \text{if } y \leq \frac{4}{3} \tag{4.23a}$$

$$P_N(y) P_N(4y(1-y)) \leq \left(P_N\left(\frac{3}{4}\right)\right)^2, \quad \text{if } \frac{4}{3} \leq y \leq 1. \tag{4.23b}$$

The discussion for general N is more difficult, and we refer to [4] for a complete proof of (4.23a), (4.23b). However, a similar result can be obtained in a simple way. To characterize the asymptotic behavior of the critical exponent when N goes to $+\infty$, one doesn't need the full force of (4.18a), (4.18b). It can also be derived from a weaker, asymptotically valid inequality, as proved by H. Volkner in [21]:

Theorem 4.2. Let $b(N)$ be the critical exponent associated with $m_0^N(\omega)$, and $\alpha(N)$ the Hölder exponent of the corresponding scaling function. Then

$$\lim_{N \to +\infty} \frac{b(N)}{N} = \frac{\log 3}{2 \log 2} \tag{4.24a}$$

and

$$\lim_{n \to +\infty} \frac{\alpha(N)}{N} = \lim_{N \to +\infty} \frac{N - b(N)}{N} = 1 - \frac{\log 3}{2 \log 2} \simeq 0.2075. \tag{4.24b}$$

Proof: This result can be viewed as a consequence of Theorem 4.1 but it can also be proved directly by using some properties of $P_N(y)$. Let us write (4.10) in the form

$$P_N(y) = \sum_{j=0}^{N-1} \binom{N-1+j}{j} \left(\frac{1}{2}\right)^j (2y)^j. \tag{4.25}$$

From (4.10) we see that $P_N\left(\frac{1}{2}\right) = 2^{N-1}$. Since P_N is an increasing function between 0 and 1, we have

$$P_N(y) \leq [\max(4y, 2)]^{N-1} = |g(y)|^{N-1}. \tag{4.26}$$

It is now trivial to verify that (4.23a) and (4.23b) are satisfied if we replace $P_N(y)$ by $g(y)$. The argument used in the proof of Theorem 4.1 leads then to

$$b(N) \leq \frac{N-1}{2 \log 2} \log \left|g\left(\frac{3}{4}\right)\right| = \frac{N-1}{2 \log 2} \log 3; \tag{4.27}$$

but from (4.17) we get

$$b(N) \geq \frac{1}{2 \log 2} \log \left|P_N\left(\frac{3}{4}\right)\right| \geq \frac{1}{2 \log 2} \log \left|\binom{2N-2}{N-1}\left(\frac{3}{4}\right)^{N-1}\right| \\ \geq \frac{N-2}{2 \log 2} \log 3. \tag{4.28}$$

This establishes the limit in (4.24a), and consequently (4.24b), since the decay index of the Fourier transform is equivalent to the Hölder exponent when both tend to $+\infty$. ∎

The use of fixed points for optimal estimates of the spectral decay is thus very efficient when one is looking for arbitrarily high regularity, since a sharp asymptotic result is obtained. For small filters, this method does not give a good result because the error on the exact regularity may have the same order as the value of the Hölder exponent itself. For such filters, methods taking advantage of the small number of taps in the filter can be used to derive more precise estimates. We now describe these methods, which are typically based on matrix computations.

§5. Regularity: Matrix-based Sharper Estimates

5a The Littlewood-Paley Approach.
We first recall some aspects of the Littlewood-Paley theory. Let $\gamma(x)$ be a real-valued symmetric function of the Schwartz class $S(\mathbb{R})$ that satisfies

$$\left. \begin{array}{ll} \widehat{\gamma}(\omega) = 0 & \text{if } |\omega| \leq \dfrac{1}{2} \text{ or } |\omega| \geq \dfrac{5}{2} \\[6pt] \widehat{\gamma}(\omega) > 0 & \text{if } \dfrac{1}{2} < |\omega| < \dfrac{5}{2} \end{array} \right\} \qquad (5.1)$$

so that the frequency axis is covered by the dyadic dilations of γ. Indeed, we have

$$0 < C_1 \leq \sum_{j=-\infty}^{+\infty} \widehat{\gamma}(2^j \omega) \leq C_2 \qquad \text{if } \omega \neq 0. \qquad (5.2)$$

Define for any f in $S'(\mathbb{R})$ the dyadic blocks $\Delta_j(f)$ by

$$\Delta_j(f) = 2^j \gamma(2^j \cdot) * f \iff \widehat{\Delta}_j(f) = \widehat{\gamma}(2^{-j} \cdot) \widehat{f}. \qquad (5.3)$$

The Littlewood-Paley theory tells us that several functional spaces can be characterized by examining only the L^p norm of these blocks. This is the case in particular for the Sobolev spaces H^s and the Hölder spaces C^α, $\alpha > 0$. To do this, it is necessary to change slightly the definition of C^α when α is an integer ; we shall say that a bounded function f is in C^n if and only if f^{n-1} belongs to the Zygmund class Λ; i.e., there exists a constant C such that, for all x and y, we have

$$\left| f^{n-1}(x+y) + f^{n-1}(x-y) - 2f^{n-1}(x) \right| \leq C|y|. \qquad (5.4)$$

With this convention, elements of the Hölder space C^α are characterized by the following conditions:

$$\|\Delta_j(f)\|_{L^\infty} \leq C 2^{-\alpha j}, \qquad \text{when } j \geq 0 ; \qquad (5.5)$$

$$f \text{ is a bounded continuous function.} \qquad (5.6)$$

Note that the choice (5.1) for γ is arbitrary and that more general functions could be chosen to divide the Fourier domain into dyadic blocks. To derive these types of estimates on the scaling function φ, we introduce a new tool.

Wavelet Bases, Approximation Theory and Subdivision Schemes

Definition 5.1. Let $L^2[0, 2\pi]$ be the space of 2π-periodic, square-integrable functions on $[0, 2\pi]$, and $C[0, 2\pi]$ the space of 2π periodic continuous functions. For any $m(\omega)$ in $C[0, 2\pi]$, we define the transition operator T_m associated with $m(\omega)$ by the prescription

$$\left. \begin{array}{l} T_m : L^2[0, 2\pi] \to L^2[0, 2\pi] \\ f \mapsto T_m f(\omega) = m\left(\frac{\omega}{2}\right) f\left(\frac{\omega}{2}\right) + m\left(\frac{\omega}{2} + \pi\right) f\left(\frac{\omega}{2} + \pi\right). \end{array} \right\} \quad (5.7)$$

This operator can be regarded as the adjoint of the "Subdivision Operator" introduced in [1]. Note that when $m(\omega)$ is a trigonometric polynomial, the study of T_m can be made in a finite-dimensional space. More precisely, if we define

$$E(N_1, N_2) = \left\{ \sum_{n=N_1}^{N_2} h_n e^{in\omega} : (h_{N_1}, \ldots, h_{N_2}) \in \mathbb{C}^{N_2 - N_1 + 1} \right\}, \quad (5.8)$$

then we have

$$(f, m) \in [E(N_1, N_2)]^2 \implies T_m f \in E(N_1, N_2). \quad (5.9)$$

This is due to the contraction $\omega \mapsto \omega/2$ which appears in the definition (5.7). If c_n is the nth Fourier coefficient of $m(\omega)$, then the matrix P of T_m with respect to the complex exponentials basis is given by

$$P_{l,n} = (2c_{2l-n}). \quad (5.10)$$

The size of this matrix in $E(N_1, N_2)$ is $L \times L$ with $L = N_2 - N_1 + 1$. This operator has been studied by J.P. Conze and A. Raugi and several ideas presented below are from their work [7,8]. We shall use it to derive Littlewood-Paley estimates for the Hölder continuity of the scaling function. For this, we need the following result:

Lemma 5.2. For all $n > 0$,

$$\int_{-\pi}^{\pi} (T_m)^n f(\omega) \, d\omega = \int_{-2^n \pi}^{2^n \pi} f(2^{-n}\omega) \prod_{k=1}^{n} m(2^{-k}\omega) \, d\omega. \quad (5.11)$$

Proof: We use induction. The result is clear for $n = 1$ since

$$\int_{-\pi}^{\pi} T_m f(\omega) \, d\omega = \int_{-\pi}^{\pi} \left[m\left(\frac{\omega}{2}\right) f\left(\frac{\omega}{2}\right) + m\left(\frac{\omega}{2} + \pi\right) f\left(\frac{\omega}{2} + \pi\right) \right] d\omega$$

$$= 2 \int_{-\pi/2}^{\pi/2} [m(\omega) f(\omega) + m(\omega + \pi) f(\omega + \pi)] \, d\omega$$

$$= 2 \int_{-\pi}^{\pi} m(\omega) f(\omega) \, d\omega = \int_{-2\pi}^{2\pi} m\left(\frac{\omega}{2}\right) f\left(\frac{\omega}{2}\right) d\omega \,].$$

Assuming (5.11) for n, we obtain by the induction hypothesis,

$$\int_{-\pi}^{\pi} (T_m)^{n+1} f(\omega) \, d\omega = \int_{-\pi}^{\pi} (T_m)^n T_m f(\omega) \, d\omega$$

$$= \int_{-2^n\pi}^{2^n\pi} \left[\prod_{k=1}^{n} m(2^{-k}\omega) \right] [m(2^{-n-1}\omega) f(2^{-n-1}\omega)$$
$$+ m(2^{-n-1}\omega + \pi) f(2^{-n-1}\omega + \pi)] \, d\omega$$

$$= 2^{n+1} \int_{-\frac{\pi}{2}}^{\frac{\pi}{2}} \left[\prod_{k=1}^{n} m(2^k \omega) \right] [m(\omega) f(\omega) + m(\omega + \pi) f(\omega + \pi)] \, d\omega$$

$$= \int_{-2^{n+1}\pi}^{2^{n+1}\pi} \left[\prod_{k=1}^{n+1} m(2^{-k}\omega) \right] f(2^{-n-1}\omega) \, d\omega \ . \quad \blacksquare$$

We now assume that $m(\omega)$ is a positive trigonometric polynomial in $E_M = E(-M, M)$ and that $m(0) = 1$ and $m(\pi) = 0$. Then m can be factored as

$$m(\omega) = \cos^{2N}\left(\frac{\omega}{2}\right) p(\omega) , \qquad (5.12)$$

where $p(\omega)$ is a trigonometric polynomial that does not vanish for $\omega = \pi$. Note that necessarily $N \leq M$. From this cancellation property, we can derive the next result.

Lemma 5.3. $\{1, \frac{1}{2}, \ldots, 2^{-2N+1}\}$ are eigenvalues of P. The row vectors $p_j = (n^j)_{n=-M,\ldots,M}$, for $0 \leq j \leq 2N-1$ generate a subspace which is left-invariant under P and contains one eigenvector for each of these $2N$ eigenvalues. Consequently, the orthogonal subspace

$$F_N = \left\{ \sum_{n=-M}^{M} h_n e^{-in\omega} \mid \sum_{n=-M}^{M} n^j h_n = 0 , \ j = 0, \ldots, 2N-1 \right\} \qquad (5.13)$$

is right invariant under T_m.

Proof: The factorization in (2.65) is equivalent to the cancellation rules

$$\sum_{n=-M}^{M} (-1)^n n^j c_n = 0 \qquad \text{for} \quad j = 0, \ldots, 2N-1 . \qquad (5.14)$$

In particular, for $j = 0$, we have

$$\sum_n c_{2n} = \sum_n c_{2n+1} = \frac{1}{2} \qquad (\text{because } m(0) = 1). \qquad (5.15)$$

This means that the sums of each column in the matrix of T_m in (5.10) are equal to 1 and that $p_0 = (1, \ldots, 1)$ is a left eigenvector for the eigenvalue 1. For $0 < j \leq 2N - 1$ we define $q_j = p_j P = (q_j^{-M}, \ldots, q_j^M)$. We have

$$q_j^l = \sum_n n^j c_{2n-l} . \qquad (5.16)$$

Thus, if l is even
$$q_j^l = \sum_n \left(n + \frac{l}{2}\right)^j c_{2n}, \tag{5.17}$$
and if l is odd
$$q_j^l = \sum_n \left(n + \frac{1}{2} + \frac{l}{2}\right)^j c_{2n+1}. \tag{5.18}$$

Using the binomial formula and the cancellation rules (5.14), we see that q_j is a linear combination of p_k for $k = 0, \ldots, j$. The coefficient of p_j is given by the last term of the binomial and is thus equal to 2^{-j}. Consequently $\{p_j\}_{j=0,\ldots,2N-1}$ is a triangulation basis for the left action of P and the eigenvalues are $\{2^{-j}\}_{j=0,\ldots,2N-1}$. ∎

We now come back to the scaling function φ, given by the infinite product
$$\widehat{\varphi}(\omega) = \prod_{k=0}^{+\infty} m(2^{-k}\omega). \tag{5.19}$$

Theorem 5.4. *Let F_N be the invariant subspace (5.13) of T_m. If λ is the largest eigenvalue of T_m restricted to F_N and if $|\lambda| < 1$, then, defining $\alpha = -\log \lambda / \log 2$ (> 0), we have,*
(i) *φ is in $C^{\alpha-\varepsilon}$ for all $\varepsilon > 0$*
(ii) *φ is in C^α if the restriction of T_m to the invariant subspace F_λ of eigenvalue λ is purely diagonal (i.e., $= \lambda I$).*
These two estimates are optimal if $\widehat{\varphi}(\omega)$ does not vanish on $[-\pi, \pi]$.

Proof: Consider the trigonometric polynomial
$$C_N(\omega) = (1 - \cos \omega)^N, \tag{5.20}$$
which clearly belongs to F_N. For all $n > 0$, we have
$$\int_{-\pi}^{\pi} (T_m)^n C_N(\omega) \, d\omega \leq (2\pi)^{1/2} \left(\int_{-\pi}^{\pi} |(T_m)^n C_N(\omega)|^2 \, d\omega \right)^{1/2}$$
$$\leq C(\lambda + \varepsilon)^n \quad \text{or} \quad C\lambda^n \quad \text{if} \quad T_m|_{F_\lambda} = \lambda I.$$

Now apply Lemma 5.2 and the inequality
$$C_N(\omega) \geq 1, \quad \text{when} \quad \frac{\pi}{2} \leq |\omega| \leq \pi, \tag{5.21}$$
to obtain
$$\int_{2^{n-1} \leq |\omega| \leq 2^n \pi} \widehat{\varphi}(\omega) \, d\omega \leq C \int_{2^{n-1}\pi \leq |\omega| \leq 2^n \pi} \prod_{k=1}^{n} m(2^{-k}\omega) \, d\omega$$
$$\leq C \int_{-2^n \pi}^{2^n \pi} C_N(2^{-n}\omega) \prod_{k=1}^{n} m(2^{-k}\omega) \, d\omega$$
$$= C \int_{-\pi}^{\pi} (T_m)^n C_N(\omega) \, d\omega.$$

Consequently, the Littlewood-Paley blocks satisfy the inequality

$$\|\hat{\Delta}_j(\varphi)\|_{L^1} \leq C 2^{-(\alpha-\varepsilon)j}, \qquad \varepsilon > 0, \ \alpha = -\frac{1}{\log 2}\log(\lambda) \qquad (5.22a)$$

$$\|\hat{\Delta}_j(\varphi)\|_{L^1} \leq C 2^{-\alpha j}, \qquad \text{if } T_m|_{F_\lambda} \text{ is purely diagonal.} \qquad (5.22b)$$

Since $\|\Delta_j(\varphi)\|_{L^\infty} \leq \|\hat{\Delta}_j\|_{L^1}$, we obtain the asserted regularity.

To prove that these estimates are optimal, we need to reverse all the inequalities which have been used. First, note that since $m(\omega)$ and $\hat{\varphi}(\omega)$ are positive, we have $\|\Delta_j(\varphi)\|_{L^\infty} = \|\hat{\Delta}_j(\varphi)\|_{L^1}$. Now let f_λ be an eigenfunction in F_λ. We have

$$\left|\int_{-2^n\pi}^{2^n\pi} f_\lambda(2^{-n}\omega) \prod_{k=1}^n m(2^{-k}\omega)\, d\omega\right| = \left|\int_{-\pi}^{\pi} (T_m)^n f_\lambda(\omega)\, d\omega\right|$$
$$= |\lambda|^n \left|\int_{-\pi}^{\pi} f_\lambda(\omega)\, d\omega\right| \geq C|\lambda|^n. \qquad (5.23)$$

Note that we have supposed that $\int_{-\pi}^{\pi} f_\lambda(\omega)\, d\omega \neq 0$. If $\int_{-\pi}^{\pi} f_\lambda(\omega)\, d\omega = 0$, then the argument has to be modified slightly ; see below (after (5.25)). Since we have supposed that $\hat{\varphi}(\omega)$ does not vanish on $[-\pi,\pi]$, we have

$$\hat{\varphi}(\omega) \geq C \prod_{k=1}^n m(2^{-k}\omega) \qquad \text{for all } n \geq 0 \text{ and } |\omega| \leq 2^n\pi. \qquad (5.24)$$

Note that this hypothesis corresponds to the condition of Theorem 2.1 with $K = [-\pi, \pi]$. In a more general setting, we could replace the integrals on $[-2^n\pi, 2^n\pi]$ by integrals on $2^n K$ and the same results would hold. Combining (5.23) and (5.24) leads to

$$\int_{-2^n\pi}^{2^n\pi} |\hat{\varphi}(\omega)|\, |f_\lambda(2^{-n}\omega)|\, d\omega \geq C\lambda^n. \qquad (5.25)$$

If $\int_{-\pi}^{\pi} f_\lambda(\omega)\, d\omega = 0$, then a slightly more sophisticated argument will suffice. Lemma 5.2 still holds if the measure $d\omega$ is replaced by any other measure of the type $g(\omega)\, d\omega$ where g is a 2π-periodic, strictly positive, continuous function. We can always choose g such that

$$\int_{-\pi}^{\pi} f_\lambda(\omega) g(\omega)\, d\omega \neq 0. \qquad (5.26)$$

Then (5.23) holds if $d\omega$ is replaced everywhere by $g(\omega)\, d\omega$. Since g is strictly positive, this modified version of (5.23) combined with (5.24), still implies (5.25).

Since f_λ has a zero of order $2N$ at the origin, the function $\gamma(x)$, defined by $\hat{\gamma}(\omega) = |f_\lambda(\omega)|\, \chi_{[-\pi,\pi]}(\omega)$, is convenient for the Littlewood-Paley analysis of Hölder regularity less than $2N$. This is the case for φ, since $2N+1$ vanishing moments would be necessary for a Hölder exponent higher than $2N$ (see [11,18,12]). Consequently (5.25) tells us that φ cannot be more regular than C^α. To prove the optimality of $C^{\alpha-\varepsilon}$ when $T_m|_{F_\lambda}$ is not purely diagonal, it suffices to replace f_λ by a function g_λ such that $T_m g_\lambda = \lambda g_\lambda + \mu f_\lambda$ with $\mu \neq 0$. This leads to

$$\int_{-2^n\pi}^{2^n\pi} |\hat{\varphi}(\omega)|\,|g_\lambda(2^{-n}\omega)|\,d\omega \geq Cn\lambda^n \tag{5.27}$$

which proves the optimality of $C^{\alpha-\varepsilon}$. ∎

REMARKS. The estimates (5.22a) and (5.22b) can be found by an equivalent technique, using the transition operator T_p corresponding to the factor $p(\omega)$ in (5.12). We simply consider the largest eigenvalue λ_p and iterate T_p on $f \equiv 1$. This leads to

$$\int_{2^{j-1}\pi \leq |\omega| \leq 2^j \pi} |\hat{\varphi}(\omega)|\,d\omega \leq C \int_{2^{j-1}\pi \leq |\omega| \leq 2^j \pi} |\omega|^{-2N} \left[\prod_{k=1}^{j} p(2^{-k}\omega)\right] d\omega$$

$$\leq C 2^{-2Nj} \int_{-\pi}^{\pi} (T_p)^j 1\, d\omega$$

$$\leq C|\lambda_p + \varepsilon|^j\, 2^{-2Nj} \quad (\text{or } C|\lambda_p|^j\, 2^{-2Nj} \text{ if } T_p|_{F_{\lambda_p}} = \lambda_p I) ,$$

and thus $\varphi \in C^{\alpha-\varepsilon}$ with $\alpha = 2N - \log|\lambda_p|/\log 2$. This estimate is in fact the same as (5.22). Indeed, if μ is an eigenvalue of T_m in F_N, then its associated eigenfunction can be written as

$$f_\mu(\omega) = \left(\sin^2\left(\frac{\omega}{2}\right)\right)^N g_\mu(\omega). \tag{5.28}$$

Replacing $m(\omega)$ by its factorized form in

$$\mu f_\mu(\omega) = f_\mu\left(\frac{\omega}{2}\right) m\left(\frac{\omega}{2}\right) + f_\mu\left(\frac{\omega}{2} + \pi\right) m\left(\frac{\omega}{2} + \pi\right), \tag{2.29}$$

we obtain, after dividing by $\left[\sin^2\left(\frac{\omega}{2}\right)\cos^2\left(\frac{\omega}{2}\right)\right]^N$,

$$\mu 2^{2N} g_\mu(\omega) = g_\mu\left(\frac{\omega}{2}\right) p\left(\frac{\omega}{2}\right) + g_\mu\left(\frac{\omega}{2} + \pi\right) p\left(\frac{\omega}{2} + \pi\right). \tag{5.30}$$

We see that the eigenvalues of T_p are given by $\mu_p = 2^{2N}\mu$. This proves the equivalence between the two techniques.

In general $m(\omega)$ is not a positive function. One can then define $M(\omega) = |m(\omega)|^2$ and use the operator T_M associated with $M(\omega)$. The result is an estimate of the L^2 norms of $\Delta_j(\varphi)$. Using the Cauchy-Schwarz inequality, we derive the following result.

Corollary 5.5. *Suppose that $M(\omega) = |m(\omega)|^2$ has a zero of order $2N$ at $\omega = \pi$. Let λ be the largest eigenvalue of T_M on F_N, and define $\alpha = -\log|\lambda|/(2\log 2)$. Then, $\varphi \in H^{\alpha-\varepsilon} \subset C^{\alpha-\frac{1}{2}-\varepsilon}$, where H^s is the Sobolev space of index s. The value α is attained if $T_M|_{F_\lambda} = \lambda I$.*

Note that the Hölder exponent cannot be optimal because we have used the Cauchy-Schwarz inequality and $\hat{\varphi}(\omega)$ is not a positive function. The Sobolev exponent, however, is optimal (see also [13] and [20]). The regularity of compactly supported wavelets was estimated with this method in [9].

A very interesting experimental fact should be mentioned here: in the case of the orthonormal wavelet bases that are constructed in [9], it seems (at least for the four smallest filters of the family) that although $m_0^N(\omega)$ (as defined in Section 4) is not a positive function, the spectral radius of T_{m_0} restricted to F_N leads to the exact Hölder exponent of φ^N. (The latter can be computed exactly by a more sophisticated method described below.) So far, we have not found any rigorous explanation of this remarkable fact.

Transition operators also play a crucial role in the biorthogonal theory with the following result:

Theorem 5.6. *Let $m_0(\omega)$ and $\tilde{m}_0(\omega)$ be a pair of dual filters, and let T_0 and $\tilde{T}_0$ be the transition operators associated with $|m_0|^2$ and $|\tilde{m}_0|^2$. Let F_N and $F_{\tilde{N}}$ be the invariant subspaces defined by Lemma 5.3. Then m_0 and $\tilde{m}_0$ generate a pair of stable biorthogonal wavelet bases if and only if*

$$\rho(T_0, F_N) < 1 \quad \text{and} \quad \rho(\tilde{T}_0, F_{\tilde{N}}) < 1, \tag{5.31}$$

where $\rho(T, F)$ is the spectral radius of T restricted to F.

The proof of this result is given in [5]. It exploits the fact that by (5.31), the functions φ and $\tilde{\varphi}$ are more than square-integrable: they belong to a Besov space $B_2^{\varepsilon,\infty}$ for some $\varepsilon > 0$. This little "extra" ingredient induces the stability of the multiscale bases. Conversely, (5.31) is a consequence of the biorthogonality relations $\langle \varphi(x-k), \tilde{\varphi}(x-l) \rangle = \delta_{k,l}$.

5b. The time-domain approach. Let $m(\omega) = \sum_{n=0}^{N} c_n e^{in\omega}$. This is a trigonometric polynomial, and we assume that $m(0) = 1$ and $m(\pi) = 0$. We do not require that $m(\omega)$ be positive. Let $\varphi(x)$ be the scaling function defined by the infinite product (3.18). It is at least a compactly supported distribution in $[0, N]$.

In the time-domain approach, we represent $\varphi(x)$ by its "vector" form,

$$w(x) : [0,1] \to \mathbb{R}^N \qquad [w(x)]_n = \varphi(x+n-1) \qquad n = 1, \ldots, N. \tag{5.32}$$

From the two-scale difference equation (1.5) we get

$$w(x) = \begin{cases} T_0 \, w(2x) & \text{if } x \leq \frac{1}{2}; \\ T_1 \, w(2x-1) & \text{if } x \geq \frac{1}{2}, \end{cases} \tag{5.33}$$

where T_0 and T_1 are $N \times N$ matrices defined by

$$(T_0)_{i,j} = c_{2i-j-1} \quad 1 \leq i,j \leq N \tag{5.34a}$$
$$(T_1)_{i,j} = c_{2i-j} \quad 1 \leq i,j \leq N. \tag{5.34b}$$

Using the notation

$$d_n(x) = n\text{-th binary digit of } x \in [0,1],$$

$$\tau(x) = \begin{cases} 2x & \text{if } x \leq \tfrac{1}{2} \\ 2x - 1 & \text{if } x \geq \tfrac{1}{2} \end{cases} \quad \text{(binary shift)},$$

we can rewrite (5.33) as a "fixed point" equation

$$w(x) = T_{d_1(x)}\, w(\tau(x)). \tag{5.35}$$

This leads to an evaluation of $w(x)$ and its derivative by an iterative process. The regularity of the result depends, of course, on the spectral properties of T_0 and T_1. Note that when $m(\omega)$ has a zero of order L (as for the transition operator studied in the previous section), the space F_L orthogonal to the vectors $p_j = (n^j)_{n=1,\ldots,N}$ for $j = 0, \ldots, L-1$ is invariant under T_0 and T_1. This method gives sharp estimates of the local regularity in x by considering the products $T_{d_1(x)} \ldots T_{d_n(x)}$ for all $n \geq 0$. The main result on global regularity proved in [11] (Theorem 3.1) is the following

Theorem 5.7. *Suppose that there exists $\rho < 1$ such that, for all binary sequences $(d_j)_{j \in \mathbb{Z}}$ and all $m > 0$, we have*

$$\|T_{d_1} T_{d_2} \ldots T_{d_m}|_{F_L}\| \leq C \rho^m. \tag{5.36}$$

Define $\alpha = -\log \rho / \log 2$. Then,
(i) *if α is not an integer, φ belongs to C^α;*
(ii) *if α is an integer, $\varphi^{\alpha-1}$ is almost Lipschitz:*

$$|\varphi^{\alpha-1}(x+t) - \varphi^{\alpha-1}(x)| \leq C|t|\,|\log|t||\quad \text{for all } (x,t).$$

REMARKS. The "generalized spectral norm"

$$\rho(T_0, T_1) = \limsup_{m \to \infty} \left[\sup_{\substack{d_j = 0 \text{ or } 1 \\ j=1,\ldots,m}} \|T_{d_1} T_{d_2} \ldots T_{d_m}|_{F_L}\|^{1/m} \right] \tag{5.37}$$

gives a sharp estimate of the global regularity. Note that it is in general superior to the spectral radius of T_0 and T_1. When N is not too large it is possible to compute the exact value of $\rho(T_0, T_1)$. For example, in the case of orthonormal wavelets, the optimal Hölder exponent was found in [11] for $N = 4, 6$ or 8. The same evaluation becomes more difficult for larger filters.

The generalization of this approach to higher dimensions (multivariate cases) is not trivial. In particular, it involves nonstandard binary expansions depending on the dilation matrix that is used.

As a conclusion of this review of regularity estimators, we could say that these three approachs are complementary: the time-domain method gives sharp results but it is only practicable for small filters; the Littlewood-Paley estimates can be derived for longer filters but they will be optimal only if $m(\omega)$ is a positive function; and finally, the Fourier approach is less precise but appropriate to asymptotic results on very large filters. Another method recently developed by O. Rioul [16] is based on $l^1(\mathbb{Z})$ norm estimates of the iterated filters and leads to interesting results; in particular, it is still manageable for larger filters than can be accommodated by the time domain method of [11].

References

1. Caveretta, A., W. Dahmen, and C. Micchelli, *Stationary Subdivision*, Memoirs of AMS Number 453, American Mathematical Society, Providence, 1991.
2. Cohen, A., Ondelettes, analyses multirésolutions et filtres miroirs en quadrature, Ann. Inst. H. Poincaré, Analyse non linéaire **7** (1990), 439–459.
3. Cohen, A., Construction de bases d'ondelettes α-Höldériennes, Revista Matemática Iberoamericana **6** (1990), 91–108.
4. Cohen, A. and J. P. Conze, Régularité des bases d'ondelettes et mesures ergodiques, Revista Matemática Iberoamericana, to appear.
5. Cohen, A. and I. Daubechies, A stability criterion for biorthogonal wavelet bases and their related subband coding schemes", preprint, AT&T Bell Laboratories, 1991.
6. Cohen, A., I. Daubechies, and J. C. Feauveau, Biorthogonal bases of compactly supported wavelets, Comm. Pure & Appl. Math., to appear.
7. Conze, J. P., Sur le calcul de la norme de Sobolev des fonctions d'échelles, Dept. de Math. Université de Rennes (France), 1990, preprint.
8. Conze, J. P. and A. Raugi, Fonction Harmonique pour un opérateur de transition et application, Dept. de Math. Université de Rennes (France), 1990, preprint.
9. Daubechies, I., Orthonormal bases of compactly supported wavelets, Comm. Pure & Appl. Math. **41** (1989), 909–996.
10. Daubechies, I., *Ten Lectures on Wavelets*, CBMS Conference Vol. 61, SIAM, Philadelphia, 1992.
11. Daubechies, I. and J. Lagarias, Two scale difference equations. Part I & II, SIAM J. Math. Anal., to appear.
12. Dyn, N. and D. Levin, Interpolating subdivision schemes for the generation of curves and surfaces, Math. Dept. Tel Aviv Univ, 1989, preprint.
13. Eirola, T., Sobolev characterization of solutions of dilation equations, Helsinki University of Technology. SIAM J. Math. Anal., submitted.

14. Mallat, S., Multiresolution approximation and wavelets orthonormal bases of $L^2(\mathbb{R})$, Trans. Amer. Math. Soc. **315** (1989) 69–87.
15. Meyer, Y. (ed.), *Ondelettes et Opérateurs*, Hermann, Paris, 1990.
16. Rioul, O., Dyadic up-scaling schemes: simple criteria for regularity, SIAM J. Math. Anal., submitted.
17. Smith, M. J. and T. P. Barnwell, Exact reconstruction techniques for tree structured subband coders, IEEE ASSP **34** (1986) 434–441.
18. Strang, G. and G. Fix, A Fourier analysis of the finite element variational method, in *Constructive Aspects of Functional Analysis*, G. Geymonant (ed.), C.I.M.E., 1973, 793–840.
19. Vetterli, M., Filter bank allowing perfect reconstruction, Signal Processing **10** (1986), 219–244.
20. Villemoes, L., Energy moments in time and frequency for two-scale difference equation solutions and wavelets, Math. Institute, Technical University of Denmark, preprint.
21. Volkner, H., On the regularity of wavelets, Dept. of Math. Univ. of Wisconsin, Milwaukee, preprint.

Albert Cohen
CEREMADE
Université Paris IX-Dauphine
Place du Maréchal de Lattre de Tassigny
75775 Paris Cedex 16
France
chavent@frulm63.bitnet

Approximation with Convex Rational Functions

B. Gao, D. J. Newman, and V. Popov

Abstract. In this paper we study the approximation of $|x|$ and of general convex functions f with continuous first derivatives by convex rational functions of degree n. For $|x|$, the uniform approximation order $e^{-c_2\sqrt{n}}$ is achieved by using an H^∞ quadrature formula. For f, we get order $n^{2-\epsilon}\|f'\|$, where $\epsilon > 0$.

§1. Introduction

The desire for co-monotone and even co-convex approximations has become very much in vogue during the last few years. The reason for this kind of approximator becomes obvious when one pictures a design for, say, a automobile *fender*.

The Tchebychev approximation *must* "zig zag" about the smooth flowing curve of the fender. The irony is that the very thing that makes for a "best possible" approximation (the alternation of signs of the maximum error) is what makes it have such an unsmooth shape. When students are shown the pictures of the true fender and the approximator, they react by describing them as "before" and "after" the collision!

So it seemed worthwhile to exchange some of the proximity for some "shape preservation." The list of names of contributors is quite long: Lorentz, Zeller, Passow, Raymon, Roulier, Iliev, Newman, to mention just a few, and quite recently, we must add, Leviatan.

The first attempt at these "form-fitting" approximations was the co-monotone polynomial approximation. The results therein were encouraging: the co-monotone polynomial can be chosen with proximity $cw(1/n)$, so the *cost* is just a constant multiple of the best possible one.

The form of this constructed polynomial, $\int (p \cdot ((f'/p) * K))$, shows the pleasant linear operator level of the problem. Here $\int$ is the integral operator,

p is the fixed polynomial bending at the bending points of the given function, K is the Jackson Kernel and f is the given function, properly modified (mollified).

But those were the good-old-days when life was linear. Life is much more complex for rational function approximation, the non-linear age.

This transition to the non-linear case is much the same as the transition from the Riemann to the Lebesgue definition of the integral. Splitting the x-axis is linear; splitting the y-axis is not.

A beautiful example of just this non-linearity is the lemma of Popov [2], which is vital to his solution of the Lip 1 conjecture for rational function approximation.[†]

Let f be an increasing function such that $f(0) = 0$ and $f(1) = 1$. Then the curve can be covered by n rectangles each of area at most $1/n^2$.

Indeed the ("book") proof of this lemma is accomplished by splitting the $(x+y)$-axis, in other words choosing the break points where $x+f(x)$ takes the values $2k/n$, $k = 0, 1, \ldots, n$. We leave it to the reader to fill in the enjoyable details.

A surprising fact about this "form-fitting" kind of approximation was observed by D. Kellman when he noticed that the classical 1964 rational function fit to $|x|$ was in fact co-monotone. This was the first super-good co-monotone approximation. The previously mentioned class-wide $\omega(1/n)$-result reflected only *ordinary* proximity. The form of the Kellman proof was also remarkable. One might say that the result proved itself, the point being that an error so small as $e^{-c\sqrt{n}}$ could not be "enormously" upset by differentiation. The crudest estimates reflect this, and we have a proof by *immense wealth*.

Guided, or perhaps misguided, by this co-monotonicity, we hoped that this same construction would give a *convex* approximator. Of course this isn't true, and even the best possible ($e^{-\pi\sqrt{n}}$) improvement due to Vjacheslovov couldn't be convex. Both of these constructions are *interpolatory* after all and so, of course, doomed not to be convex. (A curve that cuts $|x|$ five or more times cuts either x or -x three times and cutting a line three times is the very definition of non-convexity.)

§2. Approximation of $|x|$

The approximation of $|x|$ is of central importance for general classes of functions, and it behooved us to search for a convex *gigantically* close ($e^{-c\sqrt{n}}$ proximity) rational approximation to $|x|$. The existence of such an entity was not at all clear since the obvious nominees were, as previously observed, definitely not convex. After many doubts and vain attempts the affirmative answer was finally obtained. Curiously, success came, not from the $|x|$ approximation problem, but from its *dual problem*, that of quadrature formulas in H^2.

[†] Paper completed after the untimely death of Vasil Popov.

Convex Rational Approximation

The list of contributors to these quadrature formulas is quite long and includes e.g., Schwartz, Loeb, Werner, Haber, Takahashi, Mori, Stenger, Newman, and Andersen. The trick is to use

$$\tan^{-1} x = \frac{1}{2} \int_{-1}^{1} \frac{x \, dt}{1 + t^2 x^2},$$

and apply the quadrature formula to get

$$\tan^{-1} x = \frac{x}{2} \sum_{k=1}^{n} A_k \frac{1}{1 + t_k^2 x^2} + E,$$

E being the error.

This gives us an excellent ("gigantically close") rational function approximation to $\tan^{-1} x$. If we then note that

$$|x| = \frac{2}{x} \pi \tan^{-1} Nx + \mathcal{O}\left(\frac{1}{N}\right),$$

where N is *gigantic* we obtain the *gigantically* close rational function approximation to $|x|$. Here as before we use the term gigantic to mean $> e^{r\sqrt{n}}$ (where n is large but N is gigantic).

But the search was for *convex* rational functions, and this construction doesn't at first glance seem to fit the bill at all. Surely the individual terms $x/(1 + k^2 x^2)$ are not convex; but can the combination of terms in the quadrature formula be so? The answer is that this quadrature formula is not just any old linear combination, and this linear combination is convex for much the same reason that Kellman's observation worked.

The point is that the H^2 quadrature operator and integral operator are convolutions and therefore commute with differentiation. Thereby we may justify differentiating our asymptotic formula. (We still will insist to our students that differentiation of approximate formulas is a No-No, but here as in Kellman's case we find we are allowed to do it.) Omitting details, then, we pass from

$$x \tan^{-1} Nx = \sum_k A_k \frac{x^2 N}{1 + x^2 t_k^2 N^2} + \mathcal{O}\left(e^{-\sqrt{n}}\right)$$

to

$$\frac{2N}{(1 + N^2 x^2)^2} = \left(\sum A_k \frac{x^2 N}{1 + x^2 t_k^2 N^2}\right)'' + \mathcal{O}\left(N^2 e^{-\sqrt{n}}\right)$$

and conclude that,

$$\left(\sum A_k \frac{x^2 N}{1 + x^2 t_k^2 N^2}\right)'' > \frac{2}{N^3} - CN^2 e^{-\sqrt{n}}$$

which is positive as long as we choose, for example, $N = e^{\sqrt{n}/6}$.

So indeed our rational function is a lucky combination of terms which does turn out to be convex, and we have succeeded in producing a gigantically close, convex rational approximation to $|x|$. The details are scheduled to appear in Gao's thesis (Shape Preserving Approximation by Rational Functions).

§3. Approximation of Convex Functions

But now remember, this result was not obtained just for its aesthetic beauty; $|x|$ does play a central role in the general theory. The book of Petrushev and Popov [4] develops a whole collection of procedures which pivot on the function $|x|$ and its ultra-good approximation. One of the beautiful results (cf. Erdös, Túran) is about convex functions with bounded derivative, namely,

Theorem 1. *If $f(x)$ is convex with bounded derivative, then there is a rational function $r(x)$ of degree n such that $\|f(x) - r(x)\| \leq c/n^2$.*

Their proofs and methodology style themselves after Popov's proof of the Lip 1 conjecture, which doesn't seem to be able to incorporate this new convex approximation. We found, however, that if we looked instead to Newman's proof of this Lip 1 conjecture, then the new approximation can be used effectively. There was a loss of precision, however, and the best we seemed to be able to get was:

Theorem 2. *If $f(x)$ is convex with bounded derivative, then there is a rational function $r(x)$ of degree n which is convex and such that $\|f(x) - r(x)\| \leq c/n^{1.5}$.*

When this proof was shown to the great Russian mathematician A. A. Gonchar at a meeting in May, 1991 (in what was then Leningrad), he suggested an enormous improvement. In geometric language we were breaking the given $f(x)$ into a sum of triangle functions, and Gonshar suggested that a better thing to do would be to split $f(x)$ into "ski-slope" functions. Indeed these were the really appropriate building blocks, and using them instead of the triangles, we were able to get the almost best possible result:

Theorem 3. *If $f(x)$ is convex with bounded derivative, then there is a rational function $r(x)$ of degree n which is convex and such that $\|f(x) - r(x)\| \leq (c/n^2)\log^4 n$. This is the same precision as the orginal Petrushev-Popov theorem except for the power of $\log n$.*

§4. Remarks

So among other things, we have replaced the original construction of a gigantically close approximation to $|x|$ by a seemingly more commonplace object, a quadrature formula. But are we kidding ourselves? Is this H^2 quadrature formula really so commonplace? Indeed, there is a form of this formula that is postively shocking. It results from the usual one by a map taking $[-1, 1]$ onto $(-\infty, \infty)$ and thereby replacing H^2 by a certain Hilbert space of analytic functions. The form of the quadrature formula is the remarkable equality

$$\int_{-\infty}^{\infty} f(x)\,dx = h\sum_{-\infty}^{\infty} f(nh) + \mathcal{O}\left(e^{-2\pi^2/h}\|f\|\right),$$

where $\|\cdot\|$ is the norm in the new Hilbert space.

The formula looks very ordinary at first glance, the integral being equal to its Riemann sum plus an error; but look again. The error here is not the usual h^2 or h^3 or whatever. This time it's a miracle error, $e^{-2\pi^2/h}$.

Of course the complexity is hidden in the form of the $\|\cdot\|$, i.e., the definition of the Hilbert space. The details and proofs will appear in a forthcoming paper of Wubao Wang (The error estimate of H^2 quadrature), where an attempt is made to follow *Ramanujan's rule*.

Ramanujan apparently always believed that remarkably close *near* equalities are never an accident but stem from hidden exact identities. Wang has found an identity of the form

$$\sum_n \tanh^N nh \operatorname{sech}^2 nh = \frac{1}{h} e^{-2\pi^2/h} \sum_m \phi\left(m\frac{\pi^2}{h}\right) + C_N,$$

from which the aforementioned quadrature formula can be verified for the functions $\tanh^N x \operatorname{sech}^2 x$. These form a complete orthonormal system for our Hilbert space, and so may yield the hidden exact identity.

References

1. Andersson, J. E., Optimal quadrature of H_p functions, Math. Z. **172** (1980), 55–62.
2. Newman, D. J., Rational approximation to $|x|$, Michigan Math. J. **11** (1964), 11–14.
3. Newman, D. J., *Approximation with Rational Functions*, CBMS Regional Conference Series in Math. Vol. 41, Amer. Math. Soc., Providence, R. I., 1979.
4. Petrushev, P. P. and V. A. Popov, *A Rational Approximation of Real Functions*, Cambridge Univ. Press, 1987.
5. Popov, V. A., Uniform rational approximation of the class V_r and its applications, Acta. Math. Acad. Sci. Hung. **29** (1977), 119–129.
6. Vjacheslavov, N. S., On the uniform approximation of $|x|$ by rational functions, Dokl. Akad. Nauk SSSR **220** (1975), 512–515. (Russian) MR 52#1114

B. Gao and D. J. Newman
Department of Mathematics
Temple University
Broad & Montgomery
Philadelphia, PA 19122

V. Popov (deceased)

Block Structure and Recursiveness in Rational Interpolation

Martin H. Gutknecht

Abstract. We first review the Newton-Padé approximation problem and the block structure of the corresponding Newton-Padé table, which by definition contains the solutions of a double sequence of rational interpolation problems. Then we define and investigate an analogous multipoint Padé table containing the solutions of another double sequence of rational interpolation problems which include $z = \infty$ as a possibly multiple interpolation point. Finally, we derive a general recurrence formula that can be used as the basis of a stable recursive interpolation process. Briefly, it says that, for $|\mu| \leq k$, a Newton-Padé approximant of type (m,n) can be updated to one of type $(m + \mu, n + k)$ satisfying additional $\mu + k$ interpolation conditions by computing a $[\mu + 1; k]$ multipoint Padé approximant.

Introduction

Rational interpolation is a very old topic. It is a natural generalization of polynomial interpolation and can be formulated on an elementary level. Nevertheless, it leads to a nontrivial, very rich theory. This theory establishes close relationships to a number of other topics, such as continued fractions, orthogonal polynomials, moment problems, Hankel and Toeplitz matrices, tridiagonal matrices, the Lanczos process, and the conjugate and biconjugate gradient methods, to name the most important ones.

A first incentive for studying rational interpolation is its usage as a cheap alternative to rational approximation. In a large number of situations the latter is known to be very effective in the sense that a given function can be approximated up to a tiny error by a rational function of small order. Clearly, in applications such as the approximation of special functions for library routines, one should compute the best uniform approximation. In other cases, interpolation may be sufficiently good. However, the importance of rational interpolation is due to the fact that in many situations the interpolation data is really all one knows about the function to be approximated. Moreover, in many applications the rational interpolation problem does not appear explicitly, but is hidden behind the scene. Among these applications there are some where the order of the rational interpolant that is generated implicitly can become very large, say of the order $10^2 - 10^4$. To this class belong the (bi)conjugate gradient method for solving systems of linear equations, the Lanczos algorithm for tridiagonalizing a matrix and computing eigenvalue approximations, fast algorithms for solving linear systems with Hankel or Toeplitz matrices, and model identification and reduction in systems and control theory.

Most of the examples just mentioned actually refer to a very special case of rational interpolation, called *Padé approximation*, where all the data are prescribed at one point, $z = 0$ or $z = \infty$, in the form of a function value and consecutive derivatives. In other words, the initial section of a formal power series is given (or, at least, it can be constructed during the interpolation process). In the theory of Padé approximation one normally assumes that a full power series is given and considers a double sequence of rational approximants $r_{m,n}$, where m denotes the numerator and n the denominator degree. The Padé approximation $r_{m,n}$ (at $z = 0$) is chosen such that its Taylor series at 0 matches as many terms as possible in the given formal power series. A variation is two-point Padé approximation, where two formal power series, one in z and the other in $1/z$, are to be matched. It plays a particular role in general recurrences for Padé approximants, see [36].

Except for the Padé problem, the classical treatment of rational interpolation was mostly restricted to the so-called *Cauchy problem* [50], where finite values are prescribed at distinct interpolation points. Allowing multiple interpolation points, *i.e.*, prescribed derivatives up to a certain order, is, as we will see, a minor generalization if the appropriate notation is used. We then say that we have a rational interpolation or a *Newton-Padé approximation* problem. It has also been referred to as the osculatory rational interpolation problem [15,58,79], the rational Hermite interpolation problem [3,16,72], or the multipoint-Padé approximation problem [4]. It is, of course, included in our treatment, which will additionally allow us to prescribe poles.

As in Padé approximation we will consider a sequence or a double sequence of nested problems as indicated above. The term Newton-Padé approximation comes from the fact that a *formal Newton series* is to be matched instead of a formal power series. The Newton series is the key to a formally simple treatment of multiple interpolation points (*i.e.*, of prescribed function values and derivatives), but in practice it is hardly ever used. We will use the terminology *multipoint Padé approximation* for the case where additionally interpolation conditions at infinity are prescribed. (Some other authors use this designation for what we call Newton-Padé approximation.) Such problems are not just of academic interest, but are an important tool for the recursive computation of Newton-Padé approximants, as we will see in §5.

Much work on Padé approximation and rational interpolation, in general, has gone into generalizations where the coefficients of the series are vectors or matrices, *i.e.*, where one interpolates vector-valued or matrix-valued data. Vector-valued Padé approximation has a long history and comes in several flavors. Two were introduced by Hermite. One of them is now called the Padé-Hermite approximation problem; in earlier papers it has been referred to as the Latin or type 1 problem. The other is now quoted as the simultaneous or vector Padé approximation problem; earlier it has been called the German or type 2 problem. See, *e.g.*, [68,57,56] for references on the Padé-Hermite problem, and [9,70] for the simultaneous Padé approximation problem. The multipoint extension of the Padé-Hermite problem, which was pioneered by Mahler [46], is sometimes called the M-Padé problem [6,5]. Multipoint simultaneous Padé approximation is treated in, *e.g.*, [67]. Yet another form of vector rational interpolation is treated in [31,32,29]. The matrix case is of great importance in systems and control theory, see, *e.g.*, [1,23,42,65,64,34]. There has also been considerable work regarding multivariate rational approximation, but this topic will not be broached here.

Here we will restrict ourselves to the 'scalar' case, and we try to discuss the subject on a rather elementary level. For example, we let the coefficients be complex numbers. Instead they could be chosen from a more general field, and the interpolation conditions could be replaced by linear functionals.

Given the long history of the subject it may seem astonishing that just recently considerable progress has been made on the scalar univariate rational interpolation problem (or Newton-Padé approximation problem) and even on the standard Padé approximation problem, in particular regarding the *recursive computation of approximants*, and also regarding the convergence theory, which is not considered here, however. Many of the new ideas can be adapted to the vector-valued and matrix-valued case, which underlines their usefulness. To explain what they are about, we have to refer to the Padé and Newton-Padé tables. Given an infinite series of interpolation data (*i.e.*, a Taylor series or a Newton series) these tables list the double series of interpolants $\{r_{m,n}\}_{m,n=0}^{\infty}$. We let the n-axis point to the right, and the m-axis downward. Hence, m is fixed on a row, n is fixed on a column, and $m - n$ is constant on a diagonal.

One of the complications in the theory of rational interpolation and Padé approximation, in particular, is the possibility of so-called *singular blocks* in the respective table. These blocks are sets of identical entries, *i.e.*, rational functions that are for several pairs (m, n) equal to the interpolant. The occurrence of such singular blocks is paired with the existence of unattainable points; *i.e.*, with the existence of 'interpolants' that do not satisfy all conditions, but are in a certain sense best possible. These blocks have nothing to do with multiple interpolation points.

In the Padé case, all blocks are square (if $r_{m,n} \neq 0$) [26], while in the Newton-Padé case, they are a union of squares. For a long time, blocks received little attention; cf. some of the well-known books on Padé approximation [4,7]. The antipathy of many authors to blocks may be related to the often excessive use of determinants in Padé and rational interpolation theory.

We will here purposely avoid representations of interpolants by determinants. They are often just explicit expressions for the solution of a linear system, obtained by applying the classical formula involving the adjugate; in practice they are useless, and they are no longer valid when the system becomes singular, as will happen when singular blocks exist.

For the Newton-Padé table the investigation of the block structure started with Claessens's thesis, see [17]; but a complete characterization was only recently given in [37]. (While [37] was in press, Beckermann [6] established independently results on the block structure of the M-Padé table, which includes the Newton-Padé table as a special case.)

There exists a wide variety of recursive methods for computing interpolants. Most of them are based on simple relations (often called identities) among a few neighboring entries of the table. They require that all these interpolants be distinct. Identities that link entries at the border of singular blocks are often called *singular rules*. We will also refer to them as *nongeneric recurrences*, and to the recursive methods making use of them as *nongeneric algorithms*. Note that such nongeneric recurrences require exact arithmetic, because one needs to detect the exact size of singular blocks. For the Padé case, Magnus [45], McEliece and Shearer [49], Gragg and Lindquist [27], Draux [20], and Cabay and Choi [10] have presented such nongeneric recurrences, see also [39,35]. For the rational interpolation problem Werner [74] and Arndt [3] treated a special type of a nongeneric situation that appeared when they aimed at a certain reordering of the data. More general treatments of singular blocks in rational interpolation (and exact arithmetic) are due to Antoulas and Anderson [2], Gutknecht [37], Van Barel and Bultheel [66], and Beckermann [5].

However, if subject to roundoff, all these methods are numerically unstable in cases of near-singularity. One way to attain stability is to make use of the polynomial formulation of the stabilized look-ahead Lanczos algorithm defined in §9 of [35]. This algorithm produces implicitly the denominators of the well-conditioned Padé approximants along a diagonal of the Padé table. Computing additionally the numerators is nearly trivial, since the recurrences are the same. Another (weakly) stable algorithm for the recursive construction of Padé approximants along a diagonal is due to Cabay and Meleshko [12,51]. It is particularly suited for this task, and it is a generalization of classical recursive procedures. Currently this approach is being generalized to rational interpolation, see [11].

The objective of this paper is to present a general framework for the above mentioned algorithms, by reviewing the block structure of the Newton-Padé table and investigating the one of the multipoint Padé table, and by giving a general recurrence formula that can be seen to incorporate the recurrences of most of the other procedures.

While the stable recurrences of Cabay and Meleshko generate well-conditioned approximants on a particular diagonal, the general recurrence presented here allows us to move in the Padé table or the Newton-Padé table in any direction in which $m + n$ does not decrease. This recurrence makes use of rational interpolants with prescribed behavior at infinity, i.e., of multipoint Padé approximants. In the Padé case, which is treated in [36], the latter are just two-point Padé approximants.

Due to the close relationship between orthogonal polynomials, Hankel and Toeplitz matrices, and Padé approximation, the above mentioned Padé algorithm is also of relevance for the fast solution of linear systems with Toeplitz and Hankel matrices, where numerical stability is attained by related methods; see Cabay and Meleshko [12,51], Chan and Hansen [13], Freund and Zha [22], Gutknecht [36]. As already mentioned, there is also a connection to the look-ahead Lanczos algorithm of Parlett, Taylor, and Liu [55], which has recently been analyzed, revived, and extended to cover near-breakdowns; see Gutknecht [39,35], Joubert [41,40], Parlett [54], and Freund, Gutknecht and Nachtigal [21].

This paper is organized as follows. In §1 we formulate the Newton-Padé approximation problem and state a few basic properties of Newton-Padé forms and Newton-Padé approximants. In §2 we review the block structure of the Newton-Padé table. Then, in §3, we define what we call the multipoint Padé approximation problem, an analog of the Newton-Padé approximation problem with additional interpolation conditions at ∞. The block structure of the corresponding multipoint Padé table is investigated in §4. Finally, in §5, we derive a general recurrence formula for the Newton-Padé table. It is based on solving a small multipoint Padé approximation problem in each step.

§1. Newton Series, Newton-Padé Forms, Newton-Padé Approximants, and True Rational Interpolants

Given a sequence $\{z_j, \phi_j\}_{j=0}^{\infty}$ of pairs of complex numbers, we define the polynomials

$$t_0(z) := 1, \qquad t_k(z) := \prod_{j=0}^{k-1}(z-z_j), \qquad k = 1, 2, \ldots, \qquad (1.1)$$

and consider the *formal Newton series* [24,71,72]

$$f(z) := \sum_{k=0}^{\infty} \phi_k t_k(z). \qquad (1.2)$$

This is a natural generalization of a formal power series. At each of the points z_j it has a well-defined value $\sum_{k=0}^{j} \phi_k t_k(z)$. Moreover, if $\kappa(k)$ denotes the number of occurrences of z_k in $\{z_0, \ldots, z_{k-1}\}$, i.e., if

$$\kappa(k) := |\{z_j : j < k,\ z_j = z_k\}|, \qquad (1.3)$$

the formal derivatives of f (defined by termwise differentiation) up to the order

$$\sup\{\kappa(k) : k \in \mathbb{N},\ z_k = z_j\} \qquad (1.4)$$

also have well defined values at z_j, since the series for these formal derivatives reduce again to finite sums if evaluated at z_j.

The lth partial sum of the formal Newton series (1.2), that is,

$$f_l(z) := \sum_{k=0}^{l} \phi_k t_k(z), \tag{1.5}$$

is a polynomial of maximum degree l with leading coefficient ϕ_l. It is exactly the interpolation polynomial of f at the points $z_0, \ldots, z_l$:

$$\left. \frac{d^{\kappa(k)} f_l}{dz^{\kappa(k)}} \right|_{z=z_k} = \left. \frac{d^{\kappa(k)} f}{dz^{\kappa(k)}} \right|_{z=z_k}, \qquad k = 0, \ldots, l. \tag{1.6}$$

In fact, (1.5) is *Newton's interpolation formula*, and it is well known that the coefficient ϕ_k is the kth *divided difference* of f at $z_0, \ldots, z_k$, which is denoted by $[z_0, \ldots, z_k]f$:

$$\phi_k = [z_0, \ldots, z_k]f, \tag{1.7}$$

see, e.g., [19,52,62,76]. Often the kth divided difference of a function is defined as the leading coefficient of the interpolation polynomial f_k which makes use of the same points. In contrast to the usual recursive definition, confluence of points causes then no problem. We note that $[z_0, \ldots, z_k]$ is a linear functional acting on functions defined at these points. For its many properties see, e.g., [19,52,76].

Hence, given any finite or infinite sequence of Hermite interpolation data

$$\left(z_k, \frac{d^{\kappa(k)} f}{dz^{\kappa(k)}}(z_k) \right), \qquad k = 0, 1, \ldots, \tag{1.8}$$

the formal Newton series (1.2) with coefficients (1.7) satisfies all these interpolation conditions, and its lth partial sum is the interpolation polynomial satisfying the first $l+1$ of them. A nonpolynomial interpolation problem for the same $l+1$ data can therefore be reformulated as the problem of matching the first $l+1$ terms of this Newton series. However, in general there is no guarantee that the series converges at any specific point $z \notin \{z_j\}$.

Since there is a one-to-one correspondence between Newton series (1.2) and Hermite interpolation data (1.8), we use the same notation $\mathcal{N}_Z$ for the sets of these objects. The index refers to the sequence $Z := \{z_j\}_{j=0}^{\infty}$, which is assumed to be fixed. Each $\mathcal{N}_Z$ becomes a commutative algebra if addition, scalar multiplication, and multiplication are defined pointwise (using the Leibniz rule for the derivatives of the product.) There arises the question of how these operations affect the Newton series, *i.e.*, how they can be expressed in terms of the coefficients ϕ_k. To give an answer, we have to refer to the corresponding rules for divided differences: if $f, g \in \mathcal{N}_Z$, $\lambda \in \mathbb{C}$, and $l \leq k$, the linearity yields

$$[z_l, \ldots, z_k](\lambda f + g) = \lambda [z_l, \ldots, z_k]f + [z_l, \ldots, z_k]g, \tag{1.9}$$

Rational Interpolation   99

and the "Leibniz rule" for divided differences [19,52], which Opitz [53] attributes to Steffensen, says that

$$[z_l,\ldots,z_k](fg) = \sum_{j=l}^{k} ([z_l,\ldots,z_j]f)([z_j,\ldots,z_k]g) \ . \tag{1.10}$$

Actually only the case $l = 0$ of these two formulas is used for the coefficients of the resulting Newton series. But it has been noticed by Warner [72] that (1.9) and (1.10) imply that the algebra of formal Newton series is isomorphic to a commutative subalgebra of the algebra of infinite left triangular matrices. The isomorphism is induced by the mapping that associates f with

$$\mathbf{F} := \begin{bmatrix} \phi_{00} & & & \\ \phi_{10} & \phi_{11} & & \\ \phi_{20} & \phi_{21} & \phi_{22} & \\ \vdots & \vdots & & \ddots \end{bmatrix}, \qquad \text{where} \quad \phi_{kl} := [z_l,\ldots,z_k]f \ . \tag{1.11}$$

In particular, (1.10) is exactly the usual definition of the (commutative) product of two such matrices. Similarly, the $N \times N$ matrices of the form (1.11) yield a commutative algebra isomorphic to the one formed by N-term Newton series. The relevant properties of these matrices have been discovered before by Opitz [53] who calls them "Steigungsmatrizen", which should be translated as *divided differences matrices*. Note that on the diagonal these matrices contain the function values, while in the first column one finds the given divided differences $\phi_{k0} = \phi_k$. The other columns can be generated from the first one:

$$\phi_{k,l+1} = \phi_{k-1,l} + \phi_{k,l}(z_k - z_l) \ , \qquad k > l \ . \tag{1.12}$$

The sequence $\{t_k\}_{k=0}^{\infty}$ of polynomials is a basis of the above mentioned algebra $\mathcal{N}_Z$ of functions defined at the points z_l ($l = 0, 1, \ldots$), including formal derivatives up to the order (1.4). (It is a basis in the sense that each such function has a unique representation of the form (1.2); introduction of a suitable norm would turn it into a Banach space basis.) The relevance of the linear functionals $[z_0,\ldots,z_k]$, $k = 0, 1, \ldots$, which are all defined on these functions, is due to the fact that they form an adjoint basis, i.e., $\{t_k, [z_0,\ldots,z_k]\}_{k=0}^{\infty}$ is a biorthogonal family:

$$[z_0,\ldots,z_k]t_l = \delta_{kl} := \begin{cases} 0 & \text{if } k \neq l \ , \\ 1 & \text{if } k = l \ . \end{cases} \tag{1.13}$$

(This holds for $l < k$ since t_l has degree at most l, for $l > k$ since t_l has zeros at $z_0,\ldots,z_k$, and for $l = k$ since t_l has leading coefficient 1.) Note, for example, that applying (1.13) to (1.2) yields (1.7).

By $O(t_l)$ we denote any element of $\mathcal{N}_Z$ that has zeros at $z_0,\ldots z_{l-1}$, multiple ones being understood as zeros of formal derivatives. In other words, $g = O(t_l)$ if and only if the Newton series of g starts with the t_l-term:

$$g = O(t_l) \iff [z_0,\ldots,z_k]g = 0 \ , \qquad k = 0,\ldots, l-1 \ . \tag{1.14}$$

We write $g \equiv O(t_l)$ if $g = O(t_l)$ and $g \neq O(t_{l+1})$. If $l \leq 0$, the symbol $O(t_l)$ will be understood as a void condition.

The representation of data by a formal Newton series can be used for any interpolation problem where function values and, possibly, successive derivatives are prescribed in the complex plane. The extension to vector-valued or matrix-valued functions is straightforward. However, the rational interpolants, which we will discuss in a moment, are actually maps of the extended plane $\overline{\mathbb{C}} := \mathbb{C} \cup \{\infty\}$ into itself. In particular, they have poles (*i.e.*, points where the value is ∞) unless the denominator is a constant, and they are also defined at $z = \infty$. In fact, ∞ should not be considered as a special point, neither in the domain nor in the range. We will even include the constant ∞ in our set of rational functions. It therefore makes sense to allow as interpolation data prescribed poles, or, more generally, initial sections of Laurent series at poles, and initial section of the Laurent series at $z = \infty$. This view has been promoted in [38] and taken up by Van Barel and Bultheel [69] recently; see also Stahl [60]. One nice feature is then that the interpolation problem becomes invariant under Moebius transforms [38]. This is of importance for the computation of rational interpolants, because many algorithms are based on successive value transformations, a recursive reduction that amounts to applying a Moebius transformation to the range of the functions. By such a transformation a given finite function value may turn into a pole.

However, formal Newton series can only represent finite function values and derivatives at finite points z_k. In order to represent poles we therefore assume that the data is given as a formal quotient of two formal Newton series:

$$h(z) := -\frac{f(z)}{g(z)}, \qquad f(z) := \sum_{k=0}^{\infty} \phi_k t_k(z), \qquad g(z) := \sum_{k=0}^{\infty} \gamma_k t_k(z). \qquad (1.15)$$

The two series must be *relatively prime* in the sense that they do not vanish simultaneously at any z_k. Theoretically, g could be chosen to vanish wherever a pole is prescribed and to have arbitrary finite values at the other points z_k. In practice, one should rather choose f and g to have modulus at most 1 at each z_k. Obviously, there is a lot of freedom in choosing f and g. However, it is important to note that we consider formal Newton series mainly as a notational convenience and not as a computational tool. In particular, if all points z_k are distinct, it is much easier to represent the data as a set of function values than to transform these into divided differences.

The treatment of data at $z = \infty$ is deferred to §3.

Having discussed the functions of data to be approximated, we now turn to the approximants: rational functions.

Notation: $\mathcal{P}_m$ is the space of complex *polynomials* p of exact degree $\partial p \leq m$, and $\mathcal{R}_{m,n}$ is the set of *rational functions* $r = p/q$ with $p \in \mathcal{P}_m$, $q \in \mathcal{P}_n$, $q \neq 0$. If $r = p/q$ with relatively prime polynomials p and q of exact degrees ∂p and ∂q, r is said to be of *exact type* $(\partial p, \partial q)$ and to have the defect

$$\delta := \delta_{m,n} := \min\{m - \partial p, n - \partial q\} \qquad (1.16)$$

in $\mathcal{R}_{m,n}$. If $\delta > 0$, r is called *degenerate* in $\mathcal{R}_{m,n}$. In particular, if $p = 0$, $\partial p := -\infty$, and hence $r = 0$ has exact type $(-\infty, 0)$ and defect n in $\mathcal{R}_{m,n}$.

We augment $\mathcal{R}_{m,n}$ by the constant ∞, which can be represented as $1/0$ and has exact type $(0, -\infty)$ and defect m in $\mathcal{R}_{m,n}$. Let $\overline{\mathcal{R}}_{m,n} := \mathcal{R}_{m,n} \cup \{\infty\}$. Note that then $r \in \overline{\mathcal{R}}_{m,n}$ if and only if $1/r \in \overline{\mathcal{R}}_{n,m}$.

A rational function of type (m, n) has $m + n + 2$ coefficients, but since the multiplication of numerator and denominator by a common factor has no effect, there are locally at most $m + n + 1$ degrees of freedom. Hence, one might expect to be able to satisfy $m + n + 1$ interpolation conditions. However, there are exceptional sets of data, where one cannot match as many conditions. The points where the conditions are not satisfied are then called *unattainable*, (although not only these points, but all the data are special).

The rational interpolation problem is basically nonlinear. However, there is a related linear problem whose solution is the key to the nonlinear one. Given a relatively prime pair $(f, g) \in \mathcal{N}_Z \times \mathcal{N}_Z$, the double sequence of *linearized rational interpolation problems* or *Newton-Padé approximation problems* for (f, g) consists in finding, for each pair $(m, n) \in \mathbb{N} \times \mathbb{N}$ the pairs $(p, q) \in \mathcal{P}_m \times \mathcal{P}_n$ for which

$$g(z)p(z) + f(z)q(z) = O(t_{m+n+1}(z)) , \qquad (1.17)$$

or

$$\frac{d^{\kappa(k)}}{dz^{\kappa(k)}}[g(z)p(z) + f(z)q(z)]\bigg|_{z=z_k} = 0 , \qquad k = 0, \ldots, m+n . \qquad (1.17')$$

Any such pair $(p, q) \in \mathcal{P}_m \times \mathcal{P}_n$ is called an (m, n)-*Newton-Padé form* (NPF) of (f, g). The associated rational function $r = p/q$ is an (m, n)-*Newton-Padé approximant* (NPA) of (f, g) [24]. When writing $r = p/q$, we do not assume that p and q are relatively prime, but we think of r as the rational function obtained after cancellation of common factors of p and q. Therefore p/q has to be distinguished from (p, q). The pair (p, q) in (1.17) only depends on the $(m + n)$th partial sums of the Newton series f and g. Clearly, it is never uniquely determined. We will give its general form below in Theorem 1.2. As we shall see, there may be no relatively prime solution pair. The problem (1.17) is also called the *modified rational interpolation problem* [50], [72] or the *multipoint Padé approximation problem* [4]. (We will use the last notion for the more general case that includes conditions at infinity, see §3.)

The existence and uniqueness of NPAs is asserted by the following well known result [24,72,79], which is here adapted to our more general situation:

Lemma 1.1. *Given a relatively prime pair $(f, g) \in \mathcal{N}_Z \times \mathcal{N}_Z$ and $(m, n) \in \mathbb{N} \times \mathbb{N}$, there exist (m, n) Newton-Padé forms (p, q) of (f, g), and they are all equivalent in the sense that p/q defines a unique (m, n) Newton-Padé approximant $r_{m,n} \in \mathcal{R}_{m,n}$.*

Proof: According to (1.14) the condition (1.17) is equivalent to

$$[z_0, \ldots, z_k](gp + fq) = 0 , \qquad k = 0, \ldots, m+n . \qquad (1.18)$$

In view of (1.10) this means that $m+n+1$ linear functionals are acting on $(p,q) \in \mathcal{P}_m \times \mathcal{P}_n$. Hence, (1.18) is equivalent to a homogeneous linear system of $m+n+1$ equations for the $m+n+2$ coefficients of (p,q). Thus there are always nontrivial solutions.

To prove the uniqueness of the NPA, assume that (p,q) and $(\tilde{p},\tilde{q})$ are two (m,n) NPFs. Then, by (1.17),

$$g(\tilde{p}q - \tilde{q}p) = q(-f\tilde{q} + O(t_{m+n+1})) + \tilde{q}(fq + O(t_{m+n+1})) = O(t_{m+n+1})$$
$$f(\tilde{p}q - \tilde{q}p) = \tilde{p}(-gp + O(t_{m+n+1})) + p(g\tilde{p} + O(t_{m+n+1})) = O(t_{m+n+1}) \, . \tag{1.19}$$

Since f and g do not vanish simultaneously at any z_k, we conclude that $\tilde{p}q - \tilde{q}p = O(t_{m+n+1})$. In view of $\tilde{p}q - \tilde{q}p \in \mathcal{P}_{m+n}$, it follows that $\tilde{p}q - \tilde{q}p = 0$. ∎

The set of all Newton-Padé forms for a particular interpolation problem is characterized in the following fundamental theorem, which is essentially due to Maehly and Witzgall [43]. See also [37,72,78].

Theorem 1.2. *The general solution* $(p,q) \in \mathcal{P}_m \times \mathcal{P}_n$ *of (1.17) is*

$$(p,q) = (\hat{p}_{m,n}\, s_{m,n}\, w,\, \hat{q}_{m,n}\, s_{m,n}\, w) \, , \tag{1.20}$$

where $\hat{p}_{m,n}$, $\hat{q}_{m,n}$, *and* $s_{m,n}$ *are polynomials that are uniquely determined up to a common scaling factor of* $\hat{p}_{m,n}$ *and* $\hat{q}_{m,n}$, *while* $w \in \mathcal{P}_{\delta_{m,n}-\partial s_{m,n}}$ *is arbitrary.* $\hat{p}_{m,n}$ *and* $\hat{q}_{m,n}$ *are relatively prime,* $s_{m,n}$ *is a monic divisor of* t_{m+n+1} *of degree* $\partial s_{m,n} \leq \delta_{m,n}$, *and* $\delta_{m,n}$ *is the defect of* $r_{m,n} = \hat{p}_{m,n}/\hat{q}_{m,n}$ *in* $\overline{\mathcal{R}}_{m,n}$. *The zeros of* $s_{m,n}$ *are the points* z_k *where the pair* $(\hat{p}_{m,n}, \hat{q}_{m,n})$ *does not satisfy the interpolation conditions.*

Proof: Let $(p,q) \in \mathcal{P}_m \times \mathcal{P}_n$ be any NPF, and let δ be the defect of p/q in $\overline{\mathcal{R}}_{m,n}$. First, we can write $(p,q) = (\hat{p}sw, \hat{q}sw)$, where $\hat{p}$ and $\hat{q}$ are relatively prime, while s is a monic divisor of minimum degree with the property that $(\hat{p}s, \hat{q}s)$ is still an NPF. Then w contains any other common factors of p and q. Clearly, $\partial s + \partial w \leq \delta$, and s is a divisor of t_{m+n+1}, since we could divide (1.17) by any factor of s that is not a factor of t_{m+n+1} without changing the $O(t_{m+n+1})$ term. Hence, (p,q) is of the form (1.20).

In view of Lemma 1.1 $\hat{p}$ and $\hat{q}$ are unique up to scaling. The same is true for s because it can be characterized as the polynomial of minimum degree for which

$$f(z)\hat{q}(z) + g(z)\hat{p}(z) = O(t_{m+n+1}(z)/s(z)) \, ; \tag{1.21}$$

i.e., its zeros are those points where $(\hat{p}, \hat{q})$ does not satisfy the linearized interpolation problem (1.17). We call $s_{m,n} := s$ therefore the *deficiency polynomial* of the (m,n) NPA. Now, multiplying (1.21) by sw, with arbitrary $w \in \mathcal{P}_{\delta-\partial s}$, yields a solution of (1.17). As long as $w \neq 0$, any pair (p,q) of the form (1.20) is therefore an (m,n) NPF. ∎

Before we discuss some consequences of Theorem 1.2, we want to look at the nonlinear rational interpolation problem. Since formal derivatives of

Rational Interpolation

f and g up to the order indicated by (1.4) exist at z_j, the same is true for $h = -f/g$ and $1/h = -g/f$. Moreover, if at some z_j the function value and some derivatives of h and $r \in \mathcal{R}_{m,n}$ coincide, the same is true for $1/h$ and $1/r$. Therefore, in the following definition the conditions imposed on r and $1/r$ do not conflict, but are equivalent unless r is zero or infinity at an interpolation point, in which case the conditions are complementary.

Given a relatively prime pair $(f,g) \in \mathcal{N}_Z \times \mathcal{N}_Z$, the double sequence of (*nonlinear*) *true rational interpolation problems* for (f,g) consists in finding, for each pair $(m,n) \in \mathbb{N} \times \mathbb{N}$ a rational function $r \in \overline{\mathcal{R}}_{m,n}$ for which

$$h(z) - r(z) = O(t_{m+n+1}(z)), \quad \text{if} \quad h(z) \neq \infty, \qquad (1.22a)$$

$$\frac{1}{h(z)} - \frac{1}{r(z)} = O(t_{m+n+1}(z)), \quad \text{if} \quad h(z) = \infty. \qquad (1.22b)$$

By writing $h = -f/g$, $r = p/q$ and multiplying through with the denominator, we see that the existence of a solution of (1.22) implies that (p,q) is a solution of the linearized problem (1.17). Since the latter leads to a unique NPA $r_{m,n}$, it follows that, if the solution of the true rational interpolation problem exists, it is unique and equal to $r_{m,n}$. However, simple examples, such as the data $\{(z_k, h(z_k))\}_{k=0}^2 = \{(0,0), (1,0), (1,1)\}$ for $m = n = 1$, show that the NPA is not always a true interpolant. In fact, from Theorem 1.2 we know that if $\partial s_{m,n} > 0$, then $(\hat{p}_{m,n}, \hat{q}_{m,n})$ is not a solution of the linearized problem, and, hence, $r_{m,n} = \hat{p}_{m,n}/\hat{q}_{m,n}$ is also not a true interpolant. The points where $r_{m,n}$ does not interpolate up to the required order are called *unattainable*. These points are exactly the zeros of $s_{m,n}$. For this reason we called $s_{m,n}$ the deficiency polynomial in [38,37]. A Newton-Padé approximant $r_{m,n}$ that is not a true interpolant might be called a *deficient* NPA. In Padé approximation some people prefer to say that these approximants do not exist; however, these approximants are often still useful and, as we will see, the corresponding NPFs are indispensable for stable recurrences.

The connections between the true rational interpolation problem and the Newton-Padé approximation problem have been discussed by many authors, including Maehly and Witzgall [43], Meinguet [50], Stoer [61], Warner [72], Wuytack [77,79], to name some of the earlier ones. There are many ways to rephrase what we have just said about the deficiency polynomial. Here are a number of statements that follow readily from Theorem 1.2:

Corollary 1.3. *Let $r_{mn} = \hat{p}_{m,n}/\hat{q}_{m,n}$ be the (m,n) NPA in reduced form. Then, the following statements are equivalent:*

(i) $r_{m,n}$ *is a true interpolant;*
(ii) $(\hat{p}_{m,n}, \hat{q}_{m,n})$ *is an NPF;*
(iii) $\partial s_{m,n} = 0$, *i.e.*, $s_{m,n}(z) \equiv 1$;
(iv) *there exists an (m,n) NPF (p,q) such that at each interpolation point z_j $(0 \leq j \leq m+n)$ either $p(z_j) \neq 0$ or $q(z_j) \neq 0$.*

A further immediate corollary is the following characterization theorem. Note its analogy to the characterization theorem for real rational Chebyshev approximation [14].

Theorem 1.4. (NPA Characterization Theorem) *The function $r \in \overline{\mathcal{R}}_{m,n}$ with defect δ is the (m,n) NPA of $h = -f/g$ if and only if there exists a divisor $s \in \mathcal{P}_\delta$ of t_{m+n+1} such that*

$$h(z) - r(z) = O(t_{m+n+1}(z)/s(z)), \quad \text{if} \quad h(z) \neq \infty, \quad (1.23a)$$

$$\frac{1}{h(z)} - \frac{1}{r(z)} = O(t_{m+n+1}(z)/s(z)), \quad \text{if} \quad h(z) = \infty, \quad (1.23b)$$

i.e., if and only if r interpolates h in at least $m+n+1-\delta$ data points.

Proof: If $r = r_{m,n}$, then, by Theorem 1.2, (1.23) holds for $s = s_{m,n}$. Conversely, if $r = \hat{p}/\hat{q}$ and if $s \in \mathcal{P}_\delta$ satisfies (1.23), then multiplication of (1.23a) by $-gs\hat{q}$ and of (1.23b) by $-fs\hat{p}$ shows that the pair $(p,q) = (s\hat{p}, s\hat{q})$ fulfills (1.17). ∎

Unlike Padé approximation, where in the case $g(z) \equiv 1$ the Padé approximant always has the property of matching as many terms as possible of the power series of f, the analog is in general not true for the Newton-Padé approximant, unless the points are *well-ordered* for this particular approximant; see [37](p. 550).

A further, although not so immediate, consequence of Theorem 1.2 is the Block Structure Theorem of the Newton-Padé table, which we will discuss in §2.

One can display the linear system (1.18) explicitly by applying the "Leibniz rule" (1.10). Representing the polynomials p and q as finite Newton series

$$p(z) = \sum_{j=0}^{m} \pi_j t_j(z), \qquad q(z) = \sum_{j=0}^{n} \rho_j t_j(z), \quad (1.24)$$

with

$$\pi_j := [z_0, \ldots, z_j]p, \qquad \rho_j := [z_0, \ldots, z_j]q, \quad (1.25)$$

and setting, as in (1.11),

$$\phi_{kj} := [z_j, \ldots, z_k], \qquad \gamma_{kj} := [z_j, \ldots, z_k], \qquad \text{if } k \geq j, \quad (1.26)$$

and $\phi_{kj} := \gamma_{kj} := 0$ if $k < j$, we get from (1.18) the linear system

$$\sum_{j=0}^{m} \gamma_{kj} \pi_j + \sum_{j=0}^{n} \phi_{kj} \rho_j = 0, \qquad k = 0, 1, \ldots, m+n. \quad (1.27)$$

In matrix notation this becomes

$$\mathbf{G}_{m+n,m}\mathbf{p} + \mathbf{F}_{m+n,n}\mathbf{q} = \mathbf{0} \in \mathbb{C}^{m+n+1}, \quad (1.28)$$

where

$$\mathbf{p} := [\pi_0, \ldots, \pi_m]^T, \qquad \mathbf{q} := [\rho_0, \ldots, \rho_n]^T, \quad (1.29)$$

and where $\mathbf{F}_{m+n,n}$ is the $(m+n+1) \times (n+1)$ principal submatrix of $\mathbf{F}$ from (1.11), and $\mathbf{G}_{m+n,m}$ is the $(m+n+1) \times (m+1)$ principal submatrix of the divided difference matrix $\mathbf{G}$ for g. Clearly, if the $(m+n+1) \times (m+n+2)$ matrix $[\mathbf{F}_{m+n,n} \mid \mathbf{G}_{m+n,m}]$ has maximum rank $m+n+1$, then the solution $[\mathbf{p}^T \mid \mathbf{q}^T]^T$ of (1.28) is unique up to a scalar factor. Moreover, this solution is well-conditioned if an $(m+n+1) \times (m+n+1)$ submatrix of that matrix is well-conditioned. To compute the solution we need to know a component of $\mathbf{p}$ or $\mathbf{q}$ that is nonzero and can therefore be normalized to 1; then the corresponding column of the coefficient matrix can be moved to the right-hand side of (1.28).

§2. The Block Structure of the Newton-Padé Table

Associated with a double sequence of Newton-Padé approximation problems defined by (1.17) there is a *Newton-Padé table* which covers a quarter of the (m,n)-plane and contains as its (m,n)-entry the NPA $r_{m,n}$. We let the n-axis point to the right and the m-axis downward, so that the first column (indexed by 0) contains the interpolation polynomials $r_{m,0}$. For the recursive computation of rational interpolants, one often follows some path in the Newton-Padé table. For example, this path may be a diagonal, an antidiagonal, or a staircase. However, in such algorithms special measures need to be taken if two interpolants are identical on the path. Actually, for floating-point computations, it is equally important to deal appropriately with nearly identical interpolants; this is a difficulty we want to address later. But, in any case, it is important to understand in which situations identical entries can occur.

Sets of equal entries are called *blocks*; in particular, *singular blocks* if they contain more than one entry. The possible shape of a singular block is described in the *Block Structure Theorem*. A relevant partial result on the block structure of the Newton-Padé table was established by Claessens [17]; the complete Block Structure Theorem (split up in several parts) was then given in [37]. Our proof from [37] of Claessens's result is actually so short that we can repeat it here:

Lemma 2.1. Let $r = \hat{p}/\hat{q} \in \overline{\mathcal{R}}_{\bar{m},\bar{n}} \setminus \{0, \infty\}$ (with relatively prime $\hat{p}$ and $\hat{q}$) be an $(\bar{m}, \bar{n})$ NPA with defect

$$\bar{\delta} := \min\{\bar{m} - \partial\hat{p}, \bar{n} - \partial\hat{q}\}$$

and eccentricity

$$\epsilon := \max\{\bar{m} - \partial\hat{p}, \bar{n} - \partial\hat{q}\} - \bar{\delta} \ .$$

Then r is the (m,n) NPA for all (m,n) in the square

$$(m,n) = (\partial\hat{p} + \bar{\delta} + i, \partial\hat{q} + \bar{\delta} + j) \ , \qquad i,j = 0, \ldots, \epsilon \ . \tag{2.1}$$

Proof: According to Theorem 1.4, Equation (1.23) holds for $r = r_{\bar{m},\bar{n}}$ and some $s \in \mathcal{P}_{\bar{\delta}}$, e.g., $s = s_{m,n}$. Therefore, if $m + n \leq \bar{m} + \bar{n}$, and if s_1 is the greatest common divisor of s and $t_{\bar{m}+\bar{n}+1}$, then

$$h(z) - r(z) = O(t_{\bar{m}+\bar{n}+1}(z)/s_1(z)) \ , \quad \text{if} \quad h(z) \neq \infty \ , \tag{2.2a}$$

$$\frac{1}{h(z)} - \frac{1}{r(z)} = O(t_{\bar{m}+\bar{n}+1}(z)/s_1(z)) \ , \quad \text{if} \quad h(z) = \infty \ . \tag{2.2b}$$

According to (2.1) the defect δ of r in $\overline{\mathcal{R}}_{m,n}$ satisfies

$$\delta = \bar{\delta} + \min\{i,j\} \geq \bar{\delta}.$$

Hence, by Theorem 1.4, the relations (2.2) with $s \in \mathcal{P}_{\bar{\delta}} \subseteq \mathcal{P}_\delta$ imply that $r = r_{\bar{m},\bar{n}}$ is the (m,n) NPA.

If, however, $m + n > \bar{m} + \bar{n}$ $(= \partial \hat{p} + \partial \hat{q} + 2\bar{\delta} + \epsilon)$, condition (2.1) restricts $m + n$ to

$$m + n \leq \partial \hat{p} + \partial \hat{q} + 2\bar{\delta} + 2\epsilon = \bar{m} + \bar{n} + \epsilon.$$

With $s_1(z) := s(z)(z - z_{\bar{m}+\bar{n}+1}) \cdots (z - z_{m+n})$, (1.23) becomes again (2.2). Since $\partial s_1 = \partial s + m + n - (\bar{m} + \bar{n})$, but $m + n - (\bar{m} + \bar{n}) \leq \delta - \bar{\delta}$, we have $\partial s_1 \leq \delta$. Therefore, Theorem 1.4 and the relations (2.2) imply again that r is the (m,n) NPA. ∎

Lemma 2.1 states that whenever $r_{\bar{m},\bar{n}}$ ($\neq 0, \infty$) has positive eccentricity, $r_{\bar{m},\bar{n}}$ appears in the Newton-Padé table at least in the whole square defined by the upper left corner $(\partial \hat{p} + \bar{\delta}, \partial \hat{q} + \bar{\delta})$ and the corner $(\bar{m}, \bar{n})$. The upper left corner of the square lies on the diagonal which passes through $(\partial \hat{p}, \partial \hat{q})$. The lemma is readily enhanced in the following way:

Lemma 2.2. *Let $r = \hat{p}/\hat{q} \in \overline{\mathcal{R}}_{\bar{m},\bar{n}} \setminus \{0, \infty\}$ be an $(\bar{m}, \bar{n})$ NPA with defect $\bar{\delta}$, and let*

$$\hat{m} := \partial \hat{p} + \bar{\delta}, \qquad \hat{n} := \partial \hat{q} + \bar{\delta}. \tag{2.3}$$

Define Δ either as the largest integer for which r is also the $(\hat{m} + \Delta, \hat{n})$ NPA, or as the largest one for which r is still the $(\hat{m}, \hat{n} + \Delta)$ NPA. Then r is the (m, n) NPA in the whole square

$$\hat{m} \leq m \leq \hat{m} + \Delta, \qquad \hat{n} \leq n \leq \hat{n} + \Delta. \tag{2.4}$$

Among all rectangles with the same upper left corner this square is the largest one that is contained in the block of r.

Proof: By Lemma 2.1 the $(\hat{m}, \hat{n})$-entry belongs to the block of r. Applying the lemma again with $(\bar{m}, \bar{n}) := (\hat{m} + \Delta, \hat{n})$ and with $(\bar{m}, \bar{n}) := (\hat{m}, \hat{n} + \Delta)$ yields the claimed result and shows in particular that both definitions of Δ are equivalent. ∎

The following statement from [37] (Theorem 2.2) is an immediate consequence of Lemma 2.1.

Theorem 2.3. *Let $r = \hat{p}/\hat{q} \neq 0, \infty$ be an NPA of $h = -f/g$ of exact type $(\partial \hat{p}, \partial \hat{q})$. Then the block of r in the Newton-Padé table of h is a (finite or infinite) union of squares whose upper left corners lie at or below the location $(\partial \hat{p}, \partial \hat{q})$ on the diagonal passing through this location.*

However, this theorem alone does not say very much. If all the blocks were square, as is the case for the ordinary Padé table, the theorem would still hold, but would not reveal the full truth. It is therefore important that the following converse hold [37] (Theorem 2.3).

Rational Interpolation

Theorem 2.4. *Let $r = \hat{p}/\hat{q} \neq 0, \infty$ be a rational function of exact type $(\partial\hat{p}, \partial\hat{q})$, let $\{z_k\}_{k=0}^{\infty}$ be any complex sequence of distinct points, and let $\mathcal{U}$ be any union of squares with the properties described in Theorem 2.3. Then there exist complex sequences $\{\phi_k\}_{k=0}^{\infty}$ and $\{\gamma_k\}_{k=0}^{\infty}$ such that the formal Newton series f and g in (1.15) are mutually prime and such that $\mathcal{U}$ is the block of r in the Newton-Padé table of $h = -f/g$. If none of the points z_k is a pole of r, one can choose $g(z) \equiv 1$.*

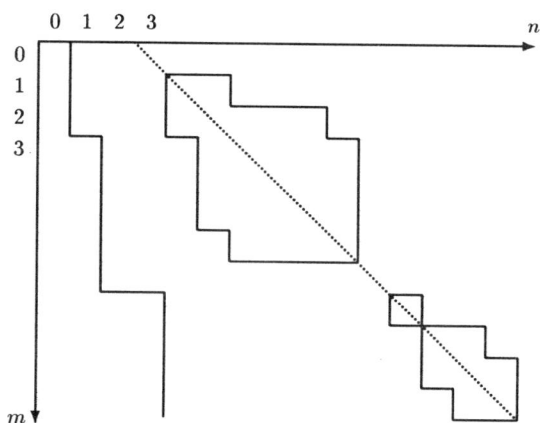

Figure 1. A Newton-Padé table with a block of the constant ∞ (at left) and another singular block consisting of three components.

In general, a singular block in the Newton-Padé table need not even be connected. It may consist of a finite or even infinite number of disjoint components, see Fig. 1. There is actually an easy interpretation for the form of these blocks. Every antidiagonal of the table is associated with an interpolation condition: e.g., the one through $m + n = k = $ const is associated with the data $(z_k, h(z_k))$ if $z_k \neq z_i$ ($\forall i < k$). If r is an NPA whose block contains at least one point of the above antidiagonal, then this block becomes broader or narrower at this antidiagonal depending on whether the new linearized interpolation condition is satisfied by r or not. For a more detailed discussion of this see [37]. Once the block gets narrower, r is certainly no longer a true interpolant. But it may even happen that an NPA r is for no index pair (m, n) a true interpolant. In fact this is true if and only if the $(\partial\hat{p}, \partial\hat{q})$-entry does not belong to the block of r.

The zero and the infinity functions have been excluded in Lemmas 2.1 and 2.2 and in Theorems 2.3 and 2.4. If one of them is an NPA, then a singular block of a different structure is associated with it, see [37] for details.

§3. Multipoint Padé Approximation with Conditions at Infinity

In this section we discuss the general scalar rational interpolation problem that includes conditions both at finite interpolation points and at infinity.

Our interest in it is not just due to a desire for generality, but due to the importance of this problem for the recursive computation of the Newton-Padé approximants (NPAs) as discussed in the two previous sections.

Again, we want to consider a double sequence of approximants. However, since conditions at infinity fix the difference between numerator and denominator degree, one of the two parameters now has a new meaning. The first, μ, determines the bias between using information at infinity and at finite interpolation points; the second, n, still denotes the nominal denominator degree of the rational function, i.e., the allowed maximum of the finite poles (with account of their multiplicity).

The data at the finite interpolation points is, for ease of notation, again assumed to be given in the Newton series form (1.15), although, as mentioned before, we suggest to use in practice the direct representation by function values and derivatives unless the interpolation points are clustered. Under the assumption that the function does not have a multiple zero at ∞, the behavior there can be prescribed by a formal Laurent series

$$\hat{h}(z) := \sum_{k=-\infty}^{\iota} \hat{\eta}_k z^k \qquad (3.1)$$

with $\hat{\eta}_\iota \neq 0$ for some $\iota \geq -1$ (read 'iota'). But here we choose in analogy to (1.15) the representation

$$\hat{h}(z) := -\frac{\hat{f}(z)}{\hat{g}(z)}, \quad \hat{f}(z) := \sum_{k=-\infty}^{\iota} \hat{\phi}_k z^k, \quad \hat{g}(z) := \sum_{k=-\infty}^{0} \hat{\gamma}_k z^k, \qquad (3.2)$$

with $\iota \geq -1$ and $\hat{\phi}_\iota \neq 0$ or $\hat{\gamma}_0 \neq 0$. In our recursive process for constructing NPAs, the data will actually occur in this form. If we had a multiple zero at ∞, we could replace h and $\hat{h}$ by $1/h$ and $1/\hat{h}$, and then take the reciprocal of the resulting rational interpolant.

Notation. $\mathcal{L}$ denotes the linear space of *formal Laurent series* $y(z) = \sum_{k=-\infty}^{\infty} \eta_k z^k$ with complex coefficients. Various subspaces of $\mathcal{L}$ are denoted by

$$\mathcal{L}_{l:m} := \{y \in \mathcal{L};\, \eta_k = 0 \text{ if } k < l \text{ or } k > m\},$$
$$\mathcal{L}_l := \mathcal{L}_{l:\infty} = \{y \in \mathcal{L};\, \eta_k = 0 \text{ if } k < l\},$$
$$\mathcal{L}_m^\star := \mathcal{L}_{-\infty:m} = \{y \in \mathcal{L};\, \eta_k = 0 \text{ if } k > m\}.$$

In particular, $\mathcal{L}_0$ and $\mathcal{L}_0^\star$ are the sets of *formal power series* in z and $1/z$, respectively. When $y \in \mathcal{L}_m^\star$, we write $y(z) = O_-(z^m)$, and if additionally $y \notin \mathcal{L}_{m-1}^\star$, we may express this as $y(z) \equiv O_-(z^m)$ or $\partial y = m$. The set of all polynomials (i.e., of formal power series with finitely many terms) is denoted by $\mathcal{P}$.

We consider now the following double sequence of *multipoint Padé approximation problems*: given $(f,g) \in \mathcal{N}_Z \times \mathcal{N}_Z$, as in (1.15), and $(\hat{f}, \hat{g}) \in$

Rational Interpolation

$\mathcal{L}_\iota^* \times \mathcal{L}_0^*$ as in (3.2), let, for each index pair $[\mu; n] \in \mathbb{Z} \times \mathbb{N}$, the nonnegative integer $m := m(\mu; n)$ be defined by

$$m := \max\{\iota - \mu - 1, \iota + n, \mu\}, \qquad (3.3)$$

and determine the pairs

$$(u, v) \in \begin{cases} \mathcal{L}_{\mu+n+1:\iota+n} \times \mathcal{P}_n = \mathcal{L}_{\iota+n-m:\iota+n} \times \mathcal{P}_n & \text{if } \mu \leq -n-1, \\ \mathcal{P}_{\iota+n} \times \mathcal{P}_n = \mathcal{P}_m \times \mathcal{P}_n & \text{if } -n-1 \leq \mu \leq \iota + n, \\ \mathcal{P}_\mu \times \mathcal{P}_n = \mathcal{P}_m \times \mathcal{P}_n & \text{if } \mu \geq \iota + n, \end{cases} \qquad (3.4)$$

which satisfy the conditions $(u, v) \neq (0, 0)$ and

$$\hat{g}(z)u(z) + \hat{f}(z)v(z) = O_-(z^\mu), \qquad (3.5a)$$
$$g(z)u(z) + f(z)v(z) = O(t_{\mu+n+1}(z)), \qquad (3.5b)$$

where $t_j(z) \equiv 1$ if $j \leq 0$. Any such pair (u, v) is here called a $[\mu; n]$ multipoint Padé form (MPF) of $(\hat{f}, \hat{g}; f, g)$, and

$$r_{\mu;n}(z) := \frac{u(z)}{v(z)} \qquad (3.6)$$

is referred to as the corresponding $[\mu; n]$ multipoint Padé approximant (MPA).

We need to comment on this definition. First, regarding the three cases in (3.4), one could subsume the middle case both in the first and the last one. However, as written, (3.4) parallels the distinction of the three cases generated by (3.3) and (3.5):

$\mu \leq -n - 1:$ Padé problem at $z = \infty$, $m = \iota - \mu - 1$,
$-n - 1 < \mu < \iota + n:$ proper multipoint Padé problem, $m = \iota + n$,
$\iota + n \leq \mu:$ Newton-Padé problem, $m = \mu$.

In the first case, where $n \leq -\mu - 1 = m - \iota$, the second line of (3.5) is vacuous and the first line implies $\iota + n - \mu = m + n + 1$ conditions at $z = \infty$ for the pair (u, v) with a total of $n + m + 2$ parameters. Hence, we have a one-point Padé approximation problem at infinity. In the third case, $\iota + n \leq \mu = m$, the first line of (3.5) is vacuous, while the second is identical to the condition (1.17) for an (m, n) NPA. Finally, in the second case, both lines of (3.5) are in effect, and we have a proper multipoint Padé problem with $\iota + n - \mu$ conditions at $z = \infty$ and $\mu + n + 1$ conditions at finite interpolation points. Hence, there is a total of $\iota + 2n + 1$ conditions for $(u, v) \in \mathcal{P}_{\iota+n} \times \mathcal{P}_n$. Problems of this kind are included in the treatment of [38]; in particular, the following analog of Lemma 1.1 can be formulated.

Lemma 3.1. Given $(\hat{f}, \hat{g}; f, g) \in \mathcal{L}_\iota^* \times \mathcal{L}_0^* \times \mathcal{N}_Z \times \mathcal{N}_Z$ (with $\hat{\gamma}_0 \neq 0$ or $\hat{\phi}_\iota \neq 0$, $\iota \geq -1$, and relatively prime f, g) and $[\mu; n] \in \mathbb{Z} \times \mathbb{N}$, there exist $[\mu; n]$ multipoint Padé forms (u, v) of $(\hat{f}, \hat{g}; f, g)$, and they are all equivalent in the sense that u/v defines a unique $[\mu; n]$ multipoint Padé approximant $r_{\mu;n}$.

The multipoint Padé approximants for a data set $(\hat{f}, \hat{g}; f, g)$ can be collected in a *multipoint Padé table* $\{r_{\mu;n} \; ; \; [\mu; n] \in \mathbb{Z} \times \mathbb{N}\}$ that covers a half-plane. We let the n-axis point to the right and the μ-axis point downward. If $\iota = 0$, the proper MPAs lie in a 90° sector with horizontal axis, cf. Fig. 2. Above this sector the table contains Padé approximants at infinity, below there are NPAs. If $\iota > 0$, the lower border of this sector lies farther down. If $\iota = -1$, it is farther up by one entry. In this case $r_{-1;0}(z) \equiv 0$, as is seen from (3.4)–(3.5).

Note that n is the nominal denominator degree of $r_{\mu;n}$, but μ is *not* the numerator degree, except in the NPA sector of the table, where $\mu = m$. In the proper MPA sector the nominal numerator degree is equal to $m = \iota + n$, and in the top sector it is seen to be $m - \iota = -\mu - 1$ if one cancels a factor z^n in $u(z)/v(z)$ and rewrites numerator and denominator as polynomials in $1/z$. The number of parameters in the numerator is always $m + 1$.

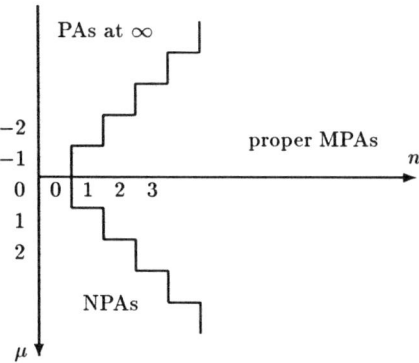

Figure 2. The case $\iota = 0$ of a multipoint Padé table with its 90° sector of multipoint Padé approximants and the two adjoining sectors of Padé approximants at infinity and Newton-Padé approximants.

We need to point out that, in contrast to our terminology, the notion of multipoint Padé approximant is often used as equivalent to Newton-Padé approximant or rational interpolant. Rational interpolation problems with data at ∞ and at several finite interpolation points have received relatively little attention so far [68,69,37,60]. The definition of a $[\mu; n]$ MPA given above is probably new. If the finite interpolation points z_k all coalesce at $z = 0$, and if $-n - 1 < \mu < \iota + n$, our multipoint Padé approximation problem becomes a two-point Padé approximation problem. Then g and f are formal power series, and the term $O(t_{\mu+n+1})$ in (3.5) becomes $O(z^{\mu+n+1})$. Such two-point Padé approximation problems for two formal power series $\hat{h} := -\hat{f}/\hat{g}$ and $h := -f/g$ in $1/z$ and z, respectively, have been investigated by McCabe and Murphy [47,48], Sidi [59], Draux [20], Cooper, Magnus and McCabe [18], and others. Except for Sidi, these authors permit no constant term in $\hat{h}$, so that $\iota = -1$ and $(u,v) \in \mathcal{P}_{n-1} \times \mathcal{P}_n$ when $|\mu+1| < n$. Our index μ is then replaced by, say, $l := \mu + 1$, so that the symmetry axis of the proper two-point Padé approximation sector of the table is at $l = 0$. For this case, McCabe and his coauthors [47,48,18] have already considered the two-point analog of our multipoint Padé table, calling it M-table in honor of Murphy. (There appear other definitions of two-point Padé approximation in the literature. Bultheel's [8] is closely related to Sidi's version. However, the one of Thron and Magnus [63,44] is further off, and the corresponding table, called the Colorado table in [18], seems of little usefulness, except for one of its diagonals, which is identical with a row of the M-table.)

In order to establish a Block Structure Theorem for multipoint Padé approximation in §4 below we need the analog of Theorems 1.2 and 1.4 for proper MPFs. Such an analog was given in [38]. But since we assumed there that the data at ∞ had another form than (3.2), we need to make some straightforward adjustments. (That other form was appropriate for the case of fixed μ, n, and m that was treated there.) We obtain the following reformulation of Theorems 2 and 3 in [38]:

Theorem 3.2. *In the case $-n - 1 < \mu < \iota + n$ of proper multipoint Padé approximation the general form of the MPFs (u,v) defined by (3.3)—(3.5) is*

$$(u,v) = (\hat{u}_{\mu;n}\, s_{\mu;n}\, w,\ \hat{v}_{\mu;n}\, s_{\mu;n}\, w)\,, \tag{3.7}$$

where $\hat{u}_{\mu;n}, \hat{v}_{\mu;n}$ are up to scaling uniquely determined relatively prime polynomials and $s_{\mu;n}$ is a monic divisor of $t_{\mu+n+1}$ of degree $\partial s_{\mu;n} \leq \min\{\delta,\nu\}$, with $\delta := \delta_{\mu;n}\ (\geq 0)$ being the defect of $r_{\mu;n} := \hat{u}_{\mu;n}/\hat{v}_{\mu;n}$ in $\overline{\mathcal{R}}_{\iota+n,n}$, and

$$\nu := \nu_{\mu;m} := \mu - \partial(\hat{g}\hat{u}_{\mu;n} + \hat{f}\hat{v}_{\mu;n}) \qquad (\geq 0) \tag{3.8}$$

being the number of extra (linearized) interpolation conditions fulfilled at ∞ by $(\hat{u}_{\mu;n}, \hat{v}_{\mu;n})$. Finally, w is an arbitrary nonzero polynomial of degree

$$\partial w \leq \min\{\delta,\nu\} - \partial s_{\mu;n}\,. \tag{3.9}$$

The zeros of the deficiency polynomial $s_{\mu;n}$ are the finite interpolation points z_k (with appropriate multiplicity) at which the pair $(\hat{u}_{\mu;n}, \hat{v}_{\mu;n})$ does not satisfy the interpolation conditions. The deficiency at ∞,

$$\hat{\sigma} := \hat{\sigma}_{\mu;n} := \begin{cases} 0 & \text{if } \nu \geq \delta, \\ \delta - \nu & \text{if } \nu < \delta, \end{cases} \tag{3.10}$$

indicates the number of (linearized) interpolation conditions at ∞ that are not fulfilled by a pair (u,v) of the form (3.7) with $\partial s_{\mu;n} + \partial w = \delta$. It satisfies

$$0 \leq \hat{\sigma} \leq \min\{n - \mu, \delta - \partial s_{\mu;n}\}\,. \tag{3.11}$$

Theorem 3.3. *(Proper MPA Characterization Theorem) Assume $-n-1 < \mu < \iota + n$. The function $w = \hat{u}/\hat{v} \in \overline{\mathcal{R}}_{\iota+n,n}$ ($\hat{u}, \hat{v}$ relatively prime) with defect δ is the $[\mu; n]$ MPA of $(\hat{f}, \hat{g}; f, g)$ if and only if there exists a polynomial divisor s of $t_{\mu+n+1}$ of degree at most δ and an integer $\hat{\sigma}$ satisfying (3.11) such that*

$$\hat{g}(z)\hat{u}(z) + \hat{f}(z)\hat{v}(z) = O_-(z^{\mu-\delta+\hat{\sigma}}), \quad (3.12a)$$
$$g(z)\hat{u}(z) + f(z)\hat{v}(z) = O(t_{\mu+n+1}(z)/s(z)), \quad (3.12b)$$

or, equivalently,

$$\frac{\hat{f}(z)}{\hat{g}(z)} + \frac{\hat{u}(z)}{\hat{v}(z)} = O_-(z^{\mu-n+\hat{\sigma}}) \quad \text{as } z \to \infty \quad \text{if } \iota \leq 0, \quad (3.13a)$$

$$\frac{\hat{g}(z)}{\hat{f}(z)} + \frac{\hat{v}(z)}{\hat{u}(z)} = O_-(z^{\mu-n+\hat{\sigma}}) \quad \text{as } z \to \infty \quad \text{if } \iota \geq 0, \quad (3.13b)$$

$$\frac{f(z)}{g(z)} + \frac{\hat{u}(z)}{\hat{v}(z)} = O(t_{\mu+n+1}(z)/s(z)) \quad \text{as } t_{\mu+n+1}(z) \to 0 \text{ and } g(z) \not\to 0, \quad (3.13c)$$

$$\frac{g(z)}{f(z)} + \frac{\hat{v}(z)}{\hat{u}(z)} = O(t_{\mu+n+1}(z)/s(z)) \quad \text{as } t_{\mu+n+1}(z) \to 0 \text{ and } f(z) \not\to 0, \quad (3.13d)$$

i.e., if and only if $(\hat{u}, \hat{v})$ fulfills at least $\iota + 2n + 1 - \delta$ of the $\iota + 2n + 1$ nonlinear interpolation conditions that are implied by (3.5) for the MPA $r_{\mu;n} = \hat{u}/\hat{v}$.

Note that the left-hand side of (3.12a) is exactly $O_-(z^{\mu-\nu})$ according to the definition (3.8) of ν. Hence, (3.12a) can be rephrased as

$$\nu \geq \delta - \hat{\sigma}. \quad (3.14)$$

On the basis of these two theorems it would be easy to formulate an analog of Corollary 1.3.

To avoid innumerable cases we formulated Theorem 3.2 only for μ between $-n-1$ and $\iota + n$. However, when the appropriate interpretations of vacuous conditions are made, it is also true for $\mu \leq -n-1$ and $\mu \geq \iota + n$. In fact, then one has a one-point Padé approximation problem at ∞ or a Newton-Padé approximation problem, respectively, and one can therefore just refer to Theorem 1.2. (In the first situation, one replaces z by $1/z$, so that all the interpolation points are moved to 0.)

It is useful to define the defect δ for all three types of MPAs. Assume r is any rational function that can be written as $r = \hat{u}/\hat{v}$ with $(\hat{u}, \hat{v})$ satisfying (3.4) and $\hat{u}, \hat{v}$ being relatively prime. In the case $\mu \leq -n-1$, where $m = \iota - \mu - 1$ and thus $\iota + n - m \leq 0$, we call $(\hat{u}, \hat{v})$ relatively prime if the polynomials $z^{\iota+m-n}\hat{u}(z)$ and $\hat{v}(z)$ are relatively prime. We define the $[\mu; n]$-defect of r by

$$\delta := \delta_{\mu;n} := \max\{\mu - \partial\hat{u}, \iota + n - \partial\hat{u}, n - \partial\hat{v}\}. \quad (3.15)$$

If $-n-1 \leq \mu \leq \iota + n$, this is just the defect of r in $\overline{\mathcal{R}}_{\iota+n,n}$. If $\mu \geq \iota + n$, so that $m = \mu$, it is the defect of r in $\overline{\mathcal{R}}_{m,n}$. Finally, if $\mu \leq -n-1$, so that $m = \iota - \mu - 1$, it is equal to the defect of the function $z \mapsto r(1/z)$ in $\overline{\mathcal{R}}_{m,n}$.

With this definition we can formulate Theorem 3.3 with no restriction on μ:

Theorem 3.4. (MPA Characterization Theorem) *Let $[\mu; n] \in \mathbb{Z} \times \mathbb{N}$, let (u, v) satisfy (3.4), and let δ be the $[\mu; n]$-defect of $r = u/v$. Then r is the $[\mu; n]$ MPA if it satisfies at least $\iota + 2n + 1 - \delta$ of the $\iota + 2n + 1$ nonlinear interpolation conditions implied by (3.5).*

§4. The Block Structure of the Multipoint Padé Table

As we mentioned earlier, the multipoint Padé table introduced in §3 is a generalization of the two-point Padé table or M-table introduced by McCabe [47]. It was shown by Cooper, Magnus, and McCabe [18] that this M-table has a very simple block structure: all singular blocks are either square or infinite, exactly as in the case of the classical Padé table. The only new feature is the possibility of a new type of an infinite block, a half-plane $\{(\mu, n) \in \mathbb{Z} \times \mathbb{N};\ n \geq n_0\}$.

Regarding the structure of the multipoint Padé table, nothing seems to have been known. We show in this section that, surprisingly, the typical singular blocks have the same form as in the Newton-Padé table. Unfortunately, the adaptation of the simple proof from [18], which establishes the square blocks of the two-point Padé table by making a connection to the ordinary Padé table, does not seem to be so easy. We have therefore chosen a completely different approach, similar to the one applied in [37] for the Newton-Padé table.

For simplicity we assume in this section that the values prescribed at ∞ are finite and nonzero, so that $\iota = 0$, $\hat{\gamma}_0 \neq 0$, and $\hat{\phi}_0 \neq 0$ in (3.1)–(3.2). We start by establishing a first analog of Lemma 2.2.

Lemma 4.1. *Let the data $(\hat{f}, \hat{g}; f, g)$ be of the form (1.15) and (3.2), with $\iota = 0$. Let $r = \hat{u}/\hat{v}$ be either a proper $[\bar{\mu}; \bar{n}]$ MPA (so that $-\bar{n} - 1 < \bar{\mu} < \bar{n}$) or a $[\bar{\mu}; \bar{n}]$ MPA that is a $(\bar{\mu}, \bar{n})$ NPA (so that $\bar{\mu} \geq \bar{n}$). Assume that $\hat{u} \in \mathcal{P}_{\bar{n}} \setminus \{0\}$ and $\hat{v} \in \mathcal{P}_{\bar{n}} \setminus \{0\}$ are relatively prime. Let $\bar{m} := \max\{-\bar{\mu} - 1, \bar{n}\}$, let $\bar{\delta}$ be the defect of r in $\mathcal{R}_{\bar{m}, \bar{n}}$, and let the integers $\check{\mu}$, $\check{\nu}$ and the monic polynomial divisor $\check{s}$ of $t_{\bar{\mu}+\bar{n}+1}$ be such that $\partial \check{s} \leq \bar{\delta}$, that $\check{\mu}$, $\check{\nu}$ and $\partial \check{s}$ are as large as possible, and*

$$\hat{g}(z)\hat{u}(z) + \hat{f}(z)\hat{v}(z) \equiv O_-(z^{\check{\mu}-\check{\nu}}), \quad (4.1a)$$

$$g(z)\hat{u}(z) + f(z)\hat{v}(z) \equiv O(t_{\bar{\mu}+\bar{n}+1}(z)/\check{s}(z)). \quad (4.1b)$$

Finally, set

$$\hat{\nu} := \bar{\delta}, \qquad \hat{\mu} := \hat{\nu} + \check{\mu} - \check{\nu}, \qquad \Delta := \check{\nu} - \hat{\nu} = \check{\mu} - \hat{\mu}. \quad (4.2)$$

(The values $\check{\mu} = \infty$ and $\check{\nu} = \infty$ are permitted.) Then r is the $[\mu; n]$ MPA in the whole square

$$\hat{\mu} \leq \mu \leq \check{\mu} = \hat{\mu} + \Delta, \qquad \hat{n} \leq n \leq \bar{n} + \Delta, \quad (4.3)$$

which may be infinite (if either $\hat{\mu} = -\infty$ or $\check{\mu} = \infty$) or a half-plane (if $\hat{\mu} = -\infty$ and $\check{\mu} = \infty$). Among all rectangles with the same upper left corner this square is the largest one that is contained in the block of r. Above the main diagonal of the square (4.3), $\hat{\sigma}$, the deficiency at ∞, is positive; on and below this diagonal, it is 0. On the antidiagonal, the deficiency polynomial is $s = \check{s}$ and has degree $\bar{\delta}$. Below the antidiagonal, it has higher degree.

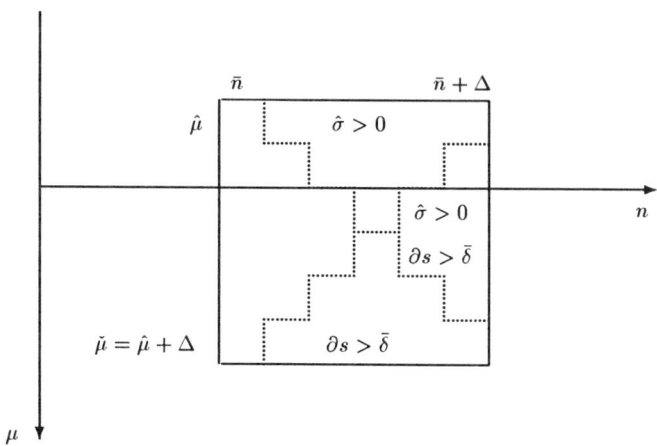

Figure 3. A basic square of a singular MPA block. The whole block is a union of such squares with a common main diagonal.

Proof: First we need to show that $\check{\mu}$, $\check{\nu}$, and $\check{s}$ are well-defined. When r is considered as the $[\bar{\mu}; \bar{n}]$-entry of the multipoint Padé table, the integers ν and $\hat{\sigma}$ of (3.8) and (3.10) take particular values $\bar{\nu}$ and $\bar{\sigma}$. If $-\bar{n} - 1 < \bar{\mu} < \bar{n}$, the deficiency polynomial from (3.7) and (3.12b) is $\bar{s} := s_{\bar{\mu};\bar{n}}$; if $\bar{\mu} \geq \bar{n}$, we let $\bar{s} := s_{\bar{\mu};\bar{n}} := s_{\bar{m},\bar{n}}$ be the deficiency polynomial from (1.20) and (1.21), where $\hat{p} := \hat{u}$, $\hat{q} := \hat{v}$. The relation (3.12b) may hold for some $\mu > \bar{\mu}$, in particular if we are willing to replace $\bar{s}$ by a polynomial s of higher degree. This degree is limited by $\bar{\delta}$, and we denote the resulting polynomial (having maximum deficiency) by $\check{s}$, and the corresponding maximum μ-value $\check{\mu}$. They yield the lower left corner of the square (4.3) if this square is finite, cf. Fig. 3. Then, due to the maximality of $\check{\mu}$ and $\check{s}$, r cannot be the $[\check{\mu} + 1; \bar{n}]$ MPA. Moreover, $\partial \check{s} = \bar{\delta}$ implies by (3.10) and (3.11) that the corresponding ν-value $\check{\nu}$ satisfies $\check{\nu} \geq \bar{\delta}$, because the $\hat{\sigma}$-value is 0. This is also true if $\check{\mu} \geq \bar{n}$.

Since the left-hand side of (3.12a) does not depend on μ and n as long as $r = \hat{u}/\hat{v}$ remains the same, the difference $\mu - \nu$ is invariant inside the block of r; cf. (3.8). Hence, if we define $\hat{\nu}$, $\hat{\mu}$, and Δ by (4.2), then $\check{\nu} \geq \hat{\nu}$, $\check{\mu} \geq \hat{\mu}$, and $\Delta \geq 0$. Moreover, since the left-hand side of (3.12a) is $O(z^{\hat{\mu}-\hat{\nu}}) = O(z^{\hat{\mu}-\bar{\delta}})$, (3.12a) is clearly satisfied for $\mu = \hat{\mu}$. If we insert in (3.12b) $\mu = \hat{\mu}$ instead of $\mu = \check{\mu}$, this condition gets relaxed and, thus, is also clearly satisfied. In general, this is even true for some $s = \hat{s}$ that is of lower degree than $\check{s}$. This makes it clear that the left edge of the square (4.3) is part of the block of r. (The block may actually extend further up.)

Assume now first that $\hat{\mu} < \bar{n}$; i.e., the upper left corner is in the PA or the proper MPA sector of the table. As we move from the left edge to the right, n and the defect δ increase steadily. (We say that a function of n or μ "increases steadily" if it increases monotonically with slope 1.) Since this allows us to increase the degree of s, there is no problem to satisfy (3.12b) as we move to the right. Also note that $\nu = \delta$ on the main diagonal of the square (4.3); cf. Fig. 3. Hence, $\hat{\sigma}$ becomes positive and increases steadily once we have crossed the diagonal of the square. The exponent $\mu - \delta + \hat{\sigma}$ in (3.12a) then remains constant as we move farther to the right. However, there is the restriction $\hat{\sigma} + \partial s \leq \delta$ from (3.11). Because $\partial s = \bar{\delta}$ on the antidiagonal of the square (4.3), there holds $\hat{\sigma} + \partial s = \Delta + \bar{\delta} = \delta$ in the upper right corner of this square. In the case under consideration this corner is in the proper MPA sector of the table. Since $\hat{\sigma} + \partial s$ increases twice as fast with n as δ, the restriction $\hat{\sigma} + \partial s \leq \delta$ is no longer fulfilled at $[\hat{\mu}; \bar{n} + \Delta + 1]$. But the whole upper edge of the square belongs to the block of r. If we move downwards from this edge, then $\hat{\sigma}$ decreases steadily, and ∂s increases at worst as fast, so that the restriction remains satisfied.

Consider next the case $\hat{\mu} \geq \bar{n}$, where the upper left corner of the square (4.3) belongs to the NPA part of the table. If the whole square is in this part, we can just refer to Lemma 2.2. But we still need to treat the situation where, in the upper right corner of the square, r is a proper MPA. Actually, this may only be formally true: in this upper right corner $\mu = \hat{\mu} \geq \bar{n}$ and $\nu = \hat{\nu} = \bar{\delta}$, so that (3.5) holds trivially for any pair $(\hat{u}s, \hat{v}s)$ with $\partial s \leq \bar{\delta}$. Since on and above the antidiagonal of the square the deficiency polynomials are limited by this degree requirement, we conclude that on the whole upper edge the linearized condition at ∞ is satisfied trivially, namely due to $\partial(\hat{g}u + \hat{f}v) \leq \mu$. However, at $[\hat{\mu}; \bar{n} + \Delta + 1]$ this is no longer true because $\partial s > \bar{\delta}$ would be needed. On the other hand, as we move down from the upper edge, $\hat{\sigma}$ decreases steadily as long as we are in the proper MPA sector, and once we get to the NPA sector, but are still above or at most on the diagonal of the square, δ increases. This allows us to increase ∂s as needed.

Altogether, we can conclude that the square (4.3) is part of the block of r. Moreover, among the rectangles with the same upper left corner, this is the largest one contained in the block. This is also true if the square extends to parts of the multipoint Padé table where one has actually a Padé approximation or a Newton-Padé approximation problem. We have only required that the left edge contains proper MPAs or NPAs. ∎

It remains to discuss the case where $\bar{\mu} \leq -\bar{n} - 1$; i.e., the left edge of the basic square contains only PAs at ∞. We have chosen to discuss this situation separately, because, first, the space $\mathcal{L}_{n-m:n} \times \mathcal{P}_n$ requires a slightly different treatment, and, second, the whole part of a singular block that lies in this sector is special, namely square (or the intersection of a square with this sector). This allows us to formulate for this part of the table a stronger result.

Lemma 4.2. *Let the data* $(\hat{f}, \hat{g}; f, g)$ *be of the form (1.15) and (3.2), with*

$\iota = 0$. Let $(\hat{u}, \hat{v}) \in \mathcal{L}_{\breve{\mu}+\bar{n}+1:\bar{n}} \times \mathcal{P}_{\bar{n}}$ be a relatively prime $[\breve{\mu}; \bar{n}]$ MPF with $\breve{\mu} \leq -\bar{n} - 1$, $\hat{u} \notin \mathcal{L}_{\breve{\mu}+\bar{n}+2:\bar{n}}$, and $[\breve{\mu}; \bar{n}]$-defect $\bar{\delta} = 0$, so that (4.1a) holds for some $\breve{\nu}$, $0 \leq \breve{\nu} \leq \infty$. Also, let $\hat{\nu}$, $\hat{\mu}$, and Δ be defined by (4.2). (In particular, $\hat{\nu} = 0$.) Then $r := \hat{u}/\hat{v}$ is the $[\mu; n]$ MPA in the whole square (4.3). In the $\mu \leq -n - 1$ part of the table (where $s(z) \equiv 1$ trivially) no other entries belong to the block of r. Above the main diagonal of the square (4.3), $\hat{\sigma}$, the deficiency at ∞, is positive; on and below this diagonal, it is 0.

Proof: Because the defect is zero and $\hat{u} \notin \mathcal{L}_{\breve{\mu}+\bar{n}+2:\bar{n}}$, the $[\breve{\mu}; \bar{n}]$-entry is at the bottom end of the left edge of the block of r; and because $\hat{\mu} = \breve{\mu} - \breve{\nu} = \breve{\mu} - \breve{\nu}$ is the exponent in (4.1a), $[\hat{\mu}; \bar{n}]$ is at the top end of the left edge. As we move from this edge to the right, (4.1a) seems to remain invariably true since n does not appear explicitly in it. But the space $\mathcal{L}_{\mu+n+1:n} \times \mathcal{P}_n$ changes, and due to $\hat{u} \notin \mathcal{L}_{\breve{\mu}+\bar{n}+2:\bar{n}}$ the pair $(\hat{u}, \hat{v})$ does no longer satisfy (3.4) if $\mu+n+1 \geq \breve{\mu}+\bar{n}+2$, i.e., if $[\mu; n]$ lies to the right of the antidiagonal of the square. In the top row, we cannot multiply $\hat{u}$ and $\hat{v}$ by a common factor since (3.5) would no longer hold. Say, k rows below, we can multiply them both by up to z^k. In this way, MPFs for the whole square are found. As long as we remain in the $\mu \leq -n - 1$ part of the table, there are no other MPFs possible that yield the same r. ∎

For the formulation of the Block Structure Theorem it is important to have a common characterization (depending only on r) for the main diagonal of the squares (4.3).

Lemma 4.3. Let $r = \hat{u}/\hat{v}$ be an MPA. Assume that $\hat{u} \neq 0$ and $\hat{v} \neq 0$ are relatively prime. Then those entries in the block of r where the $[\mu; n]$-defect δ satisfies

$$\hat{g}(z)\hat{u}(z) + \hat{f}(z)\hat{v}(z) \equiv O_-(z^{\mu-\delta}) \qquad (4.4)$$

lie on a particular diagonal of the multipoint Padé table. The diagonals of all squares of type (4.3) for the function r lie also on that diagonal of the table. The case $\mu - \delta = -\infty$ is permitted and means that the upper edge of the square (4.3) and hence also its diagonal are at infinity.

Proof: In the PA and proper MPA portions of the block of r, δ increases steadily with n, hence $\mu - \delta$ is constant on a diagonal. In the NPA sector δ increases only when both μ and n do; this also fixes a particular diagonal, however. In the upper left corner of a square (4.3) we have $\mu = \hat{\mu}$ and $\nu = \hat{\nu} = \bar{\delta} = \delta$. Hence, by definition of ν, (4.4) is satisfied. ∎

We call the constant difference $\mu - \delta$ in (4.4) the order at ∞ of $(\hat{u}, \hat{v})$ and denote it by μ_∞.

Now we are ready to formulate a first Block Structure Theorem for the multipoint Padé table.

Theorem 4.4. Let the data $(\hat{f}, \hat{g}; f, g)$ be of the form (1.15) and (3.2), with $\iota = 0$. Let $r = \hat{u}/\hat{v}$ be an MPA of $(\hat{f}, \hat{g}; f, g)$, and let μ_∞ be the order at ∞

Rational Interpolation

of $(\hat{u}, \hat{v})$. Assume that $\hat{u} \neq 0$ and $\hat{v} \neq 0$ are relatively prime. Then the block of r in the multipoint Padé table of $(\hat{f}, \hat{g}; f, g)$ is a finite or infinite union of squares whose upper left corners lie at or below the location $[\mu_\infty, \partial \hat{v}]$ on the diagonal passing through this location. If not empty, the intersection of the block with the sector $\mu < -n - 1$ is equal to the intersection of a square with this sector; the upper left corner of this square is then at $[\mu_\infty, \partial \hat{v}]$.

As for Theorem 2.3 the strength of this result is that it is best possible. By analyzing the proofs of the above lemmas and by adapting the basic ideas of the proof of Theorem 2.4 (see [37]), one can verify the following converse.

Theorem 4.5. *Let the relatively prime pair $(\hat{u}, \hat{v})$ satisfy (3.4) for $\iota = 0$ and some index pair $[\mu; n]$, and assume that n and $|\mu|$ are chosen as small as possible. (In particular, $\hat{u} \neq 0$ and $\hat{v} \neq 0$.) Moreover, assume that $\partial \hat{u} = \partial \hat{v} = n$ if $\mu \leq n$. Let $r := \hat{u}/\hat{v}$, let $\{z_k\}_{k=0}^\infty$ be any complex sequence of distinct points, and let $\mathcal{U}$ be any union of squares with the properties described in Theorem 4.4 and related to r as follows:*

(i) If $\mu \leq -n - 1$, then the $[\mu; n]$-entry is at the lower left corner of the part of $\mathcal{U}$ that lies in the PA at ∞ sector of the table.

(ii) If $-n - 1 < \mu < n$, then the first column of $\mathcal{U}$ has index at least n and contains proper MPAs; if $\mathcal{U}$ extends to the PA at ∞ sector of the table, the first column has exactly index n.

(iii) If $\mu \geq n$, then the diagonal through $[\mu; n]$ is a symmetry axis of $\mathcal{U}$, and the first column of $\mathcal{U}$ has index at least n.

Then there exist complex sequences $\{\phi_k\}_{k=0}^\infty$, $\{\gamma_k\}_{k=0}^\infty$, $\{\hat{\phi}_k\}_{k=0}^\infty$ (with $\hat{\phi}_0 \neq 0$), and $\{\hat{\gamma}_k\}_{k=0}^\infty$ (with $\hat{\gamma}_0 \neq 0$) defining the series (1.15) and (3.2) such that the formal Newton series f and g are mutually prime and $\mathcal{U}$ is the block of r in the multipoint Padé table of $(\hat{f}, \hat{g}; f, g)$.

The discussion of zero blocks (where $r(z) \equiv 0$) or infinity blocks (where $r(z) \equiv \infty$), which is again analogous to the one in [37], is left to the reader. Note, however, that the occurrence of such blocks in the PA at ∞ sector of the table is impossible when we adhere to our assumption that $\hat{\phi}_0 \neq 0$ and $\hat{\gamma}_0 \neq 0$. One could also easily formulate theorems about infinite blocks, which belong to rational functions r that interpolate infinitely many data. (This property is not sufficient however!)

§5. 'Reliable' and Stable Algorithms Based on a General Recurrence Relation

In this section we present a general recurrence relation for Newton-Padé approximants. It is modeled after the recurrence that was used by Cabay and Meleshko [51,12] for the stable computation of a diagonal sequence of Padé approximants. (Under more restrictive assumptions it was established before by Gragg, Gustavson, Warner, and Yun [25].) Here we consider Newton-Padé approximants instead of Padé approximants, and a much more general recurrence that allows us to proceed in other directions too. A similar general formula is due to Antoulas and Anderson [2], but the one given here seems to be simpler. Moreover, our interpretation of the new pair of NPAs as a combination of the old pair of NPAs and a pair of proper MPAs may be new. In the case of a diagonal sequence, the pair of proper MPAs is replaced by a pair of NPAs, so that the sequence is built up as a cascade of (small) NPAs, as in Antoulas [1]. We cannot discuss here all the details, but hope that the main idea and why it works will become clear. More information on the diagonal algorithm for NPAs will be presented in [11], while the non-diagonal Padé recurrences and the application of row-recurrences to stable fast Toeplitz solvers are discussed in [36].

According to Theorems 1.2 and 3.2 Newton-Padé forms (NPFs) and multipoint Padé forms (MPFs) are determined uniquely up to scaling if and only if in (1.20) and (3.7), respectively, the polynomial w is of degree 0, i.e., a constant. From the discussions that lead to the Block Structure Theorems in [37] and §4 above, it can be seen that w is a constant if and only if we consider in the respective table either a normal entry or a position at the border of a singular block. In view of the possible form of singular blocks the latter conditions are fulfilled if the previous or the following entry on the same diagonal is different. Hence, we are going to use pairs of NPFs (and MPFs, respectively) with the property that they are upper left and lower right neighbors, respectively, of each other, but belong to different blocks. The second member of these pairs will be called weakly regular. (We would call it regular if we knew additionally that $r = p/q$ is a true interpolant, but we will not have this knowledge in general.)

Assuming that the numerator degree $m = m(n)$ is a nondecreasing integer function of n, we just write $\{(p_n, q_n)\}$ for the sequence $\{(p_{m(n),n}, q_{m(n),n})\}$ of NPFs. For example, $m(n) = n + l$, with fixed l, for a diagonal sequence, and $m(l) = l$, with fixed l, for a row sequence. Note that m has here another meaning than in §3 and §4. For simplicity we often write m instead of $m(n)$, but we still use the longer form in definitions and other statements that will be referred to. Actually, there is normally no point in computing the full sequence $\{(p_n, q_n)\}$; only the subsequence of weakly regular NPFs, or rather, the subsequence of well-conditioned weakly regular NPFs will be constructed, together with their upper left neighbors

$$(\grave{p}_n, \grave{q}_n) := (p_{m(n)-1, n-1}, q_{m(n)-1, n-1}) \ .$$

In practice, (p_n, q_n) and $(\grave{p}_n, \grave{q}_n)$ need to be normalized, but at the moment this normalization is irrelevant.

We call the NPF $(p_n, q_n) := (p_{m(n),n}, q_{m(n),n})$ weakly regular if either $n = 0$ or

$$\frac{p_n}{q_n} \neq \frac{\grave{p}_n}{\grave{q}_n} , \quad \text{i.e.,} \quad \grave{p}_n q_n - p_n \grave{q}_n \neq 0 \in \mathcal{P} . \tag{5.1}$$

From Theorem 1.2 one can then conclude that $(\grave{p}_n, \grave{q}_n)$ does not have z_{m+n-1} or z_{m+n} as an extra interpolation point; see the discussion in §2 of [37]. If, say, z_{m+n} were one, then $(p_n, q_n) := ((z - z_{m+n-1})\grave{p}_n, (z - z_{m+n-1})\grave{q}_n)$ would be an

(m, n) NPF, possibly with z_{m+n-1} as an unattainable point, and consequently, (5.1) would not hold. By the same type of argument we can conclude that p_n and q_n cannot vanish simultaneously at z_{m+n-1} or z_{m+n}. Because, if they did, say at z_{m+n}, then we could obtain an $(m-1, n-1)$ NPF $(\grave{p}_n, \grave{q}_n)$ by cancelling the common factor $z - z_{m+n}$ in (p_n, q_n). The reverse directions of these two conclusions are verified readily too. For assume that (p_n, q_n) is not weakly regular and, say, z_{m+n} is not an extra interpolation point of $(\grave{p}_n, \grave{q}_n)$. Then, z_{m+n} is an unattainable point of p_n/q_n, and hence, by Theorem 1.2, $z - z_{m+n}$ is a factor of $s_{m,n}$ and therefore a common factor of p_n and q_n.

To reformulate these conclusions, we define the *residual* e_n of the NPF (p_n, q_n). In view of the definition (1.17) of an $(m(n), n)$ NPF we can write

$$g(z)p_n(z) + f(z)q_n(z) = t_{m(n)+n+1}(z)e_n(z) , \qquad (5.2)$$

where formally

$$e_n(z) := e_{m(n),n}(z) := \sum_{k=0}^{\infty} \epsilon_k^{(n)} t_k^{(m+n)}(z) , \qquad (5.3)$$

with

$$t_0^{(m+n)}(z) := 1 , \quad t_k^{(m+n)}(z) := \prod_{i=1}^{k}(z - z_{m(n)+n+i}) , \quad k = 1, 2, \ldots . \qquad (5.4)$$

This residual, which is itself a formal Newton series, is an important tool. In practice it suffices sometimes to compute its first or its first few coefficients or equivalent data. In any case, the Newton series in (5.3) may be replaced by some other representation of the data.

In analogy to (5.2) the residual $\grave{e}_n$ of $(\grave{p}_n, \grave{q}_n)$ is defined by

$$g(z)\grave{p}_n(z) + f(z)\grave{q}_n(z) = t_{m(n)+n-1}(z)\grave{e}_n(z) . \qquad (5.5)$$

In matrix notation (5.2) and (5.5) can be combined as

$$[g \; f]\begin{bmatrix} \grave{p}_n & p_n \\ \grave{q}_n & q_n \end{bmatrix} = t_{m(n)+n-1}[\grave{e}_n \; \tau_n e_n] , \qquad (5.6)$$

where

$$\tau_n(z) := (z - z_{m(n)+n-1})(z - z_{m(n)+n}) . \qquad (5.7)$$

What we have proved above becomes now

Lemma 5.1. *Let (p_n, q_n) be an $(m(n), n)$ NPF $(n > 0)$, and let $\grave{e}_n$ be the residual of an $(m(n)-1, n-1)$ NPF $(\grave{p}_n, \grave{q}_n)$. Then the following statements are equivalent:*

(i) (p_n, q_n) *is weakly regular, i.e.,* (5.1) *holds;*
(ii) $\grave{e}_n(z_{m(n)+n}) \neq 0$ *and* $p_n(z_{m(n)+n}) \neq 0$ *or* $q_n(z_{m(n)+n}) \neq 0$;
(iii) $\grave{e}_n(z_{m(n)+n-1}) \neq 0$ *and* $p_n(z_{m(n)+n-1}) \neq 0$ *or* $q_n(z_{m(n)+n-1}) \neq 0$.

Another simple result is

Lemma 5.2. Let (p_n, q_n) be an $(m(n), n)$ NPF $(n > 0)$; then

$$\det \begin{bmatrix} \mathring{p}_n & p_n \\ \mathring{q}_n & q_n \end{bmatrix} = \begin{cases} 0 \in \mathcal{P} & \text{if } (p_n, q_n) \text{ is not weakly regular,} \\ \delta_n t_{m(n)+n-1} & \text{if } (p_n, q_n) \text{ is weakly regular,} \end{cases} \quad (5.8)$$

where $\delta_n \neq 0$ is the leading coefficient of $\mathring{p}_n q_n - \mathring{q}_n p_n$.

Proof: The determinant is equal to $\mathring{p}_n q_n - \mathring{q}_n p_n \in \mathcal{P}_{m+n-1}$, which, by the definition of weak regularity, is identically zero if and only if the NPF (p_n, q_n) is not weakly regular. Multiplying (5.2) by $\mathring{q}_n$ and $\mathring{p}_n$, (5.5) by q_n and p_n, and subtracting pairwise we obtain

$$g(\mathring{p}_n q_n - \mathring{q}_n p_n) = O(t_{m+n-1}),$$
$$f(\mathring{p}_n q_n - \mathring{q}_n p_n) = O(t_{m+n-1}).$$

Since f and g are relatively prime by assumption, t_{m+n-1} must be a polynomial factor of $\mathring{p}_n q_n - \mathring{q}_n p_n \in \mathcal{P}_{m+n-1}$, the quotient being a scalar δ_n, which must be equal to the leading coefficient of $\mathring{p}_n q_n - \mathring{q}_n p_n$ since t_{m+n-1} is monic. ∎

To justify the usage of the pair of residuals $(e_n, \mathring{e}_n)$ as data of a Newton-Padé approximation or multipoint Padé approximation problem we will further need

Lemma 5.3. Let $Z(n) := \{z_j \in Z; j \geq m(n) + n - 1\}$, and assume that (p_n, q_n) is a weakly regular $(m(n), n)$ NPF. Then $\mathring{e}_n$ and e_n are relatively prime elements of $\mathcal{N}_{Z(n)}$.

Proof: We make here the additional assumption that the interpolation points z_j are distinct. (The general case is treated in [11].) From (5.8) we know then that the 2×2 matrix function in (5.6) is singular if and only if we evaluate it at z_j with $j \leq m + n - 2$. Hence, for $z_j \in Z(n)$ it is nonsingular, and thus $\mathring{e}_n(z_j) = e_n(z_j) = 0$ would imply $g(z_j) = f(z_j) = 0$, which is impossible since f and g are relatively prime by assumption. ∎

We will also need some information about the actual degree of the polynomials in the pairs of NPFs considered.

Lemma 5.4. If (p_n, q_n) is a weakly regular $(m(n), n)$ NPF (with $n > 0$), then
(i) $\partial q_n = n$ and $\partial \mathring{p}_n = m(n) - 1$

or

(ii) $\partial p_n = m(n)$ and $\partial \mathring{q}_n = n - 1$.

Proof: If (p_n, q_n) is a weakly regular, then in the Newton-Padé table not only the upper left neighbor of the (m, n)-entry $r_n := p_n/q_n$, but also the $(m-1, n)$-entry or the $(m, n-1)$-entry (or both) are different from r_n. If, say, the $(m-1, n)$-entry differs, then clearly $\partial p_n = m$; otherwise (p_n, q_n)

would qualify as an $(m-1, n)$ NPF. If, additionally, $\partial \mathring{q}_n < n-1$, then the pair $((z-z_{m+n-1})\mathring{p}_n, (z-z_{m+n-1})\mathring{q}_n)$ qualifies as $(m, n-1)$ NPF, so that also the left neighbor differs from r_n and, consequently, $\partial q_n = n$. Then necessarily $\partial p_n = m-1$, because otherwise we could conclude that the same pair is an $(m-1, n)$ NPF and that the $(m-1, n-1)$, the $(m, n-1)$, and the $(m-1, n)$ NPA are all the same; this would imply that they agree with the (m, n) NPA, in contrast to the weak regularity of the latter. Summarizing, we see that (i) or (ii) (or both) hold. ∎

We call (p_n, q_n) *weakly row-regular* if the $(m(n), n-1)$ and the $(m(n), n)$ NPA differ, and we call it *weakly column-regular* if the $(m(n)-1, n)$ and the $(m(n), n)$ NPA differ. According to the Block Structure Theorem, (p_n, q_n) is weakly regular if and only if it is weakly row-regular or weakly column-regular. In the first case, statement (i) of Lemma 5.4 holds; in the other case, statement (ii) holds.

Analogous definitions and statements can be made for MPFs. However, we restrict ourselves essentially to the few facts that will be needed for the general NPF recurrence formula.

We consider a sequence $\{(u_k, v_k)\} := \{(u_{\mu(k),k}, v_{\mu(k),k})\}$ of MPFs, and we let
$$(\grave{u}_k, \grave{v}_k) := (u_{\mu(k)-1, k-1}, v_{\mu(k)-1, k-1}) \ .$$
Again we are interested in those MPFs where the MPA is distinct from its upper left neighbor in the multipoint Padé table. It can be seen from Theorems 3.2 and 4.4 that this condition implies that the MPF and its neighbor are unique up to scaling, but not vice versa.

We call the MPF $(u_k, v_k) := (u_{\mu(k),k}, v_{\mu(k),k})$ *weakly regular* if

$$\frac{u_k}{v_k} \neq \frac{\grave{u}_k}{\grave{v}_k} \ , \quad \text{i.e.,} \ , \quad \grave{u}_k v_k - u_k \grave{v}_k \neq 0 \in \mathcal{L} \ . \tag{5.9}$$

The *residual* $(\hat{e}_k, e_k)$ of the multipoint Padé form (u_k, v_k) consists of a formal Laurent series $\hat{e}_k$ and a formal Newton series e_k defined implicitly by

$$\hat{g}(z) u_k(z) + \hat{f}(z) v_k(z) = z^{\mu(k)} \hat{e}_k(z) \ , \tag{5.10a}$$
$$g(z) u_k(z) + f(z) v_k(z) = t_{\mu(k)+k+1}(z) e_k(z) \ . \tag{5.10b}$$

The residual $(\grave{\hat{e}}_k, \grave{e}_k)$ of $(\grave{u}_k, \grave{v}_k)$ satisfies analogously

$$\hat{g}(z) \grave{u}_k(z) + \hat{f}(z) \grave{v}_k(z) = z^{\mu(k)-1} \grave{\hat{e}}_k(z) \ , \tag{5.11a}$$
$$g(z) \grave{u}_k(z) + f(z) \grave{v}_k(z) = t_{\mu(k)+k-1}(z) \grave{e}_k(z) \ . \tag{5.11b}$$

Again, this can be combined with (5.10):

$$\begin{bmatrix} \hat{g} & \hat{f} \\ g & f \end{bmatrix} \begin{bmatrix} \grave{u}_k & u_k \\ \grave{v}_k & v_k \end{bmatrix} = \begin{bmatrix} z^{\mu(k)-1} \grave{\hat{e}}_k & z^{\mu(k)} \hat{e}_k \\ t_{\mu(k)+k-1} \grave{e}_k & t_{\mu(k)+k+1} e_k \end{bmatrix} \ . \tag{5.12}$$

In accordance with our definition of MPAs in §3 we let u_k/v_k be a PA at ∞ if $\mu(k) \leq -k - 1$ and an NPA if $\mu(k) \geq \iota + k$; likewise, $\grave{u}_k/\grave{v}_k$ is actually a PA at ∞ if $\mu(k) - 1 \leq -(k-1) - 1 = -k$ and an NPA if $\mu(k) - 1 \geq \iota + k - 1$. However, in the following we will consider $[\mu+1; n]$ MPFs; i.e., we will have to replace $\mu(k)$ by $\mu + 1$ in all formulas.

If $(p_n, q_n) := (p_{m(n),n}, q_{m(n),n})$ is a weakly regular (m,n)-NPF, we consider the following sequence of $[\mu+1; k]$ multipoint Padé approximation problems

$$z^{-m+1}[\grave{p}_n(z)u_k^{(n)}(z) + p_n(z)v_k^{(n)}(z)] = O_-(z^{\mu+1}), \quad (5.13a)$$

$$\grave{e}_n(z)u_k^{(n)}(z) + \tau_n(z)e_n(z)v_k^{(n)}(z) = O(t_{\mu+k+2}^{(m+n-2)}(z)). \quad (5.13a)$$

The polynomial $t_{\mu+k+2}^{(m+n-2)}$ is still defined by (5.4).

In our general notation for an MPA problem, we have for problem (5.13)

$$(\hat{f}, \hat{g}; f, g) := (z^{-m+1}p_n, z^{-m+1}\grave{p}_n; \tau_n e_n, \grave{e}_n) \in \mathcal{L}_1^\star \times \mathcal{L}_0^\star \times \mathcal{N}_{Z(n)} \times \mathcal{N}_{Z(n)}. \quad (5.14)$$

(Note that g and f have here another meaning than in (5.2).) Since $\grave{p}_n \in \mathcal{P}_{m-1}$ and $p_n \in \mathcal{P}_m$, we let $\iota = 1$. We restrict $\mu + 1$ to $-k - 1 \leq \mu + 1 \leq 1 + k$, so that according to (3.4)

$$(u_k, v_k) \in \mathcal{P}_{k+1} \times \mathcal{P}_k. \quad (5.15)$$

From Lemma 5.4 we recall that $\partial(z^{-m+1}\grave{p}_n) = 0$ if (p_n, q_n) is weakly row-regular, and that $\partial(z^{-m+1}p_n) = 1$ if (p_n, q_n) is weakly column-regular. Hence, the assumption of §3 that $\hat{\gamma}_0 \neq 0$ or $\hat{\phi}_\iota \neq 0$ is fulfilled. (In §4 we assumed for simplicity that both coefficients are nonzero. However, for those results we need from there this was not crucial.)

In addition to the $[\mu+1; k]$ MPF $(u_k^{(n)}, v_k^{(n)})$ for the data (5.14) we consider a $[\mu; k-1]$ MPF

$$(\grave{u}_k^{(n)}, \grave{v}_k^{(n)}) := (u_{k-1}^{(n)}, v_{k-1}^{(n)}). \quad (5.16)$$

According to (5.12) the residuals $(\hat{e}_k^{(n)}, e_k^{(n)})$ of $(u_k^{(n)}, v_k^{(n)})$ and the residuals $(\hat{\grave{e}}_k^{(n)}, \grave{e}_k^{(n)})$ of $(\grave{u}_k^{(n)}, \grave{v}_k^{(n)})$ satisfy

$$\begin{bmatrix} z^{-m+1}\grave{p}_n & z^{-m+1}p_n \\ \grave{e}_n & \tau_n e_n \end{bmatrix} \begin{bmatrix} \grave{u}_k^{(n)} & u_k^{(n)} \\ \grave{v}_k^{(n)} & v_k^{(n)} \end{bmatrix} = \begin{bmatrix} z^\mu \hat{\grave{e}}_k^{(n)} & z^{\mu+1}\hat{e}_k^{(n)} \\ t_{\mu+k}^{(m+n-2)}\grave{e}_k^{(n)} & t_{\mu+k+2}^{(m+n-2)}e_k^{(n)} \end{bmatrix}. \quad (5.17)$$

The general recurrence formula for NPFs is based on the fact to be shown next that, given (p_n, q_n), $(\grave{p}_n, \grave{q}_n)$, and $(e_n, \grave{e}_n)$, a solution of the problem (5.13) leads in a simple way to an $(m + \mu, n + k)$ NPF of the original Newton-Padé approximation problem.

Rational Interpolation

Theorem 5.5. Let $(m,n) \in \mathbb{N} \times \mathbb{N}^+$ and $[\mu; k] \in \mathbb{Z} \times \mathbb{N}^+$ such that $-k-2 \le \mu \le k$. Let $(p_n, q_n) := (p_{m,n}, q_{m,n})$ be a weakly regular (m,n) NPF of (f,g) with residual e_n, and let $(\grave{p}_n, \grave{q}_n) := (p_{m-1,n-1}, q_{m-1,n-1})$ be an $(m-1, n-1)$ NPF with residual $\grave{e}_n$.

(i) If $(u_k^{(n)}, v_k^{(n)})$ is a $[\mu+1; k]$ MPF of $(z^{-m+1}p_n(z), z^{-m+1}\grave{p}_n(z); \tau_n e_n, \grave{e}_n)$ with residual $(\hat{e}_k^{(n)}, e_k^{(n)})$, then

$$\begin{bmatrix} p_{n+k} \\ q_{n+k} \\ t_{\mu+k+2}^{(m+n-2)} e_{n+k} \end{bmatrix} := \begin{bmatrix} \grave{p}_n & p_n \\ \grave{q}_n & q_n \\ \grave{e}_n & \tau_n e_n \end{bmatrix} \begin{bmatrix} u_k^{(n)} \\ v_k^{(n)} \end{bmatrix} \qquad (5.18)$$

yields an $(m+\mu, n+k)$ NPF (p_{n+k}, q_{n+k}) of (f,g) and the corresponding residual e_{n+k}, which is equal to $e_k^{(n)}$.

(ii) If, additionally, μ is restricted to $-k \le \mu \le k$ and $(\grave{u}_k^{(n)}, \grave{v}_k^{(n)})$ is a $[\mu; k-1]$ MPF with residual $(\grave{\hat{e}}_k^{(n)}, \grave{e}_k^{(n)})$, then

$$\begin{bmatrix} \grave{p}_{n+k} & p_{n+k} \\ \grave{q}_{n+k} & q_{n+k} \end{bmatrix} := \begin{bmatrix} \grave{p}_n & p_n \\ \grave{q}_n & q_n \end{bmatrix} \begin{bmatrix} \grave{u}_k^{(n)} & u_k^{(n)} \\ \grave{v}_k^{(n)} & v_k^{(n)} \end{bmatrix} \qquad (5.19)$$

and

$$t_{m+n+k+\mu-1}[\grave{e}_{n+k} \ \tau_n e_{n+k}] := t_{m+n-1}[\grave{e}_n \ \tau_n e_n] \begin{bmatrix} \grave{u}_k^{(n)} & u_k^{(n)} \\ \grave{v}_k^{(n)} & v_k^{(n)} \end{bmatrix} \qquad (5.20)$$

also yield an $(m+\mu-1, n+k-1)$ NPF $(\grave{p}_{n+k}, \grave{q}_{n+k})$ of (f,g) and the corresponding residual $\grave{e}_{n+k}$, which is equal to $\grave{e}_k^{(n)}$.

(iii) If, additionally, $(u_k^{(n)}, v_k^{(n)})$ is weakly regular, then (p_{n+k}, q_{n+k}) is also weakly regular.

Proof: (i) Since $(p_n, q_n) \in \mathcal{P}_m \times \mathcal{P}_n$ and $(\grave{p}_n, \grave{q}_n) \in \mathcal{P}_{m-1} \times \mathcal{P}_{n-1}$, (5.15) indicates that the pair (p_{n+k}, q_{n+k}) defined by (5.18) lies in $\mathcal{P}_{m+k} \times \mathcal{P}_{n+k}$. However, by definition of $(u_k^{(n)}, v_k^{(n)})$ as a $[\mu+1; k]$ MPF of (5.14) we have actually $p_{n+k} \in \mathcal{P}_{m+\mu}$, cf. (5.13). Moreover, by (5.6), (5.17), and (5.18),

$$[g \ f] \begin{bmatrix} p_{n+k} \\ q_{n+k} \end{bmatrix} = [g \ f] \begin{bmatrix} \grave{p}_n & p_n \\ \grave{q}_n & q_n \end{bmatrix} \begin{bmatrix} u_k^{(n)} \\ v_k^{(n)} \end{bmatrix} = t_{m+n-1}[\grave{e}_n \ \tau_n e_n] \begin{bmatrix} u_k^{(n)} \\ v_k^{(n)} \end{bmatrix}$$

$$= \left\{ \begin{matrix} t_{m+n-1} t_{k+\mu+2}^{(m+n-2)} \hat{e}_k^{(n)} = t_{m+n+k+\mu+1} e_k^{(n)} \\ t_{m+n-1} t_{k+\mu+2}^{(m+n-2)} e_{n+k} = t_{m+n+k+\mu+1} e_{n+k} \end{matrix} \right\} = O(t_{m+n+k+\mu+1}),$$

which shows that (p_{n+k}, q_{n+k}) is an $(m+\mu, n+k)$ NPF of (f,g) and that both e_{n+k} defined by (5.18) and $e_k^{(n)}$ defined by (5.17) are equal to its residual.

(ii) By applying (i) additionally to the $[\mu; k-1]$ MPF $(\grave{u}_k^{(n)}, \grave{v}_k^{(n)})$ of the data (5.14), we obtain (5.19) and (5.20).

(iii) Since (p_n, q_n) and $(u_n^{(k)}, v_n^{(k)})$ are weakly regular, the two matrices on the right-hand side of (5.19) do not vanish identically, hence the same is true for their product, and thus (p_{n+k}, q_{n+k}) is weakly regular. ∎

Viewed from the current (m,n) NPF (p_n, q_n) the newly constructed $(m+\mu, n+k)$ NPF (p_{n+k}, q_{n+k}) lies under the assumption of part (ii) in the Newton-Padé table in a 90° sector with horizontal axis, cf. Figure 4. If we would like to move instead in a 90° sector with vertical axis we just apply the same theorem, but with (f,g), (p_n, q_n), (m,n) replaced by (g,f), (q_n, p_n), (n,m).

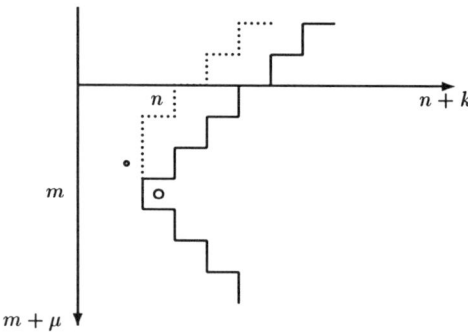

Figure 4. The index range of the new $(m+\mu, n+k)$ NPA (solid line) and its upper left neighbor (dotted line) that can be constructed according to Theorem 5.5.

Note also that, when $k = 1$, (5.19) and (5.20) allow us in particular to construct for $\mu = -1$ and $\mu = 0$ an $(m+\mu-1, n)$ NPF $(\mathring{p}_{n+1}, \mathring{q}_{n+1})$, even if we do not compute (p_{n+1}, q_{n+1}). This is an extension of part (i) of the theorem to $k = 0$, which we did not formulate under (i) because (p_{n+k}, q_{n+k}) on the left-hand side of (5.18) could be confused with (p_n, q_n) if $k = 0$. The former pair is then an $(m-1, n)$ or $(m-2, n)$ NPF, while the latter is an (m,n) NPF.

Theorem 5.5 has an interpretation in terms of continued fractions. For a diagonal sequence of weakly regular NPFs, p_n/q_n is a convergent (*i.e.*, a 'partial sum') of the *diagonal G-fraction* [37] of f/g, and $e_n/\mathring{e}_n$ is represented by the corresponding tail of the G-fraction. Hence, to get p_{n+k}/q_{n+k} we just combine a convergent of the continued fraction with a convergent of its tail.

Of course, for the recursive construction of NPAs we will apply Theorem 5.5 repeatedly. Then (5.19) provides us with a representation of two neighboring NPFs in terms of a product of 2×2 matrices

$$\begin{bmatrix} \mathring{u}_k^{(n)} & u_k^{(n)} \\ \mathring{v}_k^{(n)} & v_k^{(n)} \end{bmatrix} \quad (5.21)$$

whose elements are polynomials. In exact arithmetic we can choose k in each step as small as possible, namely just such that the determinant of the new factor is not identically zero. The formal infinite product that is then obtained by proceeding along a diagonal (*i.e.*, by choosing $\mu = k$) is a matrix representation of the diagonal G-fraction, which is itself a generalization of the classical Thiele fraction; see [11] for further details. In this case the recursive method suggested here is closely related to the algorithms of Gutknecht [37] and Van Barel and Bultheel [66,69]. A (weakly) stable algorithm of this type for diagonal sequences in the Padé table was presented by Cabay and Meleshko [12,51]. In the Padé table the diagonal G-fractions become P-fractions [45].

But Theorem 5.5 allows us also to step through the Newton-Padé table in many other ways. For example, we can follow a row of the table by choosing $\mu = 0$. In this way we could produce the product representation of a *sawtooth G-fraction*. The even part of the latter is then a *row G-fraction*. In the case of confluent interpolation points, where the Newton-Padé table becomes a Padé table, such a row G-fraction turns into a nongeneric generalization of an M-fraction [48], which one may call a *row P-fraction* [18]. By choosing $\mu = -k$ we could as well follow an antidiagonal of the Newton-Padé table, or rather, a pair of adjacent antidiagonals. Hence, in this case the algorithm has some similarity to the classical Kronecker algorithm [72], but here we obtain directly a nongeneric version, which is able to cope with singular blocks.

The algorithms which jump from one weakly regular NPF to the next are in the standard terminology 'reliable'; *i.e.*, in exact arithmetic they do not break down and produce a true interpolant if there exists one, as some of the methods described in [2,5,30,28,33,37,66,74,73,75]. The last NPF (p_n, q_n) that has to be computed need not be weakly regular, and (as in other 'reliable' algorithms) has to undergo a consistency test; *i.e.*, one has to check for common factors of p_n and q_n in order to know whether $r_n = p_n/q_n$ is a true interpolant.

However, the strength of Theorem 5.5 lies in the fact that it also leads to *stable* algorithms. In this respect it goes well beyond the above mentioned other recursive algorithms for non-normal Newton-Padé tables. For stable recurrences it is important that the factors (5.21) be not only nonsingular, but in a certain sense be well-conditioned. Hence, in general, the step sizes k have to be chosen larger than in the above construction for exact arithmetic. Recall that these factors have four polynomials as elements and are always singular for certain values of the variable z. But once we represent the polynomials in a particular basis, the 2×2 matrices (5.21) are replaced by 2×2 block matrices with scalar elements, which are the coefficients of the four polynomials in the particular basis; see [11,12,36]. (A matrix of similar type appeared in (1.28).) Hence, one could use the condition number of these typically small coefficient matrices to determine the step length k, although, of course, this condition number depends on the basis used, and the accumulation of roundoff may lead after many steps to a wrong judgement of the well-conditioning of (p_n, q_n). However, there are also other promising options for the stepsize criterion, related to Lemmas 5.1 and 5.4, see [11,12,36]. Further investigations are necessary to determine the most appropriate one.

It must be emphasized, however, that another effective strategy for stability in rational interpolation is the reordering of the interpolation points, which has some resemblance to pivoting in Gaussian elimination. This has been promoted by Graves-Morris [30,28,33] and included in Werner's algorithm [75]. It has some limitations too: first, a prescribed value for a derivative can never be taken into account before the function value and the lower derivatives at the same point z_j have been interpolated; second, function values at clustered interpolation points have to be treated like derivatives; and third, the interpolation data may actually come up only successively while the interpolation process goes on. When reordering is not applicable or not effective, the above outlined class of potentially stable 'look-ahead' method seems to be exactly the appropriate compromise. Hence, any implementation should combine the two strategies.

Finally, let us mention that Theorem 5.5 is just one formulation of a more general principle. Antoulas [1] explained recurrences for diagonal sequences of matrix Padé approximants also in this fashion, and from his and other authors' more recent work one can expect that the general principle will prevail in vector- and matrix-valued Padé approximation and rational interpolation, although a criterion for stepsize control guaranteeing stability may be more difficult to establish. Another generalization replaces the Newton-Padé table by the multipoint Padé table, for which nearly the same recursion is valid. Yet another variation of Theorem 5.5 makes use of row-regular NPAs only; for the special case of the Padé table this is also treated in [36].

References

1. Antoulas, A. C., On recursiveness and related topics in linear systems, IEEE Trans. Automatic Control, **AC-31** (1986), 1121–1135.
2. Antoulas, A. C. and B. D. O. Anderson, On the scalar rational interpolation problem, IMA J. Math. Control Inform. **3** (1986), 61–88.
3. Arndt, H., Ein verallgemeinerter Kettenbruch-Algorithmus zur rationalen Hermite-Interpolation, Numer. Math. **36** (1980), 99–107.
4. Baker, G. A., Jr. and P. Graves-Morris, *Padé Approximants (2 Vols.)*, Addison-Wesley, Reading, Mass., 1981.
5. Beckermann, B., A reliable method for computing M-Padé approximants on arbitrary staircases, J. Comput. Appl. Math. **40** (1992), 19–42.
6. Beckermann, B., The structure of the singular solution table of the M-Padé approximation problem, J. Comput. Appl. Math. **32** (1990), 3–15.
7. Brezinski, C., *Padé-Type Approximants and General Orthogonal Polynomials*, ISNM, Vol. 50, Birkhäuser, Basel, 1980.
8. Bultheel, A., *Laurent Series and their Padé Approximations*, Birkhäuser, Basel/Boston, 1987.
9. Bultheel, A. and M. Van Barel, Minimal vector Padé approximation, J. Comput. Appl. Math. **32** (1990), 27–37.
10. Cabay, S. and D.-K. Choi, Algebraic computations of scaled Padé fractions, SIAM J. Comput. **15** (1986), 243–270.

11. Cabay, S. and M. H. Gutknecht, forthcoming report.
12. Cabay, S. and R. Meleshko, A weakly stable algorithm for Padé approximants and the inversion of Hankel matrices, SIAM J. Matrix Anal. Appl., to appear.
13. Chan, T. F. and P. C. Hansen, Fortran subroutines for general Toeplitz systems, Tech. Rep. 90-21, Departement of Mathematics, University of California, Los Angeles, 1990.
14. Cheney, E. W., *Introduction to Approximation Theory*, McGraw-Hill, New York, 1966. Reprinted, Chelsea Publications, New York, 1980.
15. Claessens, G., A new algorithm for osculatory rational interpolation, Numer. Math. **27** (1976), 77–83.
16. Claessens, G., The rational Hermite interpolation problem and some related recurrence formulas, Comp. and Maths. with Appls. **2** (1976), 117–123.
17. Claessens, G., On the Newton-Padé approximation problem, J. Approx. Theory **22** (1978), 150–160.
18. Cooper, S. C., A. Magnus, and J. H. McCabe, On the non-normal two-point Padé table, J. Comput. Appl. Math. **16** (1986), 371–380.
19. de Boor, C., *A Practical Guide to Splines*, Springer, Berlin, 1978.
20. Draux, A., *Polynômes orthogonaux formels—applications*, vol. 974 of Lecture Notes in Mathematics, Springer-Verlag, Berlin, 1983.
21. Freund, R., M. Gutknecht, and N. Nachtigal, An implementation of the look-ahead Lanczos algorithm for non-Hermitian matrices, SIAM J. Sci. Stat. Comput., to appear.
22. Freund, R. and H. Zha, A look-ahead algorithm for the solution of general Hankel systems, Tech. Rep. 91.24, RIACS, NASA Ames Research Center, 1991.
23. Fuhrmann, P. A., A matrix Euclidean algorithm and matrix continued fractions, System Control Letters **3** (1983), 263–271.
24. Gallucci, M. A. and W. B. Jones, Rational approximations corresponding to Newton series (Newton-Padé approximants), J. Approx. Theory **17** (1976), 366–392.
25. Gragg, W. B., F. Gustavson, D. Warner, and D. Yun, On fast computation of superdiagonal Padé fractions, Math. Programming Stud. **18** (1982), 39–42.
26. Gragg, W. B., The Padé table and its relation to certain algorithms of numerical analysis, SIAM Rev. **14** (1972), 1–62.
27. Gragg, W. B. and A. Lindquist, On the partial realization problem, Linear Algebra Appl. **50** (1983), 277–319.
28. Graves-Morris, P. R., Efficient reliable rational interpolation, in *Padé Approximation and its Applications, Amsterdam 1980*, M. B. de Bruin and H. van Rossum (eds.), vol. 888 of Lecture Notes in Mathematics, Springer-Verlag, 1981, 28–63.
29. Graves-Morris, P. R., Generalized inverse vector valued rational interpolation, in *Padé Approximation and Its Applications, Bad Honnef*, H. Wer-

ner and H. J. Bünger (eds.), vol. 1071 of Lecture Notes in Mathematics, Springer-Verlag, 1984, 144–156.
30. Graves-Morris, P. R., Practical, reliable, rational interpolation, J. Inst. Maths. Applics. **25** (1980), 267–286.
31. Graves-Morris, P. R., Vector valued rational interpolants I, Numer. Math. **42** (1983), 331–348.
32. Graves-Morris, P. R., Vector valued rational interpolants II, IMA J. Numer. Anal. **4** (1984), 209–224.
33. Graves-Morris, P. R. and T. R. Hopkins, Reliable rational interpolation, Numer. Math. **36** (1981), 111–128.
34. Guo-liang, X. and A. Bultheel, The problem of matrix Padé approximation, in *Approximation, Optimization and Computing; Theory and Applications*, A. G. Law and C. C. Wang (eds.), Elsevier (North-Holland), 1990, 217–220.
35. Gutknecht, M. H., A completed theory of the unsymmetric Lanczos process and related algorithms, Part II, SIAM J. Matrix Anal. Appl., to appear.
36. Gutknecht, M. H., Stable row-recurrences in the Padé table and generically superfast look-ahead solvers for non-Hermitian Toeplitz systems. Forthcoming.
37. Gutknecht, M. H., Continued fractions associated with the Newton-Padé table, Numer. Math. **56** (1989), 547–589.
38. Gutknecht, M. H., The rational interpolation problem revisited, Rocky Mountain J. Math. **21** (1991), 263–280.
39. Gutknecht, M. H., A completed theory of the unsymmetric Lanczos process and related algorithms, Part I, SIAM J. Matrix Anal. Appl. **13** (1992), 594–639.
40. Joubert, W. D., Lanczos methods for the solution of nonsymmetric systems of linear equations, SIAM J. Matrix Anal. Appl., to appear.
41. Joubert, W. D., *Generalized conjugate gradient and Lanczos methods for the solution of nonsymmetric systems of linear equations*, PhD thesis, Center for Numerical Analysis, University of Texas at Austin, 1990. Tech. Rep. CNA-238.
42. Labahn, G. and S. Cabay, Matrix Padé fractions and their computation, SIAM J. Comput. **18** (1989), 639–657.
43. Maehly, H. and C. Witzgall, Tschebyscheff-Approximationen in kleinen Intervallen II, Stetigkeitssätze für gebrochen rationale Approximationen, Numer. Math. **2** (1960), 293–307.
44. Magnus, A., On the structure of the two-point Padé table, in *Analytic Theory of Continued Fractions*, W. Jones et al. (eds.), vol. 933 of Lecture Notes in Mathematics, Springer-Verlag, 1982, 176–193.
45. Magnus, A., Certain continued fractions associated with the Padé table, Math. Z. **78** (1962), 361–374.
46. Mahler, K., Perfect systems, Composito Math. **19** (1968), 95–166.
47. McCabe, J. H., A formal extension of the Padé table to include two-point Padé quotients, J. Inst. Maths Applics **15** (1975), 363–372.

48. McCabe, J. H. and J. A. Murphy, Continued fractions which correspond to power series expansions at two points, J. Inst. Maths Applics **17** (1976), 233–247.
49. McEliece, R. J. and J. B. Shearer, A property of Euclid's algorithm and an application to Padé approximation, SIAM J. Appl. Math. **34** (1978), 611–615.
50. Meinguet, J., On the solubility of the Cauchy Interpolation problem, in *Approximation Theory*, A. Talbot (ed.), Academic Press, 1970, 137–163.
51. Meleshko, R. J., *A stable algorithm for the computation of Padé approximants*, PhD thesis, Department of Computing Science, University of Alberta (Canada), 1990.
52. Milne-Thomson, L. M., *The Calculus of Finite Differences*, Macmillan, London, 1933.
53. G. Opitz, Steigungsmatrizen, Z. Ang. Math. Mech. **44** (1964), T52–T54.
54. Parlett, B. N., Reduction to tridiagonal form and minimal realizations, SIAM J. Matrix Anal. Appl. **13** (1992), 567–593.
55. Parlett, B. N., D. R. Taylor, and Z. A. Liu, A look-ahead Lanczos algorithm for unsymmetric matrices, Math. Comp. **44** (1985), 105–124.
56. Paszkowski, S., Padé-Hermite approximation: basic notions and theorems, J. Comput. Appl. Math. **32** (1987), 229–236.
57. Paszkowski, S., Recurrence relations in Padé-Hermite approximation, J. Comput. Appl. Math. **19** (1987), 99–107.
58. Salzer, H. B., Note on osculatory rational interpolation, Math. Comp. **16** (1962), 486–491.
59. Sidi, A., Some aspects of two-point Padé approximants, J. Comput. Appl. Math. **6** (1980), 9–17.
60. Stahl, H., Existence and uniqueness of rational interpolants with free and prescribed poles, in *Approximation Theory, Tampa, 1985–1986*, E. B. Saff (ed.), vol. 1287 of Lecture Notes in Mathematics, Springer-Verlag, 1987, 180–208.
61. Stoer, J., Über zwei Algorithmen zur Interpolation mit rationalen Funktionen, Numer. Math. **3** (1961), 285–304.
62. Stoer, J., *Einführung in die Numerische Mathematik I*, Springer-Verlag, Berlin, 1972.
63. Thron, W. J., Two-point Padé tables, T-fractions and sequences of Schur, in *Padé and Rational Approximation*, E. Saff and R. Varga (eds.), Academic Press, 1977, 215–226.
64. Van Barel, M. and A. Bultheel, A canonical matrix continued fraction solution of the minimal (partial) realization problem, Linear Algebra Appl. **122/123/124** (1989), 973–1002.
65. Van Barel, M. and A. Bultheel, A matrix Euclidean algorithm for minimal partial realization, Linear Algebra Appl. **121** (1989), 674–682.
66. Van Barel, M. and A. Bultheel, A new approach to the rational interpolation problem, J. Comput. Appl. Math. **32** (1990), 281–289.
67. Van Barel, M. and A. Bultheel, A new approach to the rational inter-

polation problem: the vector case, J. Comput. Appl. Math. **33** (1990), 331–346.
68. Van Barel, M. and A. Bultheel, The computation of non-perfect Padé-Hermite approximants, Numerical Algorithms **1** (1991), 285–304.
69. Van Barel, M. and A. Bultheel, A new formal approach to the rational interpolation problem, Numer. Math. **62** (1992), 87–122.
70. Van Iseghem-Le Luyer, J., *Approximants de Padé vectoriels*, PhD thesis, Université des Sciences et Techniques de Lille Flandres Artois, 1987.
71. Walsh, J. L., *Interpolation and Approximation by Rational Functions in the Complex Domain*, vol. 20 of Colloquium Publications, American Mathematical Society, Providence, R.I., 1969.
72. Warner, D. D., *Hermite Interpolation with Rational Functions*, PhD thesis, University of California at San Diego, 1974.
73. Werner, H., Ein Algorithmus zur rationalen Interpolation, in *Numerische Methoden der Approximationstheorie, Bd. 5*, L. Collatz, et al. (eds.), Birkhäuser, Basel, 1980, 319–337.
74. Werner, H., A reliable method for rational interpolation, in *Padé Approximation and its Applications*, L. Wuytack (ed.), vol. 765 of Lecture Notes in Mathematics, Springer-Verlag, 1979, 257–277.
75. Werner, H., Algorithm 51: A reliable and numerically stable program for rational interpolation in Lagrange data, Computing **31** (1983), 269–286.
76. Werner, H. and R. Schaback, *Praktische Mathematik II*, Springer-Verlag, Berlin, 1972.
77. Wuytack, L., Extrapolation to the limit by using continued fraction interpolation, Rocky Mountain J. Math. **4** (1974), 395–397.
78. Wuytack, L., On some aspects of the rational interpolation problem, SIAM J. Numer. Anal. **11** (1974), 52–60.
79. Wuytack, L., On the osculatory rational interpolation problem, Math. Comp. **29** (1975), 837–843.

Martin H. Gutknecht
Interdisciplinary Project Center for Supercomputing (IPS)
Swiss Federal Institute of Technology
ETH-Zentrum
CH-8092 Zürich
Switzerland
mhg@ips.ethz.ch

Multivariate Approximation from the Cardinal Interpolation Point of View

K. Jetter

Abstract. Cardinal interpolation provides a rich theory and powerful results in a typical situation of multivariate approximation. As the basic interpolating space we consider a principal shift-invariant subspace of $L_2(\mathbb{R}^d)$; in special examples, this space is spanned by the multi-integer translates of a box spline, or of a preconditioned radial basis function. This paper aims at reviewing some recent developments in this field. Our approach to cardinal interpolation will be formulated in the Fourier transform domain. This point of view is basically new; it is more natural and gives a more direct way of understanding some fundamental results such as, e.g., polynomial reproduction, error estimates, approximation orders for L_2-approximation, cardinal Hermite interpolation, bounds for condition numbers of collocation matrices, among others. This paper also contains a list of recent publications in the field.

§0. Introduction and Notation

This paper gives an introduction to multivariate cardinal interpolation with a special emphasis on recent research done by the present author and his co-authors. In some parts (Sections 1 - 4) the presentation will give a rather definitive view, while other parts (Sections 5 and 6) intend to provide some motivation for future research.

Cardinal interpolation was treated in the recent surveys of Powell [114], Riemenschneider [119], and the present author [73]; it is also addressed in Chui's CBMS-NSF conference monograph [33] and in Dyn's survey papers [49,50]; last but not least, Rabut's thesis [115] and his survey [117] also provide good sources for various aspects of cardinal interpolation and the Schoenberg operators. In contrast to all these works, we will give a completely new presentation (even of basic well-known properties) by working in the Fourier transform domain. As we shall see, this point of view has certain advantages over other approaches that have been considered so far.

Let us first introduce the notation to be used. We consider d-variate functions $f \in L_1(\mathbb{R}^d)$, or $f \in L_2(\mathbb{R}^d)$, or $f \in W_2^m(\mathbb{R}^d)$, where $W_2^m(\mathbb{R}^d)$ are the usual Sobolev spaces. More generally, $f \in \mathcal{S}'$, or f is a tempered distribution, if $\mathcal{S}$ denotes the Schwartz space of fast decreasing functions on $\mathbb{R}^d$. For a continuous function, we shall write $f \in C(\mathbb{R}^d)$; if in addition f has compact support then we write $f \in C_0(\mathbb{R}^d)$. If f is a function on $\mathbb{R}^d$, the notation $f_|$ denotes its restriction to $\mathbb{Z}^d$.

The *Fourier transform* of $f \in L_1(\mathbb{R}^d)$ is defined by

$$f^\wedge(\xi) := \int f(x)\, e^{-ix\cdot\xi}\, dx, \tag{0.1}$$

where $x \cdot \xi$ is the dot product in $\mathbb{R}^d$ and $\int = \int_{\mathbb{R}^d}$. This notion is generalized to $f \in \mathcal{S}'$ by requiring that Parseval's identity holds, namely,

$$\langle \psi, f^\wedge \rangle := \langle \psi^\wedge, f \rangle, \quad \text{for all} \quad \psi \in \mathcal{S}, \tag{0.2}$$

where $\langle \cdot, \cdot \rangle$ denotes the canonical sesquilinear form of the dual pairing $(\mathcal{S}, \mathcal{S}')$. By the Paley-Wiener Theorem we know that $f \in \mathcal{E}'$ (i.e., $f \in \mathcal{S}'$ has compact support) if and only if $f^\wedge$ is an entire function of exponential type.

The 2π-*periodization* of $f^\wedge$ is given by

$$f^\sim := \sum_\alpha f^\wedge(\cdot + 2\pi\alpha) ; \tag{0.3}$$

here $\sum_\alpha$ denotes summation over the entire lattice $\mathbb{Z}^d$. We note that (0.3) makes sense only under additional assumptions; typical conditions could be

(a) $f^\wedge = f_1^\wedge + f_2^\wedge \in \mathcal{S}'$ with $f_2^\wedge \in \mathcal{S}$ and $f_1^\wedge$ has compact support (e.g., f_1 is a polynomial), or

(b) $f \in W_2^m(\mathbb{R}^d)$ for "large" m; more precisely there could be assumptions in order to have Poisson's summation formula for (0.3),

$$f^\sim(\xi) = \sum_\alpha f(\alpha)\, e^{-i\alpha\cdot\xi} . \tag{0.4}$$

The *principal shift invariant space* $\mathcal{S}_\phi$ associated with $\phi \in L_2(\mathbb{R}^d)$ is defined as the L_2-closure of the space $\mathcal{S}_\phi^0$ of finite linear combinations of the set $\{\phi(\cdot - \alpha) : \alpha \in \mathbb{Z}^d\}$. Equivalently (cf. [14] as a recent important contribution along these lines), we may define these spaces in the Fourier transform domain by using

$$f^\wedge = \phi^\wedge \omega . \tag{0.5}$$

Thus, $f \in \mathcal{S}_\phi^0$ if and only if (0.5) holds for some trigonometric polynomial ω. Similarly, $f \in \mathcal{S}_\phi$ if and only if $f \in L_2(\mathbb{R}^d)$ and (0.5) holds for some 2π-periodic function ω.

In case $\phi \in \mathcal{S}'$ but $\phi \notin L_2(\mathbb{R}^d)$, we can sometimes find a 2π-periodic function such that $\phi_1^\wedge := \phi^\wedge \omega \in L_2(\mathbb{R}^d)$. For example, if ϕ is a fundamental solution of a linear partial differential operator with constant coefficients, and if ω is the symbol of a consistent discretization of this operator (with support in $\mathbb{Z}^d$), then ϕ_1 would be an analogue of what in univariate problems is called a B-spline. This method of preconditioning will be made more precise later on, and in this case we put $\mathcal{S}_\phi := \mathcal{S}_{\phi_1}$.

Here is a short outline of the paper: Section 1 gives the definition of the cardinal interpolation operator as a particular Schoenberg operator based on a fundamental interpolant of the lattice $\mathbb{Z}^d$, cf. (1.3), and analyzes its action on the corresponding principal shift-invariant space. The Toeplitz-Laurent operator approach to cardinal interpolation is also discussed, and two algorithms for computing interpolants are presented. Section 2 provides these basic examples of cardinal interpolation operators: box spline interpolation (in particular, the three- and the four-directional cases), polyharmonic spline interpolation, and general radial basis function interpolation with an emphasis on m-th order completely monotone basis functions and corresponding methods of preconditioning with powers of the discrete Laplacian. In Section 3, error estimates in terms of L_2-Sobolev-seminorms are considered, and as a part of this consideration, the Strang-Fix conditions are treated. Section 4 shows the relation between cardinal interpolation and L_2-approximation from shift-invariant spaces. A new type of cardinal Hermite interpolation is introduced in Section 5; and condition numbers for (not necessarily cardinal) interpolation are derived in Section 6 for the case where the basis function is a special kind of positive definite function.

This paper largely contains results that have not appeared before in their present form. The presentation is based on the author's recent research and his specific point of view on cardinal interpolation problems. Only Section 7 deviates from this personal point of view by giving some references and hints to other approaches.

The author's present understanding of multivariate cardinal interpolation certainly reflects many discussions with his colleagues and co-authors. Here, in particular, J. Stöckler (and the discussions during a joint research stay at the Oberwolfach institute in March 1991) had a major influence on the material presented here; also, the discussions with C. Chui, S. Riemenschneider, Z. Shen, N. Sivakumar, and J. Ward during two research visits at Texas A&M University and the University of Alberta are gratefully acknowledged.

§1. Cardinal Interpolation Defined

For a tempered (i.e., slowly growing) and continuous function ϕ we define the corresponding *Schoenberg operator* T_ϕ by the equation

$$T_\phi(f)^\wedge := \phi^\wedge f^\sim . \tag{1.1}$$

Since we are multiplying distributions, the domain of this operator depends very much upon the properties of $\phi^\wedge$. For example, if $\phi^\wedge \in L_2(\mathbb{R}^d)$ and

$\phi^\wedge f^\sim \in L_2(\mathbb{R}^d)$ as well, then $T_\phi(f) \in S_\phi$. In particular, if $\phi \in C_0(\mathbb{R}^d)$ and $f^\sim(\xi) = \sum f(\alpha) e^{-i\alpha \cdot \xi}$ with $f_| \in \ell_1(\mathbb{Z}^d)$, then

$$T_\phi(f) = \sum_\alpha f(\alpha) \phi(\cdot - \alpha) .$$

The special case in which $\phi = M_m =$ the m-th order cardinal B-spline, (i.e., $M_m^\wedge(t) := (\text{sinc } \frac{t}{2})^m$ with sinc $t := t^{-1} \sin t$), was originally considered by Schoenberg.

The Schoenberg operator T_ϕ is said to be *local* if ϕ has compact support (equivalently, $\phi^\wedge$ is an entire function of exponential type).

The *commutator* (of ϕ and f) was introduced in [35] by the definition

$$[\phi|f] := T_f(\phi) - T_\phi(f)$$

in case ϕ has compact support. By putting

$$[\phi|f]^\wedge := f^\wedge \phi^\sim - \phi^\wedge f^\sim \qquad (1.2)$$

we get a more general definition (for example, this notion can be applied to radial basis functions as well, a fact which was not considered so far). We shall use this later for analyzing the error of cardinal interpolation. Another useful fact is given by what is called the "flip property" in [19]:

Theorem 1.1. *If $\phi \in L_2(\mathbb{R}^d)$ and $f \in S_\phi$, then $[\phi|f] = 0$.*

Proof: Since $f \in S_\phi$ we have $f^\wedge = \phi^\wedge \omega \in L_2(\mathbb{R}^d)$, where ω is a 2π-periodic function, and hence $f^\sim = \phi^\sim \omega$. Substituting this into (1.2) gives the desired property. ∎

The *cardinal interpolation operator* $\mathcal{J}_\phi$ associated with ϕ is the Schoenberg operator corresponding to Λ_ϕ where

$$\Lambda_\phi^\wedge := \frac{\phi^\wedge}{\phi^\sim} = \frac{\phi^\wedge}{\sum_\alpha \phi^\wedge(\cdot + 2\pi\alpha)} ; \qquad (1.3a)$$

thus we define

$$\mathcal{J}_\phi(f)^\wedge := \Lambda_\phi^\wedge f^\sim . \qquad (1.3b)$$

Of course, $\phi^\wedge$ and $\phi^\sim$ should be functions in the classical sense in order for (1.3a) to be meaningful (cf. the examples in Section 2), and the set of zeros of $\phi^\sim$ should be assumed to be a set of Lebesgue measure zero.

It has been observed by several authors (dealing with specific special cases) that Λ_ϕ is often a *fundamental interpolant* as follows:

Theorem 1.2. *Let $\Lambda_\phi^\wedge$ belong to $L_1(\mathbb{R}^d)$. Then we have*

$$\Lambda_\phi(\alpha) = \begin{cases} 1, & \alpha = 0 \\ 0, & 0 \neq \alpha \in \mathbb{Z}^d \end{cases}. \tag{1.4}$$

For the proof it is sufficient to note that $\Lambda_\phi^\sim \equiv 1$ and hence for $\alpha \in \mathbb{Z}^d$:

$$\Lambda_\phi(\alpha) = \frac{1}{(2\pi)^d} \int \Lambda_\phi^\wedge(\xi) e^{i\alpha \cdot \xi} d\xi = \frac{1}{(2\pi)^d} \int_C \Lambda_\phi^\sim(\xi) e^{i\alpha \cdot \xi} d\xi = \delta_{0,\alpha},$$

where $\delta_{0,\alpha}$ is the Kronecker delta. More generally, since $\mathcal{J}_\phi(f)^\sim = \Lambda_\phi^\sim f^\sim = f^\sim$, we can conclude that $\mathcal{J}_\phi(f)^\sim$ and $f^\sim$ have the same Fourier coefficients; hence, with an application of the Poisson summation formula (under certain additional mild assumptions), we have the *interpolation property*

$$\mathcal{J}_\phi(f)(\alpha) = f(\alpha), \qquad \alpha \in \mathbb{Z}^d. \tag{1.5}$$

Let us show that $\mathcal{J}_\phi$ is a projection onto S_ϕ:

Theorem 1.3. *If $\phi \in L_2(\mathbb{R}^d)$ and $f \in S_\phi$, then $f = \mathcal{J}_\phi(f)$.*

This follows from Theorem 1.1, since

$$f^\wedge - \mathcal{J}_\phi(f)^\wedge = f^\wedge - \frac{\phi^\wedge}{\phi^\sim} f^\sim = \frac{[\phi|f]^\wedge}{\phi^\sim} \tag{1.6}$$

and since we have assumed that the zero set of $\phi^\sim$ is small.

If $\phi \notin L_2(\mathbb{R}^d)$, we can often create an L_2-function by differencing ϕ; in other words, we multiply $\phi^\wedge$ by, say, trigonometric polynomials τ

$$\phi_1^\wedge := \tau \phi^\wedge.$$

Since $\phi_1^\wedge / \phi_1^\sim = \phi^\wedge / \phi^\sim$, it is obvious that if $\phi_1^\wedge / \phi_1^\sim \in L_1(\mathbb{R}^d)$ we can replace (1.3a) by

$$\Lambda_\phi^\wedge := \frac{\phi_1^\wedge}{\phi_1^\sim},$$

and hence, Theorem 1.3 may be applied to S_{ϕ_1}. This idea of *preconditioning* of ϕ is rather old. In problems of scattered data interpolation it goes back to Dyn and Levin [53]; later it was applied to radial basis functions by Jackson [71], and by Buhmann [21]. The following preconditioner is due to Rabut [115] (application to polyharmonics) and to Chui, Jetter and Ward [36] (cf. Theorem 2.3 below); here $\tau = \sigma^\mu$ for some $\mu \in \mathbb{R}_+$, where

$$\sigma(\xi) = \sum_{i=1}^{d} 2(1 - \cos \xi_i) = \sum_{i=1}^{d} \left(2 \sin \frac{\xi_i}{2}\right)^2 \tag{1.7}$$

is the symbol of the discrete Laplacian.

If $\phi_| \in \ell_1(\mathbb{Z}^d)$ we have the following *Toeplitz-Laurent operator approach to cardinal interpolation*. Consider the operator

$$\Xi_\phi : \ell_2(\mathbb{Z}^d) \to \ell_2(\mathbb{Z}^d) , \qquad a \mapsto \phi_|\tilde{*}a, \tag{1.8}$$

where $\tilde{*}$ denotes the discrete convolution, i.e., $(a\tilde{*}b)_\alpha = \sum_\beta a_\alpha b_{\beta-\alpha}$. The interpolation property (1.5) can often be interpreted as the following problem (for more details refer to the approach to cardinal interpolation taken in [73]): For given data $b = (f(\alpha))_{\alpha \in \mathbb{Z}^d}$ find another sequence a such that

$$\Xi_\phi(a) := \phi_|\tilde{*}a = b . \tag{1.9}$$

Using the notation of singular equations, the problem is called *normal* if Ξ_ϕ is invertible, and *non-normal* otherwise. We have this well-known result:

Theorem 1.4. *Cardinal interpolation with $\phi_| \in \ell_1(\mathbb{Z}^d)$ is normal if and only if the symbol*

$$\sigma_\phi(\xi) := \sum_\alpha \phi(\alpha) e^{-i\alpha \cdot \xi} \tag{1.10}$$

does not vanish on $\mathbb{R}^d$.

Proof: For $a \in \ell_2(\mathbb{Z}^d)$ and $a^\sim(\xi) = \sum_\alpha a_\alpha e^{-i\alpha \cdot \xi}$, we have, by Parseval's identity,

$$\sum_\alpha \overline{a}_\alpha (\phi_|\tilde{*}a)_\alpha = \frac{1}{(2\pi)^d} \int_C \sigma_\phi(\xi) |a^\sim(\xi)|^2 \, d\xi . \tag{1.11}$$

If $\sigma_\phi(\xi_0) = 0$ for some $\xi_0 \in C$ then, for any $\varepsilon > 0$ we can find $a \in \ell_2(\mathbb{Z}^d)$ with $|a|_2 = 1$ such that the right-hand side of (1.11) in absolute value is less than ε, and hence Ξ_ϕ is not invertible. If $\sigma_\phi(\xi) \neq 0$ for all ξ, then by Wiener's Lemma we have $\lambda \in \ell_1(\mathbb{Z}^d)$, where

$$\lambda_\alpha := \frac{1}{(2\pi)^d} \int_C \frac{e^{i\alpha \cdot \xi}}{\sigma_\phi(\xi)} d\xi , \qquad \alpha \in \mathbb{Z}^d , \tag{1.12}$$

and (1.9) is solved by setting $a = \lambda \tilde{*} b$. ∎

We note that in many cases (cf. the examples below), we have

$$\sigma_\phi = \phi^\sim . \tag{1.13}$$

Sufficient conditions for this identiy to hold are that $\phi^\wedge$ is continuous and the series (0.4) defining $\phi^\sim$ is uniformly convergent; with these assumptions (1.13) is just Poisson's summation formula.

Let us finally point out a *method for computing cardinal interpolants*. Usually in the literature the following algorithm is considered: Find a from b in (1.9) and then evaluate the series $\sum_\alpha a_\alpha \phi(\cdot - \alpha)$. This method is only stable when the problem is normal. The following algorithm can be applied whenever $f^\sim(\xi) = \sum_\alpha f(\alpha) e^{-i\alpha \cdot \xi}$ and the sequence $f_|$ decays fast enough:

Algorithm 1.

Step 1: Evaluate $f^\sim(\xi) = \sum_\alpha f(\alpha)e^{-i\alpha\cdot\xi}$.

Step 2: Evaluate $\mathcal{J}_\phi(f)^\wedge = \Lambda_\phi^\wedge f^\sim$.

Step 3: Perform an inverse Fourier transform.

This algorithm was tested on various examples where $\Lambda_\phi^\wedge$ is readily available. In Steps 1 and 3, FFT-methods can be applied (for details see [75,78]).

In contrast to this, *the Chui-Diamond Neumann series approach* [34] could be applied as well. In the Fourier transform domain this method is realized by truncations of the series

$$\frac{1}{\phi^\sim} \sim \sum_{k=0}^\infty (1-\phi^\sim)^k \qquad (1.14)$$

in order to get the following recurrence.

Algorithm 2.

Step 1: Evaluate $f_0^\sim := f^\sim$ and $f_k^\sim := f_0^\sim + (1-\phi^\sim)f_{k-1}^\sim$ for $k=1,\ldots,m$.

Step 2: Evaluate $\mathcal{J}_\phi^{(m)}(f)^\wedge := \phi^\wedge f_m^\sim$.

Step 3: Perform an inverse Fourier transform.

It should be noted that in place of (1.14) other series expansions (such as Faber series, which are more adapted to the range of $\phi^\sim$, [39]) can also be used.

§2. Examples of Cardinal Interpolation Operators

2.1. Box Spline Interpolation.
For a directional matrix

$$X = (\xi_1, \ldots, \xi_N), \quad 0 \neq \xi_i \in \mathbb{Z}^d, \quad \text{rank } X = d, \qquad (2.1)$$

the corresponding (centered) *box spline* M_X is defined by

$$M_X^\wedge(\xi) := \prod_{i=1}^N \frac{2\sin\frac{1}{2}\xi_i\cdot\xi}{\xi_i\cdot\xi} = \prod_{i=1}^N \operatorname{sinc}\frac{1}{2}\xi_i\cdot\xi. \qquad (2.2)$$

The rank condition in (2.1) ensures that $M_X^\wedge$ (and hence M_X) is an L_2-function. Moreover, $M_X^\wedge \in L_1(\mathbb{R}^d)$ if and only if rank $X_i = d$, $i=1,\ldots,N$, where X_i is obtained from X by deleting column i. The support of M_X is given by

$$\operatorname{supp} M_X = \sum_{i=1}^N \left[-\frac{1}{2}, +\frac{1}{2}\right]\xi_i. \qquad (2.3)$$

Further properties of M_X are given in [73]; in particular, (1.13) holds for $\phi = M_X(\cdot + \tau)$, where τ is an arbitrary shift. Hence the symbol (1.10) takes the form

$$\sigma_\phi(\xi) = \phi^\sim(\xi) = \sum_\alpha M_X(\alpha + \tau)e^{-i\alpha\cdot\xi} . \tag{2.4}$$

A number of normal problems were listed in [73, Section 4]. Meanwhile, the *three-directional case* is (almost) completely understood. Here $d = 2$, and

$$X = (\underbrace{e_1, \ldots, e_1}_{k}, \underbrace{e_2, \ldots, e_2}_{\ell}, \underbrace{e_1 + e_2, \ldots, e_1 + e_2}_{m}) \tag{2.5a}$$

with $e_1 = (1,0)^T$ and $e_2 = (0,1)^T$, and we use the notation

$$M_{k,\ell,m} := M_X . \tag{2.5b}$$

The following result can be found [15] in case $\tau = 0$, [128] in case $k = \ell = m \in 2\mathbb{N}$, [37,40] in case $k = \ell = m \in \mathbb{N}$, and [10] in full generality:

Theorem 2.1. *Cardinal interpolation with $\phi = M_{k,\ell,m}(\cdot + \tau)$ and $\tau \in [-\frac{1}{2}, +\frac{1}{2}[^2$ is normal if (and in case $k = \ell = m$ only if) τ is from the hexagonal shift region*

$$\Omega = \left\{ (\tau_1, \tau_2); \ |\tau_1| < \frac{1}{2}, \ |\tau_2| < \frac{1}{2}, \ |\tau_1 - \tau_2| < \frac{1}{2} \right\}.$$

We also note that by (1.12) the fundamental sequence λ, and hence also the fundamental box spline

$$\Lambda_\phi(x) = \sum_\alpha \lambda_\alpha M_{k,\ell,m}(x + \tau - \alpha) ,$$

decay exponentially if $\tau \in \Omega$, but the rate of decay becomes worse as τ approaches $\partial\Omega$.

The *four-directional box spline*

$$M_{k,\ell,m,n} := M_Y \tag{2.6a}$$

is determined by the directional matrix

$$Y = (X | \underbrace{e_1 - e_2, \ldots, e_1 - e_2}_{n}) \tag{2.6b}$$

where X is as in (2.5a). Here it is easy to see that $\phi^\sim(\pi, \pi) = 0$ for any $\phi = M_{k,\ell,m,n}(\cdot + \tau)$, so that cardinal interpolation is never normal. Nevertheless, some interesting facts could be obtained for $\phi = M_{k,\ell,m,n}$:
(i) The submodule idea (cf. [119, Section 2]) shows that cardinal interpolation with

$$\psi(x_1, x_2) = 2\,\phi(x_1 + x_2, x_1 - x_2) \tag{2.7}$$

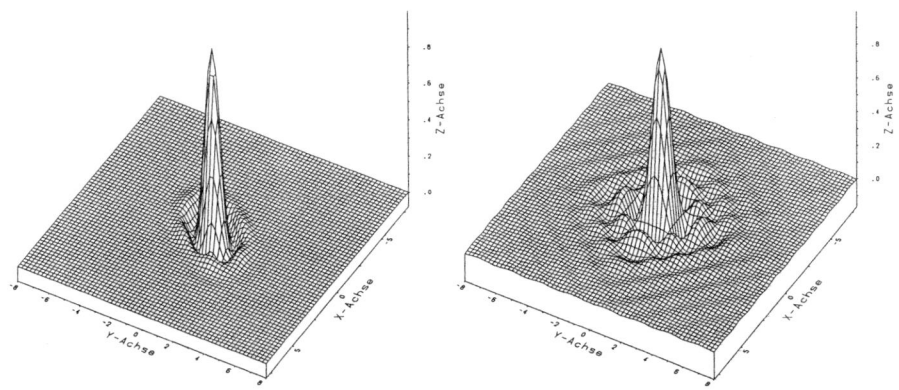

Figures 1 and 2. Fundamental interpolants for the box splines $M_{2,2,2}$ (left) and $M_{2,2,2,2}$ (right).

is indeed normal, and hence Λ_ψ decays exponentially. Interpolation on submodules is also considered in [139].

(ii) In case $k = \ell$ and $m = n$, the results of [78] show that $\phi^\wedge / \phi^\sim \in L_1(\mathbb{R}^d)$. Hence, the definition (1.3) of $\mathcal{J}_\phi$ is applicable, and Algorithm 1 gives a stable method for computing interpolants. In this case Λ_ϕ decays, by the Riemann-Lebesgue theorem, but we don't have exponential decay. Explicit results along these lines, based on a generalization of (1.12), were obtained in [86].

2.2. Polyharmonic Spline Interpolation. Cardinal interpolation with polyharmonic splines was considered in [64], [99]. Let Δ be the d-variate Laplacian,

$$\Delta = \sum_{i=1}^{d} \frac{\partial^2}{\partial x_i^2} \, ,$$

and let ϕ be the fundamental solution of $(-\Delta)^k$ where

$$2k > d \, , \tag{2.8}$$

(i.e., $(-\Delta)^k \phi = \delta$, where δ is the Dirac distribution). Thus,

$$\phi^\wedge(\xi) = |\xi|^{-2k} \, . \tag{2.9}$$

Now assumption (2.8) ensures that

$$\Lambda_\phi^\wedge(\xi) = \frac{|\xi|^{-2k}}{\sum_\alpha |\xi + 2\pi\alpha|^{-2k}} \in L_1(\mathbb{R}^d) \, , \tag{2.10}$$

and we have the following result from [99]:

Theorem 2.2. *The function $\Lambda_\phi^\wedge$ defined by (2.8) - (2.10) is analytic in a tube $\mathbb{R}_\epsilon^d = \{x + iy : x, y \in \mathbb{R}^d, |y| < \epsilon\}$, for some $\epsilon > 0$. Consequently, Λ_ϕ decays exponentially.*

While it is essential to have $k \in \mathbb{N}$ in this theorem, (2.10) is valid for any $k \in \mathbb{R}_+$ satisfying (2.8); e.g., in case $d = 2$ we could take $k = \frac{3}{2}$. For $d = k = 2$ we get the thin plate spline. Correspondingly, in Algorithm 1 only Step 2 must be adjusted.

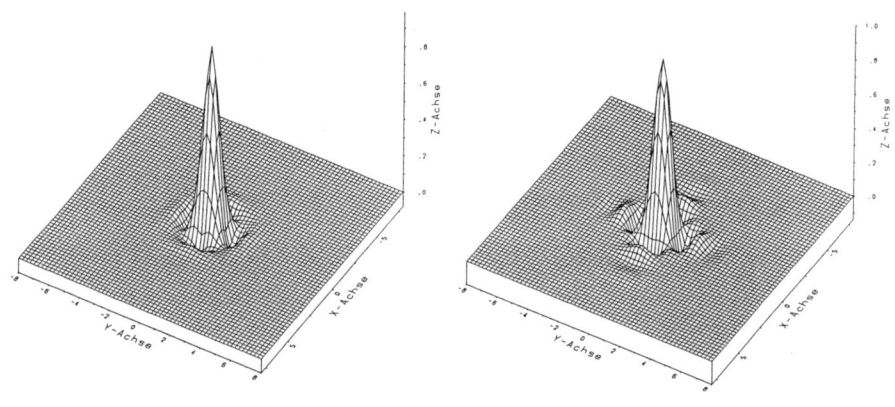

Figures 3 and 4. Fundamental interpolants for the thin plate spline (left) and the Hardy multiquadric (right).

We could also base our definition of Λ_ϕ on Rabut's polyharmonic B-spline [115]
$$B_k := (-\nabla)^k \phi$$
where ∇ is the discrete Laplacian; equivalently, by using the symbol (1.7), one can prove that the Fourier transform of B_k is given by

$$B_k^\wedge(\xi) = \left(\frac{\sum_{i=1}^d 2(1 - \cos \xi_i)}{|\xi|^2} \right)^k . \qquad (2.11)$$

This formula applies again to any $k \in \mathbb{R}_+$, and (2.8) implies that $B_k^\wedge \in L_1(\mathbb{R}^d) \cap C(\mathbb{R}^d)$. It is obvious that the fundamental function (2.10) then can be written as

$$\Lambda_\phi^\wedge = \frac{B_k^\wedge}{B_k^\sim} . \qquad (2.12)$$

The advantage of (2.12) over (2.10) is that $B_{k|} \in \ell_1(\mathbb{Z}^d)$ while $\phi_|$ is a sequence of polynomial growth; from this, the Toeplitz-Laurent operator approach to cardinal interpolation with polyharmonic splines becomes apparent.

2.3. Radial Basis Function Interpolation.

The previous example is a particularly interesting special case of a radial basis function; another case is the Hardy multiquadric

$$\phi(x) = \sqrt{c^2 + |x|^2},$$

where c is a fixed real parameter. Such radial basis functions show interesting features for interpolation in general [114] and cardinal interpolation in particular [23,28].

By definition, a *radial basis function* ϕ is defined through a slowly increasing univariate continuous function

$$g : [0, \infty) \to \mathbb{R} \tag{2.13a}$$

by putting

$$\phi(x) := g(|x|^2), \quad x \in \mathbb{R}^d. \tag{2.13b}$$

For example, in the case of Hardy multiquadrics we have $g(t) = \sqrt{c^2 + t}$ while in the polyharmonic case we have

$$g(t) = \text{const} \begin{cases} t^{k-d/2}, & \text{for } d \text{ odd} \\ t^{k-d/2} \log t, & \text{for } d \text{ even} \end{cases}$$

with some (positive) constant. Typical for these examples is the fact that some derivative of g is completely monotone.

We thus define the class CM_m of *m-th order completely monotone radial basis functions* by

$$\begin{aligned} \phi \in CM_m :&\Leftrightarrow g \in C[0, \infty) \cap C^\infty(0, \infty) \text{ and} \\ &(-1)^j g^{(j)}(t) \geq 0 \text{ for } j \geq m. \end{aligned} \tag{2.14}$$

By Bernstein's Theorem [140, Chapter IV], for $\phi \in CM_m$ we have

$$(-1)^m g^{(m)}(t) = \int_0^\infty e^{-\tau t} \, d\mu(\tau), \tag{2.15}$$

where $d\mu$ is a non-negative Borel measure. For example, in the above special cases, Equation (2.15) reduces to the following:

$$\frac{1}{t} = \int_0^\infty e^{-\tau t} \, d\tau \quad \text{and} \quad \frac{1}{\sqrt{c^2 + t}} = \int_0^\infty e^{-\tau t} \frac{e^{-c^2 \tau}}{\sqrt{\pi \tau}} \, d\tau, \quad t > 0.$$

This has led in [36] to the introduction of the following class of radial basis functions:

Definition. We write $\phi \in CM_m(\gamma, \beta)$ if $\phi \in CM_m$ and the if measure in (2.15) is of form

$$d\mu(t) = \gamma(t) \, t^{-\beta} \, dt, \tag{2.16a}$$

where $0 \leq \gamma \in C[0, \infty)$, γ is bounded, $\gamma(0) > 0$, $\beta \in \mathbb{R}_+$, and

$$\beta + m - 1 > 0. \tag{2.16b}$$

It was observed in [21] and elsewhere that (1.3a) can be taken as a definition of fundamental interpolants for radial basis functions if $\phi^\wedge$ is the Fourier transform of $\phi \in \mathcal{S}'$. Following [36], we can also apply an analogue of Equation (2.12) in order to obtain *cardinal interpolation by preconditioned radial basis functions*:

Theorem 2.3. Let $\phi \in CM_m(\gamma, \beta)$ and let $k \in \mathbb{N}$, $k \geq \max\{m, \delta\}$, where $\delta := \frac{d}{2} + \beta + m - 1$. With σ given by (1.7), we have

$$\psi^\wedge(\xi) := (\sigma(\xi))^k \phi^\wedge(\xi) = 4^\delta \pi^{d/2} \frac{(\sigma(\xi))^k}{|\xi|^{2\delta}} \int_0^\infty \gamma\left(\frac{|\xi|^2 \tau}{4}\right) \frac{e^{-1/\tau}}{\tau^{\delta+1}} \, d\tau \,, \quad (2.17)$$

and hence $\Lambda_\phi^\wedge := \psi^\wedge/\psi^\sim \in L_p(\mathbb{R}^d) \cap C(\mathbb{R}^d)$, $p \geq 1$. If $k = \delta$, the Toeplitz-Laurent operator Ξ_ψ defined by $\psi_|$ is positive definite, and in particular invertible; i.e., cardinal interpolation with ψ is normal.

There is a similar theorem for $k \in \frac{1}{2}\mathbb{N}$; (cf. [36]). We also note that if $k = \delta$, the expression (2.17) contains the function in Equation (2.11) as a factor. This shows that ψ is the convolution of Rabut's polyharmonic B-spline with a distribution.

2.4. Conditionally Positive Definite Functions. It was observed in [64,100] that the polyharmonic cardinal interpolating spline minimizes a certain "energy norm." This generalizes the well-known minimal property of Duchon's thin plate splines, and is part of the variational theory that was established as early as 1983 by Madych and Nelson in an unpublished manuscript [96]. Let us give the essentials about this energy norm from the current point of view.

The class CPD_m of continuous functions $\phi \in \mathcal{S}'$ which are *conditionally positive definite of order* m is defined by

$$\phi \in CPD_m :\Longleftrightarrow \int \phi(x) \, (\psi * \psi^*)(x) \, dx \geq 0 \quad \text{for all} \quad \psi \in \mathcal{D}_m \,; \quad (2.18)$$

here, $\psi^*(x) := \overline{\psi(-x)}$ is the involution of ψ,

$$\mathcal{D}_m := \left\{ \psi \in C_0^\infty(\mathbb{R}^d) : \int x^\alpha \psi(x) \, dx = 0 \quad \text{for} \quad |\alpha| < m \right\} \quad (2.19)$$

and $(f * g)(x) := \int f(t) g(x-t) \, dt$ is the convolution of f and g.

It is proved in [101] that the class CM_m defined in (2.14) is a subclass of CPD_m; the interesting additional fact here is that the integral in (2.18) defines a proper *energy norm* on $\mathcal{D}_m$ via

$$\|\psi\|_\phi^2 := \int \phi(x) \, (\psi * \psi^*)(x) \, dx \,. \quad (2.20)$$

In order to give an alternate expression of this norm we use the following canonical decomposition of $\phi \in CPD_m$.

Theorem 2.4. (cf. [98, Theorem 2.1]) If $\phi \in CPD_m$ then $\phi^\wedge = \phi_1^\wedge + \phi_2^\wedge$, where

(i) $\phi_1^\wedge = d\mu$ is a positive Borel measure on $\mathbb{R}^d \setminus \{0\}$ satisfying $\int (|\xi| / (1 + |\xi|))^{2m} \, d\mu < \infty$, and

(ii) $\phi_2^\wedge = p(D)\delta$ where $p(D) = \sum_{|\gamma|\leq 2m} a_\gamma D^\gamma$, a linear partial differential operator of order $2m$ (with constant coefficients) in which the leading coefficients satisfy

$$\sum_{|\alpha|=m} \sum_{|\beta|=m} c_\alpha\, a_{\alpha+\beta}\, \overline{c}_\beta \geq 0$$

for any complex numbers c_α.

Using this result, we infer via Parseval's identity that (2.20) takes the following alternate form in terms of the Fourier transform of ψ:

$$(2\pi)^d \|\psi\|_\phi^2 = \int |\psi^\wedge(\xi)|^2 d\mu(\xi) + \sum_{|\alpha|=m} \sum_{|\beta|=m} a_{\alpha+\beta} \frac{D^\alpha \psi^\wedge(0)}{\alpha!} \frac{\overline{D^\beta \psi^\wedge(0)}}{\beta!} \quad (2.21)$$

for $\psi \in \mathcal{D}_m$.

Let us finally show how the definition (2.18) is related to the assertion that

$$\sum_{k,\ell=1}^N c_k\, \phi(x_k - x_\ell)\, \overline{c}_\ell \geq 0 \qquad (2.22a)$$

for all distinct points $x_1, \ldots, x_N$ in $\mathbb{R}^d$, and all constants $c_1, \ldots, c_N$ satisfying

$$\sum_{k=1}^N c_k\, x_k^\alpha = 0 \quad \text{for all} \quad |\alpha| < m\,, \qquad (2.22b)$$

where $N \in \mathbb{N}$ is arbitrary. In order to see this we follow [98, Section 6] and let $\tau(\xi) = \sum_{k=1}^N c_k e^{-i x_k \cdot \xi}$. Then by (2.22b) we have

$$\tau(\xi) = \mathcal{O}(|\xi|^m) \quad \text{as} \quad |\xi| \to 0\,.$$

If we choose $g \in C_0^\infty(\mathbb{R}^d)$ with $\int g(\xi) d\xi = 1$, and $g_\epsilon(x) := \epsilon^{-d} g(x/\epsilon)$ for $\epsilon > 0$, then $\psi_\epsilon \in \mathcal{D}_m$, where

$$\psi_\epsilon^\wedge(\xi) := \tau(\xi) g_\epsilon^\wedge(\xi);$$

hence by (2.18), it follows that

$$0 \leq \int \phi(x)\, (\psi_\epsilon * \psi_\epsilon^*)(x)\, dx = \iint \phi(x-y)\, \psi_\epsilon(x)\, \overline{\psi_\epsilon(y)}\, dx\, dy$$

and taking the limit as $\epsilon \to 0$ yields (2.22a), due to the continuity of ϕ.

§3. Polynomial Reproduction and Approximation Orders

It was observed in [35] that *cardinal interpolation by 3-directional box splines provides optimal approximation orders*. By this we mean the following: Given the shift invariant space S_ϕ, the approximation order m is the largest non-negative number such that

$$\inf\{\|f - s_h\| : s_h \in S_{\phi,h}\} = \mathcal{O}(h^m) \quad \text{as} \quad h \to 0 \tag{3.1}$$

for sufficiently smooth f (e.g., $f \in C_0^\infty(\mathbb{R}^d)$) where $S_{\phi,h}$ is the scaled version of S_ϕ,

$$s_h \in S_{\phi,h} \iff s_h(\cdot/h) \in S_\phi \tag{3.2}$$

for the scale parameter $h > 0$.

We are going to use L_2-norms $\|\cdot\| = |\cdot|_2$, and the approximation order (3.1) will follow from the *error estimate for cardinal interpolation*

$$|f - \mathcal{J}_\phi(f)|_2 \leq \text{const} \, |f|_{m,2} \,, \tag{3.3}$$

where $m > d/2$, and the Sobolev seminorm of order m, in terms of the Fourier transform, is given by

$$|f|_{m,2}^2 := \frac{1}{(2\pi)^d} \int |\xi|^{2m} |f^\wedge(\xi)|^2 \, d\xi \,. \tag{3.4}$$

When $m \in \mathbb{N}$, estimate (3.3) shows that $\mathcal{J}_\phi$ reproduces the space Π_{m-1} of polynomials of degree $m-1$. This property can be verified through the *Strang-Fix conditions* for the fundamental interpolant (1.3a). These read as follows:

$$D^\alpha \Lambda_\phi^\wedge(2\pi\beta) = 0 \quad \text{for} \quad 0 \neq \beta \in \mathbb{Z}^d \quad \text{and} \quad |\alpha| < m \tag{3.5a}$$

and

$$D^\alpha(1 - \Lambda_\phi^\wedge)(0) = 0 \quad \text{for} \quad |\alpha| < m \,. \tag{3.5b}$$

Here we have to assume that $\Lambda_\phi^\wedge \in C^m(2\pi\mathbb{Z}^d)$; i.e., $\Lambda_\phi^\wedge \in C^m$ in a neighborhood of $2\pi\alpha$, for all $\alpha \in \mathbb{Z}^d$. The central result is

Theorem 3.1. *Let $m \in \mathbb{N}$ and $\Lambda_\phi^\wedge \in C^m(2\pi\mathbb{Z}^d)$. Then $\mathcal{J}_\phi(f) \in \Pi_{m-1}$ for all $f \in \Pi_{m-1}$ if and only if (3.5a) holds; in particular $\mathcal{J}_\phi(f) = f$ for $f \in \Pi_{m-1}$ if and only if both (3.5a) and (3.5b) hold.*

Proof: This follows immediately from (1.6), which reads as follows

$$\left(f - \mathcal{J}_\phi(f)\right)^\wedge = (1 - \Lambda_\phi^\wedge)f^\wedge - \Lambda_\phi^\wedge \sum_{\alpha \neq 0} f^\wedge(\cdot + 2\pi\alpha) \,. \tag{3.6}$$

Now $f \in \Pi_{m-1}$ if and only if $f^\wedge = \sum_{|\beta|<m} a_\beta D^\beta \delta$ (and in particular, the support of $f^\wedge$ is $\{0\}$). From this the theorem becomes obvious. ∎

Approximation from Cardinal Interpolation 145

If $\phi^\sim$ is smooth at the origin and $\phi^\sim(0) \neq 0$ we can write (3.5) as the Strang-Fix condition for the basis function ϕ itself; viz. (3.5a) is equivalent to
$$D^\alpha \phi^\wedge(2\pi\beta) = 0 \quad \text{for} \quad 0 \neq \beta \in \mathbb{Z}^d \quad \text{and} \quad |\alpha| < m, \tag{3.7a}$$
and (3.5b) is equivalent to
$$D^\alpha(\phi^\sim - \phi^\wedge)(0) = 0 \quad \text{for} \quad |\alpha| < m. \tag{3.7b}$$
This is the usual way of writing the (normalized) Strang-Fix conditions, where the condition $\phi^\wedge(0) = 1$ is also assumed.

There is another advantage of the preconditioned ψ in Theorem 2.3 over the radial basis function ϕ: In (2.17), $\psi^\wedge$ and $\psi^\sim$ have the required smoothness properties while $\phi^\wedge$ is singular at the origin. Also, in order to check (3.7a) only the zero properties of σ^k have to be analyzed.

In order to give an *error estimate* of type (3.3) for arbitrary $m \in \mathbb{R}_+$, we employ a strong version of (3.5); viz. we assume that, for some $c \in \ell_2(\mathbb{Z}^d)$, we have these estimates
$$|1 - \Lambda_\phi^\wedge(\xi)| \leq c_0 |\xi|_2^m \tag{3.8a}$$
$$|\Lambda_\phi^\wedge(\xi + 2\pi\alpha)| \leq c_\alpha |\xi|_2^m, \quad 0 \neq \alpha \in \mathbb{Z}^d, \tag{3.8b}$$
holding almost everywhere on $C = [-\pi, \pi]^d$. Then

Theorem 3.2. *If Λ_ϕ satisfies (3.8) almost everywhere on C, then estimate (3.3) holds for all $f \in L_2(\mathbb{R}^d)$ with $\operatorname{supp} f^\wedge \subseteq C$.*

Proof: This is another consequence of (3.6). Putting $g_0^\wedge := (1 - \Lambda_\phi^\wedge)f^\wedge$ and $g_\alpha^\wedge := -\Lambda_\phi^\wedge f^\wedge(\cdot + 2\pi\alpha)$ for $\alpha \neq 0$, we find that the family $(g_\alpha^\wedge)_{\alpha \in \mathbb{Z}^d}$ is orthogonal, and hence
$$|(f - \mathcal{J}_\phi(f))^\wedge|_2^2 = \sum_\alpha |g_\alpha^\wedge|_2^2 \leq \sum_\alpha c_\alpha^2 \int_C |\xi|^{2m} |f^\wedge(\xi)|^2 \, d\xi$$
where in the latter inequality we have used an obvious substitution. From this the following constant is obtained in (3.3):
$$\text{const} = \sqrt{\sum_\alpha c_\alpha^2}. \quad \blacksquare \tag{3.9}$$

The restriction to band-limited functions in this theorem addresses a particularly important special case of (3.3) from which more general estimates could be obtained by density arguments.

In the case of *normal cardinal interpolation* with compactly supported ϕ, estimate (3.3) for $f \in W_2^m(\mathbb{R}^d)$ was proved in [35]. Some other estimates are given by the Cambridge group (cf. [114]) in terms of the Chebyshev norm $|\cdot|_\infty$. Binev [9] gives an estimate in terms of the τ-moduli that applies to box spline interpolation. Madych and Nelson [98] give estimates in terms of the lub-norm $|f|_\phi$ associated with the energy norm (2.20). This applies to $\phi \in CPD_m$ and $f \in \mathcal{S}'$ satisfying
$$|\langle f, \psi \rangle| \leq \text{const} \, \|\psi\|_\phi \quad \text{for all} \quad \psi \in \mathcal{D}_m,$$
and $|f|_\phi$ is the infimum of all constants for which this estimate holds true.

§4. Cardinal Interpolation and L_2-approximation

The subject of L_2-approximation from shift invariant spaces was recently treated in the general setting by de Boor, DeVore and Ron [14]. If $\phi \in L_2(\mathbb{R}^d)$ then the L_2-projection to the principal shift invariant space S_ϕ is defined by

$$\mathcal{P}_\phi : L_2(\mathbb{R}^d) \to S_\phi, \qquad f \mapsto \mathcal{P}_\phi(f), \tag{4.1}$$

where

$$\mathcal{P}_\phi(f)^\wedge := \frac{[f^\wedge, \phi^\wedge]}{[\phi^\wedge, \phi^\wedge]} \phi^\wedge \; ;$$

here we employ the following notation for $f, g \in L_2(\mathbb{R}^d)$:

$$[f^\wedge, g^\wedge] := (f * g^*)^\sim = \sum_\alpha f^\wedge(\cdot + 2\pi\alpha)\overline{g^\wedge(\cdot + 2\pi\alpha)}, \tag{4.2}$$

and we put

$$\mathcal{P}_\phi(f)^\wedge(\xi) := 0 \quad \text{if} \quad [\phi^\wedge, \phi^\wedge](\xi) = 0. \tag{4.3}$$

This latter condition is vacuous if the function $[\phi^\wedge, \phi^\wedge]$ (which is equal to $(\phi * \phi^*)^\sim$) has a "small" zero set; in other cases the range of the operator $\mathcal{P}_\phi$ is restricted. We also note that (4.2) defines an $L_1(C)$ function, for any $f, g \in L_2(\mathbb{R}^d)$.

In order to show the connection of (4.1) to cardinal interpolation it is useful duce the *smoothing (convolution) operator*

$$f \mapsto \Sigma_\phi(f) \quad \text{for} \quad f \in L_2(\mathbb{R}^d) \tag{4.4a}$$

defined by

$$\Sigma_\phi(f)^\wedge := (f * \phi^*)^\wedge = f^\wedge \overline{\phi^\wedge}. \tag{4.4b}$$

It is then a trivial observation that (in case $\phi^\wedge$ is nonzero almost everywhere)

$$|f|_2^2 = \frac{1}{(2\pi)^d} \int |f^\wedge(\xi)|^2 \, d\xi = \frac{1}{(2\pi)^d} \int |\phi^\wedge(\xi)|^{-2} |\Sigma_\phi(f)^\wedge(\xi)|^2 \, d\xi. \tag{4.5}$$

Theorem 4.1. For $\phi \in L_2(\mathbb{R}^d)$ let $\mathcal{P}_\phi$ be the L_2-projector given by (4.1), and Σ_ϕ the smoothing operator defined by (4.4). Let

$$\Phi := \phi * \phi^* = \Sigma_\phi(\phi) \tag{4.6}$$

and let $\mathcal{J}_\Phi$ be the associated cardinal interpolation operator of (1.3). Then

$$\mathcal{J}_\Phi(\Sigma_\phi(f)) = \Sigma_\phi(\mathcal{P}_\phi(f)) \quad \text{for} \quad f \in L_2(\mathbb{R}^d). \tag{4.7}$$

Proof: Using the definitions of these operators, we get

$$\mathcal{J}_\Phi(\Sigma_\phi(f))^\wedge = \frac{\Phi^\wedge}{\Phi^\sim} \Sigma_\phi(f)^\sim = \frac{|\phi^\wedge|^2}{[\phi^\wedge, \phi^\wedge]} [f^\wedge, \phi^\wedge]$$
$$= \mathcal{P}_\phi(f)^\wedge \overline{\phi^\wedge} = \Sigma_\phi(\mathcal{P}_\phi(f))^\wedge. \blacksquare$$

The *fundamental interpolant* Λ_Φ of Φ as defined in (4.6) plays the role of a key function in L_2-approximation. From

$$\Lambda_\Phi^\wedge = \frac{\Phi^\wedge}{\Phi^\sim} = \frac{|\phi^\wedge|^2}{\sum_\alpha |\phi^\wedge(\cdot + 2\pi\alpha)|^2}, \quad (4.8)$$

we see that

$$0 \le \Lambda_\Phi^\wedge(\xi) \le 1 \quad \text{for all} \quad \xi \in \mathbb{R}^d \quad (4.9)$$

and $\Lambda_\Phi^\wedge \in L_1(\mathbb{R}^d)$. Using Parseval's identity, we get the following representation

$$|f - \mathcal{P}_\phi(f)|_2^2 = \frac{1}{(2\pi)^d} \int_C (1 - \Lambda_\Phi^\wedge(\xi)) |f^\wedge(\xi)|^2 \, d\xi \quad (4.10)$$

subject to the condition that $f \in L_2(\mathbb{R}^d)$ with supp $f^\wedge \subseteq C = [-\pi, +\pi]^d$. This result should be compared with Theorem 3.2. It is also an intermediate step in proving the following result on approximation orders in [14]; (cf. also de Boor's contribution to this volume):

Theorem 4.2. *Scaled versions of S_ϕ provide L_2-approximation order $m \in \mathbb{R}_+$ if and only if*

$$1 - \Lambda_\Phi^\wedge(\xi) \le \text{const } |\xi|^{2m}$$

almost everywhere on C.

The results in this section can be extended to cases where $\phi \notin L_2(\mathbb{R}^d)$, e.g., by using the method of preconditioning as in Theorem 2.3, letting $k = \delta$. Then Theorem 4.1 holds with ϕ replaced by $\psi = (-\nabla)^k \phi$, and Φ replaced by

$$\Psi = \psi * \psi^* = \Sigma_\psi(\psi) \ . \quad (4.11)$$

For $f \in L_2(\mathbb{R}^d)$ Equation (4.5) also yields

$$|f|_2^2 = \frac{1}{(2\pi)^d} \int |\psi^\wedge(\xi)|^{-2} |\Sigma_\psi(f)^\wedge(\xi)|^2 \, d\xi = \frac{1}{(2\pi)^d} \int |\phi^\wedge(\xi)|^{-2} |g^\wedge(\xi)|^2 \, d\xi$$

with $g^\wedge = f^\wedge \overline{\phi^\wedge}$.

§5. Cardinal Hermite Interpolation

Cardinal Hermite interpolation with box splines, as dealt with by Riemenschneider and Scherer [120,121], is described in [119]. Only very recently, [76], a new approach to multivariate cardinal Birkhoff interpolation was considered that is applicable in particular to the examples of Section 2. We describe here only first order Hermite interpolation and, in contrast to [76], we again define our operators in the Fourier transform domain in order to be consistent.

For $\phi \in S' \cap C^1(\mathbb{R}^d)$ we require that

$$\Lambda_\phi^\wedge = \frac{\phi^\wedge}{\phi^\sim} \in L_1(\mathbb{R}^d) \quad \text{and} \quad |\xi| \, \Lambda_\phi^\wedge \in L_1(\mathbb{R}^d) \ . \quad (5.1)$$

We also use the notation

$$f_j := \frac{\partial}{\partial x_j} f \quad \text{for} \quad f \in C^1(\mathbb{R}^d), \tag{5.2}$$

and assume that $\phi_j^\sim / \phi^\sim$, $j = 1, \ldots, d$, are bounded functions. Let us consider the functions:

$$\Lambda_j^\wedge := \frac{1}{2\pi i} \frac{\phi^\wedge(\cdot - 2\pi e_j) - \phi^\wedge}{\phi^\sim}, \quad j = 1, \ldots, d, \tag{5.3a}$$

and

$$\Lambda_0^\wedge := \Lambda_\phi^\wedge - \sum_{j=1}^d \Lambda_j^\wedge \frac{\phi_j^\sim}{\phi^\sim}, \tag{5.3b}$$

where $e_1, \ldots, e_d$ denote the canonical unit vectors. These functions define the fundamental solutions for first order Hermite interpolation:

Theorem 5.1. For $j = 1, \ldots, d$ and $\alpha \in \mathbb{Z}^d$:

$$\Lambda_j(\alpha) = 0 \quad \text{and} \quad \frac{\partial}{\partial x_k} \Lambda_j(\alpha) = \delta_{jk} \delta_{0\alpha}, \quad k = 1, \ldots, d, \tag{5.4}$$

and

$$\Lambda_0(\alpha) = \delta_{0\alpha} \quad \text{and} \quad \frac{\partial}{\partial x_k} \Lambda_0(\alpha) = 0, \quad k = 1, \ldots, d. \tag{5.5}$$

The proof of this result is not difficult. We note that the fundamental function Λ_j takes the form

$$\Lambda_j(x) = (e^{2\pi i e_j \cdot x} - 1) \Lambda_\phi(x), \quad j = 1, \ldots, d. \tag{5.6}$$

Now, $\Lambda_\phi \in C^1(\mathbb{R}^d)$ by (5.1), and straightforward differentiation establishes (5.4). In order to verify (5.5) it is probably better to use the inverse Fourier transform.

Equation (5.6) shows that in this type of Hermite interpolation, the interpolation space is an extension of S_ϕ by adding *oscillating factors*. Surprisingly, this simple idea settles the somewhat critical question of how the interpolation space should be defined in Hermite problems. (The natural first impulse of using the sum of the spaces S_ϕ and S_{ϕ_j}, $j = 1, \ldots, d$, leads to a singular problem.)

Based on the functions (5.3) the *cardinal Hermite interpolation operator* $\mathcal{H}_\phi$ is defined by

$$\mathcal{H}_\phi(f)^\wedge := \Lambda_0^\wedge f^\sim + \sum_{j=1}^d \Lambda_j^\wedge f_j^\sim. \tag{5.7}$$

The identity

$$\mathcal{H}_\phi(f)^\wedge = \mathcal{J}_\phi(f)^\wedge + \sum_{j=1}^d \Lambda_j^\wedge \left\{ f_j^\sim - \frac{\phi_j^\sim}{\phi^\sim} f^\sim \right\} \tag{5.8}$$

shows that $\mathcal{H}_\phi$ is, in a certain sense, a modification of the cardinal Lagrange interpolation operator $\mathcal{J}_\phi$ as defined in (1.3b). But (5.8) can be also used to prove the analogue of Theorem 1.3:

Approximation from Cardinal Interpolation 149

Theorem 5.2. *With the above assumptions, if $\phi \in L_2(\mathbb{R}^d)$, then $f = \mathcal{H}_\phi(f)$ for $f \in S_\phi$.*

Proof: Since $f^\wedge = \omega \phi^\wedge$, where ω is a 2π-periodic function, we have $f_j^\wedge = i\xi_j f^\wedge = \omega i\xi_j \phi^\wedge = \omega \phi_j^\wedge$ for $j = 1, \ldots, d$. Hence $f^\sim = \omega \phi^\sim$ and $f_j^\sim = \omega \phi_j^\sim$, and

$$f_j^\sim - \frac{\phi_j^\sim}{\phi^\sim} f^\sim = 0, \qquad j = 1, \ldots, d. \qquad \blacksquare$$

Another application of (5.8) gives an error bound of type (3.3) for Hermite interpolation. The following result is also from [76]:

Theorem 5.3. *For $\phi \in C_0^1(\mathbb{R}^d)$ satisfying the Strang-Fix conditions (3.7) of order $m > d/2+1$ assume that $\phi^\sim = \sum_\alpha \phi(\alpha) e^{-i\alpha \cdot \xi}$. If cardinal interpolation with ϕ is normal, then*

$$|f - \mathcal{H}_\phi(f)|_2 \le \text{const } |f|_{m,2}$$

for all $f \in W_2^m(\mathbb{R}^d)$.

§6. Condition Numbers for Non-Cardinal Interpolation

A *non-cardinal type of interpolation* that is successfully used in various applications proceeds as follows: Given the basis function $\phi \in S' \cap C(\mathbb{R}^d)$, and the set

$$X = \{x_1, \ldots, x_N\} \tag{6.1}$$

of distinct knots $x_i \in \mathbb{R}^d$, find $p \in \text{span}\{\phi(\cdot - x_i), i = 1, \ldots N\}$ interpolating some given data at $x_1, \ldots, x_N$. This leads to a linear system of equations with collocation matrix

$$A := (\phi(x_k - x_\ell))_{k,\ell=1}^N. \tag{6.2}$$

We want to point out some results on *bounds for the condition number*

$$\text{cond}(A) = \|A\|_2 \, \|A^{-1}\|_2 \tag{6.3}$$

(where $\|\cdot\|_2$ is the spectral norm) in terms of constants depending on properties of the set (6.1) only.

As we have seen in Section 2.4, many basis functions ϕ yield conditionally positive definite matrices. This was used in papers by Ball, Narcowich, Sivakumar and Ward [3,4,104-107] where bounds on $\|A^{-1}\|_2$ relative to the subspace of $\mathbb{R}^N$ given by the restrictions (2.22b) were derived. In order to get rid of these restrictions, the paper [12] assumes that

$$\phi^\wedge \in L_1(\mathbb{R}^d) \cap C(\mathbb{R}^d) \quad \text{and} \quad \phi^\wedge(-\xi) = \phi^\wedge(\xi) \in \mathbb{R} \quad \text{for all} \quad \xi \in \mathbb{R}^d. \tag{6.4}$$

In the case of the class $CM_m(\gamma, \beta)$, (6.4) can be obtained by the method of preconditioning. Let us emphasize that, although $CM_m(\gamma, \beta)$ is a subclass of CPD_m, all examples that have shown favorable properties in numerical applications belong to this more restrictive class.

From (6.4) we see that the matrix A is real symmetric. It is often definite, as can be verified from the identity, for any real numbers $c_1, \ldots, c_N$,

$$\sum_{k,\ell=1}^{N} c_k \phi(x_k - x_\ell) c_\ell = \frac{1}{(2\pi)^d} \int \phi^{\wedge}(\xi) |\sum_{k=1}^{N} c_k e^{-ix_k \cdot \xi}|^2 \, d\xi \, . \tag{6.5}$$

A typical result along these lines is

Theorem 6.1. *If, in addition to (6.4), we have $\phi^{\wedge} \geq 0$ and $\phi^{\wedge}(0) > 0$, then the collocation matrix A given by (6.2) is always positive definite.*

The assumptions of this theorem are satisfied, in particular, in box spline interpolation (provided that the multiplicities k, ℓ, m, n of Section 2 are even numbers), and by the class $CM_m(\gamma, \beta)$ in case preconditioning with $k = \delta$ is used.

In order to bound the condition number of A we would like to bound (6.5) in terms

$$\lambda(q) \sum_{k=1}^{N} |c_k|^2 \leq \sum_{k,\ell=1}^{N} c_k \phi(x_k - x_\ell) c_\ell \leq \Lambda(q) \sum_{k=1}^{N} |c_k|^2 \tag{6.6}$$

where q is a parameter that is somehow characteristic of the set X. In [12] the ℓ_∞-distance

$$q := \min\{|x_k - x_\ell|_\infty \, ; \, k, \ell = 1, \ldots, N \, , \, k \neq \ell\} \tag{6.7}$$

is used (in contrast to the usually employed ℓ_2-distance), and the following decay property of $\phi^{\wedge}$ is assumed

$$|\phi^{\wedge}(\xi)| \leq M_0 \quad \text{for} \quad |\xi|_\infty \leq q\pi \, , \tag{6.8}$$
$$|\phi^{\wedge}(\xi)| \leq M_1 |\xi|_\infty^{-n} \quad \text{for} \quad |\xi|_\infty \geq q\pi \, ,$$

where $n \in \mathbb{N}$ and M_0, M_1 are positive constants. We have the following result for bivariate interpolation $(d = 2)$ from [12]:

Theorem 6.2. *Assume that in addition to (6.4), (6.7) and (6.8)*

$$\phi^{\wedge}(\xi_1, \xi_2) \geq K \cos \frac{\xi_1}{3q} \cos \frac{\xi_2}{3q} \quad \text{for} \quad -\frac{3}{2}\pi q \leq \xi_1, \xi_2 \leq \frac{3}{2}\pi q \, . \tag{6.9}$$

Then estimate (6.6) holds with

$$\lambda(q) = \rho K \frac{9q^2}{\pi^2} \quad \text{and} \tag{6.10}$$

$$\Lambda(q) = (2 - \rho) \frac{36q^2}{\pi^2} \left\{ M_0 + \frac{8M_1}{(q\pi)^n} \left(1 + \frac{2n-1}{3^n(n-2)} \right) \right\} ,$$

where $\rho = 2 - (2 - \frac{\pi}{3\sqrt{3}})^2 \sim .0528$.

For example, in the case of Rabut's thin plate spline (2.11) with $k = 2$ (and $d = 2$) we get for $q = 1$ the bound $\text{cond}(A) < 2150$.

Since several groups are still working on this important problem it is to be expected that further results will emerge in the near future.

It is much easier to bound $\text{cond}(A)$ in case $X \subseteq \mathbb{Z}^d$. Here, we assume that in addition to (6.4), we have

$$\phi_| = (\phi(\alpha))_{\alpha \in \mathbb{Z}^d} \in \ell_1(\mathbb{Z}^d) . \tag{6.11}$$

Then, for any sequence $c = (c_\alpha)_{\alpha \in \mathbb{Z}^d}$ with compact support, (6.5) takes the form (cf. [6,66])

$$\sum_\alpha c_\alpha (\phi_| \tilde{*} c)_\alpha = \frac{1}{(2\pi)^d} \int_C \phi^\sim(\xi) | \sum_\alpha c_\alpha e^{-i\alpha \cdot \xi}|^2 \, d\xi , \tag{6.12}$$

and Theorem 6.1 can be replaced by

Theorem 6.3. *Suppose that X is a subset of $\mathbb{Z}^d$, and ϕ is a function satisfying (6.4), (6.11) such that for all $\xi \in \mathbb{R}^d$*

$$\phi^\sim(\xi) = \sum_\alpha \phi^\wedge(\xi + 2\pi\alpha) = \sum_\alpha \phi(\alpha) e^{-i\alpha \cdot \xi} .$$

Then

$$\lambda \sum_{\alpha \in X} |c_\alpha|^2 \leq \sum_{\alpha, \beta \in X} c_\alpha \phi(\alpha - \beta) c_\beta \leq \Lambda \sum_{\alpha \in X} |c_\alpha|^2 \tag{6.13}$$

for any ℓ_2-sequence $(c_\alpha)_{\alpha \in \mathbb{Z}^d}$, where

$$\lambda := \min \phi^\sim(\xi) \quad \text{and} \quad \Lambda := \max \phi^\sim(\xi) . \tag{6.14}$$

An application of this theorem to the examples of Section 2 shows that for preconditioned $\phi \in CM_m(\gamma, \beta)$ with $k = 6$, or for a 3-directional box spline (and $d = 2$)

$$\text{cond}(\phi(\alpha - \beta))_{\alpha, \beta \in X} \leq \frac{\Lambda}{\lambda}, \tag{6.15}$$

(in case $X \subseteq \mathbb{Z}^d$) where the constants $0 < \lambda < \Lambda$ are independent of X. In case of the 4-directional box spline an estimate of type (6.15) holds for any finite set $X \subseteq \mathbb{Z}^d$ (with $\Lambda = 1$), but $\lambda \to 0$ as X approaches the full lattice $\mathbb{Z}^d$.

We note that the assumption $X \subseteq \mathbb{Z}^d$ can be extended to any regularly spaced set

$$X \subseteq x_0 + q\mathbb{Z}^d , \quad x_0 \in \mathbb{R}^d , \quad q \in \mathbb{R}_+ \tag{6.16}$$

by scaling the function ϕ; i.e., in place of ϕ one has to consider $\phi_q = \phi(q \cdot)$. Adjusting the basis function to the mesh width (6.7) as in (6.16) is often done, for example, in multiquadric computations.

§7. Additional Notes

We have given only selected references in Sections 1 - 6. Also, the more recent aspects of wavelets or multiscale analysis and its relation to cardinal interpolation were not discussed here since C. Chui's contribution to this volume will address this important field. (We also refer to the papers [41,42,81,87,92,94,102,103,122,136].) This section will provide more references; it intends to indicate how the work of other authors fits in our presentation of cardinal interpolation.

Schoenberg operators T_ϕ (in terms of infinite series) play an essential role in univariate and multivariate spline theory. In the usual L_∞-estimates for quasi-interpolants (which are special Schoenberg operators) it is important for ϕ to have compact support or, at least, to have reasonable decay at infinity. Such quasi-interpolants are often used for deriving approximation order results.

The commutator operator was introduced (for $\phi \in C_0(\mathbb{R}^d)$) in 1984 by Chui, Jetter and Ward [35] for the estimation of the error of cardinal interpolation as given in (1.6); for their estimate they employed the Bramble-Hilbert Lemma. The result of Theorem 1.1 is not explicitly mentioned there although it is implicit; this was presented later in the papers [13,18].

The representation (1.3a) for the fundamental interpolant is, of course, well-known for all the examples in Section 2; it even appears in early work of Schoenberg on univariate cardinal interpolation (which is beautifully surveyed in [126]). Also, Theorem 1.2 is not original, but probably the approach to the interpolation property (1.5) shows some new aspect.

Preconditioning plays an important role in the more numerically oriented work of Dyn, Levin and Rippa [53-57]. The discrete Laplacian as used in this paper is, of course, based on the regular grid. Baxter's method of preconditioning (as presented in his talk at this Austin conference) constructs $\psi \in S_\phi$ so that $\psi^\wedge$ is a reasonable approximation to the characteristic function of C.

A nice presentation of (univariate) Toeplitz forms is given in [66]. Discrete convolutions are also used as filters in signal analysis. Here the zeros of the symbol σ_ϕ (also called the transfer-function) result in distortions of the signal. A good reference for this is the paper [72]. From the point of view of numerical computations the condition of Theorem 1.4 is the proper condition of stability. It also appears in a modified version in the notion of frames in wavelet theory. (See Chui's article in this volume.) While the notion of "linear independence" (see for example, [82,83,129]) leads to interesting analytical questions, it is too weak to be relevant for computations.

The examples in Section 2 are selected according to what basis functions have shown favorable performance in numerical computations (cf. [62,63,2]). Certainly, the most complete treatment of box splines can be found in the recent book [17] of de Boor, Höllig and Riemenschneider. Polyharmonic splines are an extension of Duchon's thin plate splines [44-48,109]; their theory was further advanced by Madych and Nelson [94,96-100] and by Rabut [115-117]. The Hardy multiquadric, in some sense, is a regularization of some (fractional) polyharmonic (cf. the recent survey paper [68]); its application to univariate and multivariate interpolation has been very much advanced by the Cambridge group (cf. [7,8,69-71,110-114]) with Buhmann [20-31] dealing profoundly with many aspects of cardinal interpolation involving radial basis functions. Certainly, Micchelli's paper [101] (cf. [96], too) can be viewed as a break-through; it is related to the fundamental work of Schoenberg (cf. [108,125]), as is the notion of m-th order completely monotone functions, or of m-th order conditionally positive definite functions.

The Strang-Fix conditions [61,137] and their relation to approximation orders is nicely surveyed in recent papers by Light and others ([67,89-91]). Error estimates of type (3.3) give, in general, lower bounds for approximation orders by scaled shift-invariant spaces. Concerning these approximation orders, we refer to the papers by Jia and Lei [79,80], Dyn et al. [52], de Boor et al. [14,19] and Ron and Sivakumar [123].

Section 4 is based on the fundamental paper by de Boor, DeVore and Ron [14], but Theorem 4.1 (extending a well-known property of univariate cardinal spline interpolation) is new. Of course, it should be compared with the minimal property of cardinal interpolation by polyharmonic splines as given by Goodman and Lee [64] and Madych and Nelson [100].

Section 5 deals with a very special case of the results by Jetter, Riemenschneider and Shen [76] on a type of cardinal Birkhoff interpolation, but the presentation in the Fourier transform domain is again new. There is some evidence that this approach could also lead to a new type of wavelets.

Bounding the condition numbers for multivariate (scattered or non-scattered data) interpolation is one of the major numerical questions. So far the following groups are involved with this difficult problem: Ball, Narcowich, Sivakumar and Ward [3,4,104-107], Baxter [5,6], Madych [95], Schaback and Wu [124,141]. Our presentation here is from Binev and Jetter [12]; using the preconditioned functions avoids involving the polynomial part in the collocation matrix. The estimate in Theorem 6.2 is based on a type of estimate in univariate problems which can be found in Zygmund's book [142]. The mesh constant q as given in (6.7) should result in a dilation of the basis function; numerical experiments with multiquadrics along these lines are reported by Carlson et al. in [32,84,85], among others.

Acknowledgements. This research was partially supported by NATO grant # CRG 900 158.

References

1. Abramowitz, M. and I. A. Stegun, *Handbook of Mathematical Functions*, Dover Publications, 1970.
2. Arge, E., M. Daehlen, and A. Tveito, Box spline interpolation: a computational study, preprint, 1991.
3. Ball, K., Eigenvalues of Euclidean distance matrices, J. Approx. Theory **68** (1992), 74–82.

4. Ball, K., N. Sivakumar, and J. D. Ward, On the sensitivity of radial basis interpolation to minimal data separation distance, Constr. Approx., to appear.
5. Baxter, B. J. C., Conditionally positive functions and p-norm distance matrices, Constr. Approx. **7** (1991), 427–440.
6. Baxter, B. J. C., Norm estimates for inverses of Toeplitz distance matrices, Report #NA16, DAMTP, University of Cambridge, 1991.
7. Beatson, R. K. and M. J. D. Powell, Univariate multiquadric approximation: quasi-interpolation to scattered data, Report #NA7, DAMTP, University of Cambridge, 1990.
8. Beatson, R. K. and M. J. D. Powell, Univariate interpolation on a regular finite grid by a multiquadric plus a linear polynomial, Report #NA2, DAMTP, University of Cambridge, 1991.
9. Binev, P. G., Error estimate for box spline interpolation, in *Constructive Theory of Functions, Varna 1987* R. Maleev et al. (eds.), Publ. House Bulg. Acad. Sciences, Sofia, 1988, 50–55.
10. Binev, P. G. and K. Jetter, Cardinal interpolation with shifted 3-directional box splines, Proc. Royal Soc. Edinburgh, to appear.
11. Binev, P. G. and K. Jetter, Euler splines from 3-directional box splines, in *Constructive Theory of Functions, Varna 1991*, K. Ivanov et al. (eds.), to appear.
12. Binev, P. G. and K. Jetter, Estimating the condition number for multivariate interpolation problems, in *Numerical Methods in Approximation Theory Vol. 9*, D. Braess and L. L. Schumaker (eds.), Birkhäuser, Basel, 41–52.
13. de Boor, C., The polynomials in the linear span of integer translates of a compactly supported function, Constr. Approx. **3** (1987), 199–208.
14. de Boor, C., R.A. DeVore, and A. Ron, Approximation from shift-invariant subspaces of $L_2(\mathbb{R}^d)$, CMS Tech. Report #92-2, University of Wisconsin, Madison, 1991.
15. de Boor, C., K. Höllig, and S. D. Riemenschneider, Bivariate cardinal interpolation by splines on a three-direction mesh, Ill. J. Math. **29** (1985), 533–566.
16. de Boor, C., K. Höllig, and S. D. Riemenschneider, Fundamental solutions for multivariate difference equations, Amer. J. Math. **111** (1989), 403–415.
17. de Boor, C., K. Höllig, and S. D. Riemenschneider, *Box Splines*, Springer-Verlag, Berlin, 1992.
18. de Boor, C. and A. Ron, The exponentials in the span of the integer translates of a compactly supported function, C.S. Tech. Report #887, University of Wisconsin, Madison, 1989.
19. de Boor, C. and A. Ron, Fourier analysis of the approximation power of principal shift-invariant spaces, CMS Tech. Report #92-01, University of Wisconsin, Madison, 1991.
20. Buhmann, M. D., Convergence of univariate quasi-interpolation using multiquadrics, IMA J. Num. Anal. **8** (1988), 365–383.

21. Buhmann, M. D., Multivariable interpolation using radial basis functions, Ph.D. Dissertation, University of Cambridge, May 1989.
22. Buhmann, M. D., Multivariate interpolation in odd dimensional Euclidean spaces using multiquadrics, Constr. Approx. **6** (1990), 21–34.
23. Buhmann, M. D., Multivariate cardinal interpolation with radial basis functions, Constr. Approx. **6** (1990), 225–255.
24. Buhmann, M. D., Cardinal interpolation with radial basis functions: an integral transform approach, Report #NA8, DAMTP, University of Cambridge, 1989.
25. Buhmann, M. D., On quasi-interpolation with radial basis functions, Report #NA3, DAMTP, University of Cambridge, 1991.
26. Buhmann, M. D. and N. Dyn, Error estimates for multiquadric interpolation, in *Curves and Surfaces* P. J. Laurent, A. Le Méhauté and L. L. Schumaker (eds.), Academic Press, New York, 1991, 51–58.
27. Buhmann, M. D. and N. Dyn, Spectral convergence of multiquadric interpolation, Report #NA10, DAMTP, University of Cambridge, 1991.
28. Buhmann, M. D. and C. A. Micchelli, Multiply monotone functions for cardinal interpolation, Adv. Appl. Math. **12** (1991), 358–386.
29. Buhmann, M. D. and C. A. Micchelli, On radial basis approximation on periodic grids, Report #NA9, DAMTP, University of Cambridge, 1991.
30. Buhmann, M. D. and C. A. Micchelli, Multiquadric interpolation improved, Comput. Math. Appl., to appear.
31. Buhmann, M. D. and M. J. D. Powell, Radial basis functions on an infinite regular grid, in *Algorithms for Approximation II*, J. C. Mason and M. G. Cox (eds.), Chapman and Hall, London, 1989, 146–169.
32. Carlson, R. E. and T. A. Foley, The parameter r^2 in multiquadric interpolation, Comput. Math. Appl. **21** (1991), 29–42.
33. Chui, C. K., *Multivariate Splines*, CBMS-NSF Reg. Conf. Series in Appl. Math., vol. 54, SIAM, Philadelphia, 1988.
34. Chui, C. K. and H. Diamond, A natural formulation of quasi-interpolation by multivariate splines, Proc. Amer. Math. Soc. **99** (1987), 643–646.
35. Chui, C. K., K. Jetter, and J. D. Ward, Cardinal interpolation with multivariate splines, Math. Comp. **48** (1987), 711–724.
36. Chui, C. K., K. Jetter, and J. D. Ward, Cardinal interpolation with differences of tempered functions, Comput. Math. Appl., to appear.
37. Chui, C. K., J. Stöckler, and J. D. Ward, Invertibility of shifted box spline interpolation operators, SIAM J. Math. Anal. **22** (1991), 543–553.
38. Chui, C. K., J. Stöckler, and J. D. Ward, Polynomial expansions for cardinal interpolants and orthonormal wavelets, in *Curves and Surfaces*, P. J. Laurent, A. LeMéhauté and L. L. Schumaker (eds.), Academic Press, New York, 1991, 83–90.
39. Chui, C. K., J. Stöckler, and J. D. Ward, A Faber series approach to cardinal interpolation, Math. Comp., to appear.
40. Chui, C. K., J. Stöckler, and J. D. Ward, Singularity of cardinal interpolation with shifted box splines, CAT Report #217, Texas A&M University, 1990.

41. Chui, C. K. and J. Z. Wang, An analysis of cardinal spline-wavelets, J. Approx. Theory, to appear.
42. Chui, C. K. and J. Z. Wang, Computational and algorithmic aspects of cardinal spline wavelets, CAT Report #235, Texas A&M University, 1990.
43. Daubechies, I., S. Jaffard, and J.-L. Journé, A simple Wilson orthonormal basis with exponential decay, SIAM J. Math. Anal. **22** (1991), 554-572.
44. Duchon, J., Fonctions-spline du type "plaque mince" en dimension 2, Séminaire d'analyse numérique #231, Grenoble, 1975.
45. Duchon, J., Fonctions-spline à énergie invariante par rotation, Rapport de recherche #27, mathématiques appliqueés, Grenoble, 1976.
46. Duchon, J., Interpolation des fonctions de deux variables suivant le principe de la flexion des plaques minces, R.A.I.R.O. Analyse numérique **10** (1976), 5–12.
47. Duchon, J., Splines minimizing rotation-invariant semi-norms in Sobolev spaces, in *Constructive Theory of Functions of Several Variables*, W. Schempp and K. Zeller (eds.), Springer Lecture Notes in Math., vol. 571, 1977, 85–100.
48. Duchon, J., Sur l'erreur d'interpolation des fonctions de plusieurs variables par les D^m-splines, R.A.I.R.O. Analyse numérique **12** (1978), 325–334.
49. Dyn, N., Interpolation of scattered data by radial functions, in *Topics in Multivariate Approximation*, C. K. Chui, L. L. Schumaker and F. I. Utreras (eds.), Academic Press, New York, 1987, 47–61.
50. Dyn, N., Interpolation and approximation by radial and related functions, in *Approximation Theory VI*, C. K. Chui, L. L. Schumaker and J. D. Ward (eds.), Academic Press, New York, 1989, 211–234.
51. Dyn, N., T. Goodman, and C. A. Micchelli, Positive powers of certain conditionally negative definite matrices, Proc. A Koninkl. Nederl. Akad. Wetensch. **89** (1986), 163–178.
52. Dyn, N., I. R. A. Jackson, D. Levin, and A. Ron, On multivariate approximation by integer translates of a basis function, Israel J. Math., to appear.
53. Dyn, N. and D. Levin, Bell-shaped basis functions for surface fitting, in *Approximation Theory and its Applications*, Z. Ziegler (ed.), Academic Press, New York, 1981, 113–129.
54. Dyn, N. and D. Levin, Construction of surface spline interpolants of scattered data over finite domains, R.A.I.R.O. Analyse numérique **16** (1982), 201–209.
55. Dyn, N. and D. Levin, Iterative solution of systems originating from integral equations and surface interpolation, SIAM J. Numer. Anal. **20** (1983), 377–390.
56. Dyn, N., d. Levin, and S. Rippa, Surface interpolation and smoothing by "thin plate" splines, in *Approximation Theory IV*, C. K. Chui, L. L. Schumaker, J. D. Ward (eds.), Academic Press, New York, 1983, 445–449.

57. Dyn, N., D. Levin, and S. Rippa, Numerical procedures for surface fitting of scattered data by radial functions, SIAM J. Sci. Stat. Comput. **7** (1986), 639–659.
58. Dyn, N., W. A. Light, and E. W. Cheney, Interpolation by piecewise linear radial basis functions, I, J. Approx. Theory **59** (1989), 202–223.
59. Dyn, N. and C. A. Micchelli, Interpolation by sums of radial functions, Numer. Math. **58** (1990), 1–9.
60. Dyn, N. and A. Ron, Local approximation by certain spaces of exponential polynomials, approximation order of exponential box splines, and related interpolation problems, Trans. Amer. Math. Soc. **319** (1990), 381–403.
61. Fix, G. and G. Strang, Fourier analysis of the finite element method in Ritz-Galerkin theory, Stud. Appl. Math. **48** (1969), 265–273.
62. Franke, R., Scattered data interpolation: test of some methods, Math. Comp. **38** (1982), 181–200.
63. Franke, R., Recent advances in the approximation of surfaces from scattered data, Naval postgraduate school technical report, Monterey, 1986.
64. Goodman, T. N. T. and S. L. Lee, Cardinal interpolation by D^m-splines, Proc. Royal Soc. Edinburgh **94A** (1983), 149–161.
65. Goodman, T. N. T. and A. A. Taani, Cardinal interpolation by symmetric exponential box splines on a three-direction mesh, Proc. Edinburgh Math. Soc. **33** (1990), 251–264.
66. Grenander, U. and G. Szegö, *Toeplitz Forms and Their Applications*, 2nd ed., Chelsea Publ. Comp., New York, 1984.
67. Halton, E. J. and W. A. Light, On local and controlled approximation order, Math. Tech. Rep. #27, Lancaster University, 1990.
68. Hardy, R. L., Theory and applications of the multiquadric-biharmonic method, Comput. Math. Appl. **19** (1990), 163–208.
69. Jackson, I. R. A., Convergence properties of radial basis functions, Constr. Approx. **4** (1988), 243–264.
70. Jackson, I. R. A., An order of convergence for some radial basis functions, IMA J. Numer. Anal. **9** (1989), 567–587.
71. Jackson, I. R. A., Radial basis functions: a survey and new results, in *The Mathematics of Surfaces III*, D. C. Handscomb (ed.), Oxford University Press, Oxford, 1989, 115–133.
72. Janssen, A. J. E. M., The Zak transform: A signal transform for sampled time-continuous signals, Philips J. Res. **43** (1988), 23–69.
73. Jetter, K., A short survey on cardinal interpolation by box splines, in *Topics in Multivariate Approximation*, C. K. Chui, L. L. Schumaker and F. I. Utreras (eds.), Academic Press, New York, 1987, 125–139.
74. Jetter, K., The Bernoulli spline and approximation by trigonometric blending polynomials, Results in Mathematics **16** (1989), 243–252.
75. Jetter, K. and P. Koch, Methoden der Fourier-Transformation bei der kardinalen Interpolation periodischer Daten, in *Multivariate Approximation IV*, C. K. Chui, W. Schempp and K. Zeller (eds.), ISNM vol. 90, Birkhäuser-Verlag, Basel, 1989, 201–208.

76. Jetter, K., S. D. Riemenschneider, and Z. W. Shen, Hermite interpolation on the lattice $\mathbb{Z}^d$, preprint, 1992.
77. Jetter, K., S. D. Riemenschneider, and N. Sivakumar, Schoenberg's exponential Euler spline curves, Proc. Royal Soc. Edingburgh **118A** (1991), 21–33.
78. Jetter, K. and J. Stöckler, Algorithms for cardinal interpolation using box splines and radial basis functions, Numer. Math. **60** (1991), 97–114.
79. Jia, R. Q., A characterization of the approximation order of translation invariant spaces of functions, Proc. Amer. Math. Soc. **111** (1991), 61–70.
80. Jia, R. Q. and J. J. Lei, Approximation by multiinteger translates of functions having global support, J. Approx. Theory, to appear.
81. Jia, R. Q. and C. A. Micchelli, Using the refinement equations for the construction of pre-wavelets II: powers of two, in *Curves and Surfaces*, P. J. Laurent, A. LeMéhauté and L. L. Schumaker (eds.), Academic Press, New York, 1991, 209–246.
82. Jia, R. Q. and C. A. Micchelli, On linear independence of integer translates of a finite number of functions, Proc. Royal Soc. Edinburgh, to appear.
83. Jia, R. Q. and N. Sivakumar, On the linear independence of integer translates of box splines with rational directions, Linear Algebra Appl. **135** (1990), 19–31.
84. Kansa, E. J., Multiquadrics - A scattered data approximation scheme with applications to computational fluid-dynamics-I: Surface approximations and partial derivative estimates, Comput. Math. Appl. **19** (1990), 127–145.
85. Kansa, E. J. and R. E. Carlson, Inproved accuracy of multiquadric interpolation using variable shape parameters, Comput. Math. Appl., to appear.
86. Lai, M. J., On fundamental solutions for multivariate singular interpolation, preprint, 1991.
87. Lemarie, P. G., Some remarks on wavelet theory and interpolation, Report #91-13, Université de Paris-Sud, Mathématiques, 1991.
88. Lenze, B. and F. Locher, On ridge-type functions with elliptic contour lines, in *Multivariate Approximation and Interpolation*, W. Haußmann and K. Jetter (eds.), ISNM vol. 94, Birkhäuser-Verlag, Basel, 1990, 193–204.
89. Light, W. A., Recent developments in the Strang-Fix theory for approximation orders, in *Curves and Surfaces*, P. J. Laurent, A. LeMéhauté and L. L. Schumaker (eds.), Academic Press, New York, 1991, 285–292.
90. Light, W. A., Some aspects of radial basis function approximation, preprint, 1991.
91. Light, W. A. and E. W. Cheney, Quasi-interpolation with translates of a function having non compact support, Constr. Approx. **8** (1992), 35–48.
92. Lorentz, R. A. and W. R. Madych, Wavelets and generalized box splines, Arbeitspapiere der GMD #563, GMD Sankt Augustin, 1991.

93. Madych, W. R., Cardinal interpolation with polyharmonic splines, in *Multivariate Approximation Theory IV*, C. K. Chui, W. Schempp and K. Zeller (eds.), ISNM vol. 90, Birkhäuser-Verlag, Basel, 1989, 241–248.
94. Madych, W. R., Polyharmonic splines, multiscale analysis, and entire functions, in *Multivariate Approximation and Interpolation*, W. Haußmann and K. Jetter (eds.), ISNM vol. 94, Birkhäuser-Verlag, Basel, 1990, 205–216.
95. Madych, W. R., Error estimates for interpolation by generalized splines, in *Curves and Surfaces*, P. J. Laurent, A. Le Méhauté and L. L. Schumaker (eds.), Academic Press, New York, 1991, 297–306.
96. Madych, W. R. and S. A. Nelson, Multivariate interpolation: a variational theory, unpublished manusucript, 1983.
97. Madych, W. R. and S. A. Nelson, Multivariate interpolation and conditionally positive definite functions, Approx. Theory Appl. **4** (1988), 77–89.
98. Madych, W. R. and S. A. Nelson, Multivariate interpolation and conditionally positive definite functions II, Math. Comp. **54** (1990), 211–230.
99. Madych, W. R. and S. A. Nelson, Polyharmonic cardinal splines, J. Approx. Theory **60** (1990), 141–156.
100. Madych, W. R. and S. A. Nelson, Polyharmonic cardinal splines: a minimization property, J. Approx. Theory **63** (1990), 303–320.
101. Micchelli, C. A., Interpolation of scattered data: distance matrices and conditionally positive definite functions, Constr. Approx. **2** (1986), 11–22.
102. Micchelli, C. A., Using the refinement equation for the construction of prewavelets, Numerical Algorithms **1** (1991), 75–116.
103. Micchelli, C. A., C. Rabut, and F. I. Utreras, Using the refinement equation for the construction of pre-wavelets III: elliptic splines, Numerical Algorithms **1** (1991), 331–351.
104. Narcowich, F. J. and J. D. Ward, Norms of inverses and condition numbers for matrices associated with scattered data, J. Approx. Theory **64** (1991), 69–94.
105. Narcowich, F. J. and J. D. Ward, Norm estimates for inverses of scattered-data interpolation matrices associated with completely monotonic radial functions, CAT Report #197, College Station.
106. Narcowich, F. J. and J. D. Ward, Norm estimates for the inverses of a general class of scattered data radial basis interpolation matrices, J. Approx. Theory, to appear.
107. Narcowich, F. J. and J. D. Ward, Norms of inverses for matrices associated with scattered data, in *Curves and Surfaces*, P. J. Laurent, A. LeMéhauté and L. L. Schumaker (eds.), Academic Press, New York, 1991, 341–348.
108. von Neumann, J. and I. J. Schoenberg, Fourier integrals and metric geometry, Trans. Amer. Math. Soc. **50** (1941), 226–251.
109. Paihua, L., Methodes numériques pour l'obtention de fonctions-spline du type plaque mince en dimension 2, Séminaire d'analyse numérique #273, Grenoble, 1977.

110. Powell, M. J. D., Radial basis functions for multivariable interpolation: a review, in *Algorithms for Approximation*, J. C. Mason and M. G. Cox (eds.), Oxford University Press, Oxford, 1987, 143–167.
111. Powell, M. J. D., Radial basis function approximations to polynomials, in *Numerical Analysis 1987*, D. E. Griffiths and G. A. Watson (eds.), Longman Scientific & Technical, Burnt Mill, 1988, 223–241.
112. Powell, M. J. D., Univariate multiquadric approximation: reproduction of linear polynomials, in *Multivariate Approximation and Interpolation*, W. Haußmann and K. Jetter (eds.), Birkhäuser-Verlag, Basel, 1990, 227–240.
113. Powell, M. J. D., Univariate multiquadric interpolation: some recent results, in *Curves and Surfaces*, P. J. Laurent, A. LeMéhauté and L. L. Schumaker (eds.), Academic Press, New York, 1991, 371–382.
114. Powell, M. J. D., The theory of radial basis function approximation in 1990, in *Advances in Numerical Analysis, volume II: Wavelets, Subdivision Algorithms, and Radial Basis Functions*, W. A. Light (ed.), Oxford University Press, 1992, 105–210.
115. Rabut, Ch., *B*-splines polyharmoniques cardinales: interpolation, quasi-interpolation, filtrage, Thèse, Université Paul Sabatier, Toulouse, 1990.
116. Rabut, Ch., How to build quasi-interpolants: Application to polyharmonic *B*-splines, in *Curves and Surfaces*, P. J. Laurent, A. LeMéhauté and L. L. Schumaker (eds.), Academic Press, New York, 1991, 391–402.
117. Rabut, Ch., An introduction to Schoenberg's approximation, Comput. Math. Appl., to appear.
118. Richards, J. I. and H. K. Youn, *Theory of Distributions: A Non-Technical Introduction*, Cambridge University Press, Cambridge, 1990.
119. Riemenschneider, S. D., Multivariate cardinal interpolation, in *Approximation Theory VI*, C. K. Chui, L. L. Schumaker and J. D. Ward (eds.), Academic Press, New York, 1989, 561–580.
120. Riemenschneider, S. D. and K. Scherer, Cardinal Hermite interpolation with box splines, Constr. Approx. **3** (1987), 223–238.
121. Riemenschneider, S. D. and K. Scherer, Cardinal Hermite interpolation with box splines II, Numer. Math **58** (1991), 591–602.
122. Riemenschneider, S. D. and Z. Shen, Box splines, cardinal series, and wavelets, in *Approximation Theory and Functional Analysis*, C.K. Chui (ed.), Academic Press, New York, 1991, 133–149.
123. Ron, A. and N. Sivakumar, The approximation order of box spline spaces, Proc. Amer. Math. Soc., to appear.
124. Schaback, R., Lower bounds for norms of inverses of interpolation matrices for radial basis functions, preprint, 1991.
125. Schoenberg, I. J., Metric spaces and completely monotone functions, Ann. of Math. **39** (1938), 811–841.
126. Schoenberg, I. J., *Cardinal Spline Interpolation*, CBMS Reg. Conf. Series, vol. 12, SIAM, Philadelphia, 1973.
127. Sivakumar, N., Studies in box splines, Ph.D. Thesis, Edmonton, 1990.
128. Sivakumar, N., On bivariate cardinal interpolation by shifted splines on a three-direction mesh, J. Approx. Theory **61** (1990), 178–193.

129. Sivakumar, N., Concerning the linear dependance of integer translates of exponential box splines, J. Approx. Theory **64** (1991), 95–118.
130. Sivakumar, N., On univariate cardinal interpolation by shifted splines, Rocky Mount. J. Math. **19** (1989), 481–489.
131. Smith, P. W. and J. D. Ward, Quasi-interpolants from spline interpolation operators, Constr. Approx. **6** (1990), 97–110.
132. Stöckler, J., Interpolation mit mehrdimensionalen Bernoulli-Splines and periodischen Box-Splines, Dissertation, Duisburg, 1988.
133. Stöckler, J., Cardinal interpolation with translates of shifted bivariate box-splines, in *Mathematical Methods in Computer Aided Geometric Design*, T. Lyche and L. L. Schumaker (eds.), Academic Press, Boston, 1989, 583–592.
134. Stöckler, J., Minimal properties of periodic box-spline interpolation on a three direction mesh, in *Multivariate Approximation Theory IV*, C. K. Chui, W. Schempp and K. Zeller (eds.), Birkhäuser-Verlag, Basel, 1989, 329–336.
135. Stöckler, J., Multivariate Bernoulli splines and the periodic interpolation problem, Constr. Approx. **7** (1991), 105–122.
136. Stöckler, J., Multivariate wavelets, in *Wavelets - A Tutorial in Theory and Applications*, C. K. Chui (ed.), Academic Press, Boston, 1992, 325–355.
137. Strang, G. and J. Fix, A Fourier analysis of the finite element variational method, in *Constructive Aspects of Functional Analysis*, G. Geymonet (ed.), C.I.M.E., 1973, 793–840.
138. Sun, X., On the solvability of radial function interpolation, in *Approximation Theory VI* C. K. Chui, L. L. Schumaker and J. D. Ward (eds.), Academic Press, New York, 1989, 643–646.
139. Sun, X., Cardinal and scattered-cardinal interpolation by functions having non-compact support, Comput. Math. Appl., to appear.
140. Widder, D. V., *The Laplace Transform*, Princeton University Press, Princeton, 8th printing, 1972.
141. Wu, Z. and R. Schaback, Local error estimates for radial basis function interpolation of scattered data, preprint, 1990.
142. Zygmund, A., *Trigonometric Series*, Vol. I, 2nd ed., Cambridge University Press, Cambridge, 1959.

Kurt Jetter
FB Mathematik
Universität Duisburg
4100 Duisburg 1
Germany

Ridge Functions, Sigmoidal Functions and Neural Networks

Will Light

Abstract. This paper considers mainly approximation by ridge functions. Fix a point $a \in \mathbb{R}^n$ and a function $g : \mathbb{R} \to \mathbb{R}$. Then the function $f : \mathbb{R}^n \to \mathbb{R}$ defined by $f(x) = g(ax)$, $x \in \mathbb{R}^n$, is a ridge or plane wave function. A sigmoidal function is a particular example of the function g which closely resembles 1 at ∞ and 0 at $-\infty$. This paper discusses approximation problems involving general ridge functions and specific research connected with sigmoidal functions. The type of problems discussed lead naturally to a consideration of neural networks, particularly multi-layered feedforward networks. Most important is the existence of constructive proofs of the fact that networks of this type can approximate a given continuous function to any desired accuracy. A mathematician's view of these networks may be found in Section 5.

§1. Introduction

There has been much attention paid recently to the development of simple strategies for approximating a function $f : D \to \mathbb{R}$, where D is some suitable domain in $\mathbb{R}^n$. The motivation for such studies comes from a variety of sources — for example, scattered data interpolation, or approximation involving a large value of the parameter n. One approach is to construct an approximating subspace of functions $f_1, \ldots, f_m$ where each f_i is a composition $f_i = h \circ g_i$, with $g_i : \mathbb{R}^n \to \mathbb{R}$ and $h : \mathbb{R} \to \mathbb{R}$. The best known example of this is the radial basis function approximation: here to obtain g_i, one fixes an $x_i \in D$ and sets $g_i(x) = \|x - x_i\|_2$, $x \in \mathbb{R}^n$. Then, for example, the multiquadric radial functions are obtained by taking $h(t) = \sqrt{t^2 + c}$, $c > 0$, so that

$$f_i(x) = \sqrt{\|x - x_i\|_2^2 + c}, \qquad x \in \mathbb{R}^n.$$

Ridge functions follow a similar scheme. One fixes a point $x_i \in D$ and defines $g_i(x) = x x_i$, $x \in \mathbb{R}^n$. Then by composition with $h : \mathbb{R} \to \mathbb{R}$ one obtains

$$f_i(x) = h(x x_i), \qquad x \in \mathbb{R}^n.$$

(Note here that xx_i represents the usual scalar product of x with x_i.) A sigmoidal function is a function $\sigma : \mathbb{R} \to \mathbb{R}$ such that $\lim_{t \to \infty} \sigma(t) = 1$ and $\lim_{t \to -\infty} \sigma(t) = 0$. Such functions (or shifts of them) are common choices for the function h. For example, given $f \in C(D)$, one might try to determine constants a_j, $j = 1, \ldots, m$ and points x_j, $j = 1, \ldots, m$ (either in D or $\mathbb{R}^n$) such that f is well-approximated by a function of the form

$$x \longmapsto \sum_{j=1}^{m} a_j \sigma(xx_j) .$$

If one were to allow shifts of the sigmoidal function, then one might in addition seek constants $\theta_1, \ldots, \theta_m$ such that

$$x \longmapsto \sum_{j=1}^{m} a_j \sigma(xx_j + \theta_j)$$

is a good approximation to f. The difficulties with this sort of approximation problem are similar to those with radial functions. In particular, the approximating functions show rather bad behaviour at infinity even for very good choices of the function h or σ. We will attempt to give an introduction to this area, providing some results of recent research, and indicate the connection with the topic of neural networks. In particular, we will address the question of density of ridge functions, constructive techniques for obtaining approximations and questions related to the feasibility of certain interpolation problems.

Let us conclude this section with some simple considerations. Fix a point $a \in \mathbb{R}^n$ and a function $h : \mathbb{R} \to \mathbb{R}$. Define $f : \mathbb{R}^n \to \mathbb{R}$ by $f(x) = h(ax)$, $x \in \mathbb{R}^n$. Throughout this exposition we shall write ax for the scalar product of the vectors a and x. Now suppose for the moment that $n = 2$ and $a = (\alpha_1, \alpha_2)$. Then the vector $b = (-\alpha_2, \alpha_1)$ satisfies $ab = 0$ and so

$$f(x) = h(ax) = h(a(x + \lambda b)) = f(x + \lambda b) , \quad \text{for all } x \in \mathbb{R}^n, \lambda \in \mathbb{R} .$$

Thus the function f is constant along the line $\{x + \lambda b : \lambda \in \mathbb{R}\}$, for any value of $x \in \mathbb{R}^n$. The graph of such a function is a "ruled" surface, and it is this ruled nature that make ridge functions difficult to use in any standard way in approximation theory. As an example we quote the following elementary result.

Proposition 1.1. *The space $L^1(\mathbb{R}^2)$ contains no ridge function except 0.*

Proof: Let f be a nonzero measurable ridge function on $\mathbb{R}^2$. By using a rotation if necessary, we can assume that $f(x) = h(e_1 x)$, where $h : \mathbb{R} \to \mathbb{R}$ and $e_1 = (1, 0)$. Since $f \neq 0$, we have $|f(x)| > \epsilon > 0$ on some set A of positive measure. If P denotes the usual projection map from $\mathbb{R}^2$ onto the first coordinate axis, then $f(x) = h(Px)$, and consequently $|h(t)| > \epsilon$ on the set $P(A)$, which has positive measure. It follows that $|f(x)| > \epsilon$ on $P(A) \times \mathbb{R}$, a set of infinite measure. Hence $f \notin L^1(\mathbb{R}^2)$. ∎

The concept of a ridge function admits the following straightforward generalization. Let X be a Banach space, and let $\phi \in X^*$, where X^* is the dual space of X. Let $h : \mathbb{R} \to \mathbb{R}$. Then the pair $\{\phi, h\}$ defines a ridge function f, where $f(x) = h(\phi(x)) = (h \circ \phi)(x)$, $x \in X$. Sometimes this level of generality will be achievable in establishing results, but in most cases we will confine our treatment to $X = \mathbb{R}^n$.

It is not our intention in this account to provide an historical survey. Given the present literature from neural networks alone, this would be a massive undertaking. Our aim is more modest, and we set out to give an introductory account of the ideas which are of interest to approximation theorists.

§2. Simple Density Arguments for Ridge Functions

In this section we consider X, a Banach space, and $C(X)$, the space of real-valued, continuous mappings on X. In the space $C(X)$, we will use the topology of compact convergence. For K a compact set in X we define the semi-norm

$$\|f\|_K = \sup_{x \in K} |f(x)|, \qquad (f \in C(X)).$$

A net $\{f_\nu\} \subset C(X)$ converges to f if $\|f_\nu - f\|_K \to 0$ for all compact sets $K \subset X$. (This notion of convergence defines the topology of compact convergence.) Now let G be a given subset of $C(\mathbb{R})$ and Φ a subset of X^*. We define $A(G, \Phi)$ to be the subset of $C(X)$ defined by

$$\{g \circ \phi : g \in G \text{ and } \phi \in \Phi\}.$$

In Sun and Cheney [43], one finds a number of results describing when $A(G, \Phi)$ is fundamental in $C(X)$. (Note that a subset of a Banach space is fundamental if its linear span is dense.) We describe some of those results now. It was pointed out by Cybenko [9] prior to the work of Sun and Cheney that the Stone-Weierstrass theorem was a powerful tool in establishing such results, and indeed, this theorem forms a major part of the theory.

Theorem 2.1. (Diaconis and Shahshahani [14]). Let X be a Banach space. The set of ridge functions $\{\exp \circ \phi : \phi \in X^*\}$ is fundamental in $C(X)$.

Proof: Let S denote the linear span of the set $\{\exp \circ \phi : \phi \in X^*\}$. Then S is an algebra in $C(X)$. To see this, it is enough to observe that for $x \in X$

$$[(\exp \circ \phi)(\exp \circ \psi)](x) = \exp(\phi(x)) \exp(\psi(x))$$
$$= \exp(\phi(x) + \psi(x))$$
$$= [\exp \circ (\phi + \psi)](x).$$

Also, if we choose ϕ_0 to be the zero functional, then $(\exp \circ \phi_0)(x) = 1$ for all $x \in X$. Hence S contains constants. Finally, S separates points of X: if $x, y \in X$ and $x \neq y$, then the Hahn-Banach theorem guarantees the existence of $\phi \in X^*$ such that $\phi(x - y) \neq 0$. Thus

$$1 \neq \exp[\phi(x - y)] = \exp[\phi(x)] \exp[-\phi(y)] = \exp[\phi(x)]/\exp[\phi(y)],$$

and so $(\exp \circ \phi)(x) \neq (\exp \circ \phi)(y)$.

Unfortunately, one cannot quite apply the classical Stone-Weierstrass Theorem (see Cheney [6]) at this point. However, it suffices to prove each basic neighborhood of a fixed point f in $C(X)$ intersects S. A basic neighborhood of f corresponding to the compact set $K \subset X$ and $\epsilon > 0$ has the form

$$B_{K,\epsilon} = \{g \in C(X) : \|f - g\|_K < \epsilon\}.$$

Now restrict f and all members of S to the compact set K. Then $\{h|K : h \in S\}$ is still an algebra containing constants and separating points of K. Hence $\{h|K : h \in S\}$ is dense in $C(K)$ by the Stone-Weierstrass Theorem, and consequently S intersects $B_{K,\epsilon}$ as required. ∎

Theorem 2.2. (Sun and Cheney [43]). *Let G be a fundamental set in $C(\mathbb{R})$. Let Φ be a subset of the continuous dual X^* of a Banach space X. If the set*

$$\Phi_0 = \{\phi/\|\phi\| : \phi \in \Phi, \ \phi \neq 0\}$$

is dense in $S(X^*)$ (the unit sphere of X^*), then the set

$$A(G, \Phi) = \{g \circ \phi : g \in G, \ \phi \in \Phi\}$$

is fundamental in $C(X)$.

Proof: Fix $f \in C(X)$, and let K be a compact subset of X. Take $\epsilon > 0$. By **2.1**, there exists $h_i \in C(\mathbb{R})$ and $\psi_i \in X^*$ such that

$$\left\| f - \sum_{i=1}^m h_i \circ \psi_i \right\|_K < \epsilon/3. \tag{1}$$

(Note that in fact **2.1** tells us that we can let $h_i(t) = a_i e^t$, for $a_i \in \mathbb{R}$, but this observation turns out not to be very helpful!)

We can assume $\|\psi_i\| = 1, 1 \leq i \leq m$. Set $M = \sup_{x \in K} \|x\|$, and select $\delta > 0$ so that when $|s|, |t| \leq M$ and $|s - t| < \delta$, we have $|h_i(s) - h_i(t)| < \epsilon/3m$ for $1 \leq i \leq m$. Since Φ_0 is dense in $S(X^*)$ we may select $\phi_i \in \Phi$ such that $\|\phi_i/\|\phi_i\| - \psi_i\| < \delta/M$ for $1 \leq i \leq m$. Put $\lambda_i = 1/\|\phi_i\|$, $1 \leq i \leq m$, and $\mu = \max_i \|\phi_i\|$. Since G is fundamental in $C(\mathbb{R})$ we can find coefficients $a_{ij} \in \mathbb{R}$ and functions $g_{ij} \in G$, $1 \leq i \leq m$, $1 \leq j \leq N$ so that

$$\left| h_i(\lambda_i t) - \sum_{j=1}^N a_{ij} g_{ij}(t) \right| < \epsilon/3m, \quad \text{for } 1 \leq i \leq m \text{ and } |t| \leq M. \tag{2}$$

Now write, for $x \in K$,

$$|f(x) - \sum_{i=1}^{m}\sum_{j=1}^{N} a_{ij}g_{ij}(\phi_i(x))| \leq |f(x) - \sum_{i=1}^{m} h_i(\psi_i(x))| +$$

$$+ \left|\sum_{i=1}^{m} h_i(\psi_i(x)) - \sum_{i=1}^{m} h_i(\lambda_i \phi_i(x))\right|$$

$$+ \left|\sum_{i=1}^{m} h_i(\lambda_i \phi_i(x)) - \sum_{i=1}^{m}\sum_{j=1}^{N} a_{ij}g_{ij}(\phi_i(x))\right|$$

$$=: I_1 + I_2 + I_3 \ .$$

By (1), $I_1 < \epsilon/3$. Now observe that

$$|\psi_i(x) - \lambda_i \phi_i(x)| \leq \|x\| \|\psi_i - \lambda_i \phi_i\| < M\delta/M = \delta \ .$$

It follows from the choice of δ that

$$I_2 \leq \sum_{i=1}^{m} |h_i(\psi_i(x)) - h_i(\lambda_i(\phi_i(x)))| \leq \sum_{i=1}^{m} \epsilon/3m = \epsilon/3 \ .$$

Finally, by (2),

$$I_3 \leq \sum_{i=1}^{m} |h_i(\lambda_i \phi_i(x)) - \sum_{j=1}^{N} a_{ij} g_{ij}(\phi_i(x))| \leq \sum_{i=1}^{m} \epsilon/3m = \epsilon/3 \ .$$

Thus,

$$\left\|f - \sum_{i=1}^{m}\sum_{j=1}^{N} a_{ij} g_{ij} \circ \phi_i\right\|_K < \epsilon \ ,$$

and this is enough to show that any basic neighborhood of f intersects the linear span of $A(G, \Phi)$. ∎

As a concrete example of this theorem we state the following immediate corollary.

Corollary 2.3. *Let P be the set of all monomials on $\mathbb{R}$. Then the set of functions*

$$\{x \longmapsto p(zx) : z \in \mathbb{Z}^n, \ x \in \mathbb{R}^n, \ p \in P\}$$

is fundamental in $C(\mathbb{R}^n)$.

Thus any $f \in C(\mathbb{R}^n)$ can be represented to within any required tolerance $\epsilon > 0$ by a linear combination of functions of the form $x \longmapsto p(zx)$, so that

$$\sup_{x \in K} |f(x) - \sum_{i=0}^{m} \alpha_i p(z_i x)| < \epsilon$$

where $\alpha_i \in \mathbb{R}$, $z_i \in \mathbb{Z}^n$ and K is a compact subset of $\mathbb{R}^n$.

In the cases considered so far, the greatest generality is achieved by Sun and Cheney [43]. There is a very recent paper by Lin and Pinkus [26] that tackles amongst other things the following problem. Let $1 \leq m \leq n-1$. Let Ω be a set of $n \times m$ real matrices, and let $g \in C(\mathbb{R}^n)$. Then for a given $A \in \Omega$, the function
$$x \longmapsto g(Ax)$$
is a continuous mapping from $\mathbb{R}^m \to \mathbb{R}$. Set
$$M(\Omega) = \{x \longmapsto g(Ax) : A \in \Omega, g \in C(\mathbb{R}^n)\} .$$
We now ask when $M(\Omega)$ is fundamental in $C(\mathbb{R}^n)$. The characterization is provided by reference to a special set $L(\Omega)$ whose definition is now given.

Definition 2.4. *Given a set Ω of $n \times m$ real matrices and $A \in \Omega$, we define $L(A)$ to be the subspace of $\mathbb{R}^m$ spanned by the n rows of A. The set Ω is then partitioned into equivalence classes by the relation $A \sim B$ if $L(A) = L(B)$. Also, we set*
$$L(\Omega) = \bigcup_{A \in \Omega} L(A) .$$

Let H_k denote the set of homogeneous polynomials of total degree k on $\mathbb{R}^m$. Thus a typical member p of H_k has the form
$$p(v) = \sum_{|\alpha|=k} c_\alpha v^\alpha , \qquad v \in \mathbb{R}^m .$$
Finally, denote by H the set of all homogeneous polynomials $\mathbb{R}^m$.

We need an elementary result concerning decompositions of elements in $\mathbb{R}^m$ before coming to the main result.

Lemma 2.5. *Let A be a real $n \times m$ matrix with $\dim L(A) = k$. Let P denote the orthogonal projection of $\mathbb{R}^m$ onto $L(A)$. Let $E : L(A) \to \mathbb{R}^k$ and $F : L(A)^\perp \to \mathbb{R}^{m-k}$ be the usual change of basis mappings. Then every $x \in \mathbb{R}^m$ can be written as $x = (x', x'')$ where $x' = EPx$ and $x'' = F(I-P)x$. Furthermore, there is an $n \times k$ matrix B such that $Ax = Bx'$ for all $x \in \mathbb{R}^m$.*

Proof: The fact that each $x \in \mathbb{R}^m$ has the appropriate decomposition is elementary. Furthermore, $L(A)^\perp$ is the kernel of the matrix A. Thus
$$Ax = A[Px + (I-P)x] = APx = AE^{-1}EPx = AE^{-1}x' .$$
Now AE^{-1} is the appropriate $n \times k$ matrix to choose for B. ∎

Theorem 2.6. *The set $M(\Omega)$ is fundamental in $C(\mathbb{R}^n)$ if and only if there is no non-trivial polynomial in H which vanishes identically on $L(\Omega)$.*

Proof: Assume that p is a non-trivial member of H which vanishes on $L(\Omega)$. Then p has the form
$$p(v) = \sum_{|\alpha|=k} c_\alpha v^\alpha , \qquad v \in \mathbb{R}^m .$$

Let ϕ be a non-trivial, infinitely continuously differentiable function from $\mathbb{R}^m$ to $\mathbb{R}$ which has compact support. Let

$$D^\alpha = \frac{\partial^{|\alpha|}}{\partial x_1^{\alpha_1}, \ldots, \partial x_m^{\alpha_m}}, \qquad \alpha = (\alpha_1, \ldots, \alpha_m) \in \mathbb{Z}_+^m .$$

Define

$$\psi(x) = \sum_{|\alpha|=k} c_\alpha (D^\alpha \phi)(x), \qquad x \in \mathbb{R}^m .$$

Then $\psi \in C^\infty(\mathbb{R}^m)$, ψ is non-trivial, and has compact support. By the standard theory of Fourier transforms (see [33]) we have $\widehat{\psi} = i^k \widehat{\phi} p$. We shall show that

$$\int_{\mathbb{R}^m} g(Ax)\psi(x)\,dx = 0$$

for all $A \in \Omega$ and $g \in C(\mathbb{R}^n)$, thus exhibiting a non-trivial linear functional which annihilates $M(\Omega)$. It then follows from the Hahn-Banach theorem that $M(\Omega)$ cannot be fundamental in $C(\mathbb{R}^n)$. To see this, take $A \in \Omega$ and let $k = \dim L(A) \leq m$. Decompose any $x \in \mathbb{R}^m$ into $x = (x', x'')$ using **2.5**. Now take any vector $a \in \mathbb{R}^k$ and write $v_a = (a,0) \in \mathbb{R}^m$. Then, under the identification of **2.5** again, v_a can be considered as an element of $L(A)$. Hence, adopting the obvious notation,

$$\begin{aligned}
0 &= i^k \widehat{\phi}(v_a) p(v_a) \\
&= \widehat{\psi}(v_a) \\
&= \frac{1}{(2\pi)^{m/2}} \int_{\mathbb{R}^m} \psi(x) e^{-i v_a x}\, dx \\
&= \frac{1}{(2\pi)^{m/2}} \int_{\mathbb{R}^k} \int_{\mathbb{R}^{m-k}} \psi(x', x'') e^{-i a x'}\, dx''\, dx' \\
&= \frac{1}{(2\pi)^{m/2}} \int_{\mathbb{R}^k} h(x') e^{-i a x'}\, dx' = \widehat{h}(a),
\end{aligned}$$

where we have made the obvious definition for h. Note that h is again a compactly supported function in $C^\infty(\mathbb{R}^k)$, so that h does indeed have a Fourier transform. It follows that $h \equiv 0$. Now using **2.5** again,

$$\begin{aligned}
\int_{\mathbb{R}^m} g(Ax)\psi(x)\,dx &= \int\int_{\mathbb{R}^k \times \mathbb{R}^{m-k}} g(Bx')\psi(x', x'')\,dx''\,dx' \\
&= \int_{\mathbb{R}^k} g(Bx') h(x')\,dx' = 0 .
\end{aligned}$$

For the reverse implication, suppose that no non-trivial member of H_k vanishes identically on $L(\Omega)$. We will show that $H_k \subset \operatorname{span} M(\Omega)$.

Let $d_k = \dim H_k$. Then there exist $b_1, \ldots, b_{d_k} \in L(\Omega)$ such that if $p \in H_k$ and $p(b_i) = 0$, $i = 1, \ldots, d_k$, then $p \equiv 0$. Each b_i can be written as $b_i = y_i A_i$ for some $y_i \in \mathbb{R}^n$ and $A_i \in \Omega$. Thus we can write

$$(b_i x)^k = (y_i A_i x)^k =: g_i(A_i x) ,$$

showing that $(b_i x)^k \in M(\Omega)$. We now claim that $\{(b_i x)^k\}_{i=1}^{d_k}$ spans H_k. It suffices to show that these functions form a linearly independent set. If not, there exist λ_i, $i = 1, \ldots, d_k$, such that

$$\sum_{i=1}^{d_k} \lambda_i (b_i x)^k = 0, \qquad x \in \mathbb{R}^n,$$

but not all λ_i are zero. Then for $\alpha \in \mathbb{Z}_+^n$, $|\alpha| = k$,

$$0 = D^\alpha \left(\sum_{i=1}^{d_k} \lambda_i (b_i x)^k \right) = k! \sum_{i=1}^{d_k} \lambda_i b_i^\alpha .$$

The $\{b_i^\alpha\}_{i=1,|\alpha|=k}^{d_k}$ form a $d_k \times d_k$ square matrix, which the above equality shows to be singular. Thus there exist $\{\mu_\alpha\}_{|\alpha|=k}$ such that

$$\sum_{|\alpha|=k} \mu_\alpha b_i^\alpha = 0, \qquad i = 1, 2, \ldots, d_k,$$

and not all μ_α are zero. Set $p(x) = \sum_{|\alpha|=k} \mu_\alpha x^\alpha$. Then $p \in H_k$, $p \not\equiv 0$ and $p(b_i) = 0$, $i = 1, \ldots, d_k$. This contradiction shows that $H_k \subset \text{span } M(\Omega)$, and so $H \subset \text{span } M(\Omega)$. The Weierstrass theorem completes the proof. ∎

§3. Sigmoidal Functions

A sigmoidal function is a function $\sigma : \mathbb{R} \to \mathbb{R}$ such that $\lim_{t \to \infty} \sigma(t) = 1$ and $\lim_{t \to -\infty} \sigma(t) = 0$. Such functions arise naturally in neural network theory as the activation function of a neural node. See Section 5 for a brief explanation of this. Within the theory of neural networks an important question is whether for a given continuous sigmoidal function σ, the set

$$\{x \longmapsto \sigma(yx + \theta) : y \in \mathbb{R}^n, \theta \in \mathbb{R}\} \qquad (x \in \mathbb{R})$$

is fundamental in $C(\mathbb{R}^n)$. This result was first established by Cybenko [9], although his arguments were arranged for showing density in $C(I_n)$, where $I_n = [0, 1]^n$. An analogous result for the L_p metrics can be found in [40]. The paper of Cybenko has been superceded by a paper of Sun and Cheney [43] and by two papers by Chui and Li, [7], [8]. The papers [43] and [7] deal with the fundamentality of the set

$$\{x \longmapsto \sigma(zx + k) : z \in \mathbb{Z}^n \text{ and } k \in \mathbb{Z}\} \qquad (x \in K)$$

in the space $C(K)$. Here K is as usual a compact set. The real benefit over Cybenko's result is that the fundamental set is now considerably smaller.

Then there is work by Mhaskar and Micchelli [29] on k^{th} degree sigmoidal functions. A kth degree sigmoidal function is a function $\sigma : \mathbb{R} \to \mathbb{R}$ such that

$$\lim_{t \to \infty} \frac{\sigma(t)}{t^k} = 1 \, , \quad \lim_{t \to -\infty} \frac{\sigma(t)}{t^k} = 0$$

and σ is bounded by a polynomial of degree at most k on $\mathbb{R}$.

A related question is the construction of algorithms to generate accurate approximations. Advances in this area are of recent vintage. The second paper by Chui and Li contains a constructive proof of the density result, using Bernstein polynomial operators. The paper by Chen, Chen and Liu [5] and Maskar and Micchelli also provide constructive proofs, drawing upon ideas from the area of spline functions in the univariate case and Fourier series in the multivariate case. Finally two papers by Xu, Light and Cheney [44,45], also provided constructive proofs and algorithms.

We begin by establishing the density result of Chui and Li, since it is a small step away from **2.2**, the result of Sun and Cheney. The following result is in fact a straightforward extension of one in [7].

Theorem 3.1. *Let σ be a continuous sigmoidal function. Then the set of functions*

$$\{t \longmapsto \sigma(zt + \ell) : z, \ell \in \mathbb{Z}\} \quad (t \in \mathbb{R})$$

is fundamental in $C(\mathbb{R})$.

Proof: Fix $f \in C(\mathbb{R})$ and let S denote the linear span of the set of functions $\{t \longmapsto \sigma(zt + \ell) : z, \ell \in \mathbb{Z}, t \in \mathbb{R}\}$. We intend to follow the technique used in the proof of **2.1** – showing that every basic neighbourhood of f intersects S. Suppose this is not the case. Then there exists an $\epsilon > 0$ and a compact set K in $\mathbb{R}$ such that

$$B_{K,\epsilon} = \{g \in C(\mathbb{R}) : \|f - g\|_K < \epsilon\}$$

has $B_{K,\epsilon} \cap S$ empty. The compact set K lies in some closed interval $I = [a, b]$. We will actually establish that the restriction of functions in S to I is dense in $C(I)$, which will in turn establish that the restriction of functions in S to K is dense in $C(K)$, and so must intersect $B_{K,\epsilon}$. Let E denote the restriction to I of functions in S. It will help at this stage to assume a and b are rational numbers. Let μ be a functional in $[C(I)]^*$ such that $\mu(e) = 0$ for all $e \in E$. This functional is associated (via the Riesz Representation theorem) with a regular Borel measure, which we again denote by μ. Then

$$\int_I \sigma(zt + \ell) \, d\mu(t) = 0 \, ,$$

for all $z, \ell \in \mathbb{Z}$. Now fix $\ell, p, q \in \mathbb{Z}$ with $q > 0$ and $p/q \in I$. Define $r : I \to \mathbb{R}$ by

$$r(t) = \begin{cases} 1 & p/q < t \leq b \\ 0 & a \leq t < p/q \\ \sigma(\ell) & t = p/q \, . \end{cases}$$

Consider the expression

$$\sigma\bigl(nq(t - p/q) + \ell\bigr), \quad \text{for } t \in I.$$

As $n \to \infty$, this expression converges pointwise to r on I. Then by the Lebesgue Dominated Convergence theorem,

$$\begin{aligned}
0 &= \lim_{n\to\infty} \int_I \sigma\bigl(nq(t-p/q)+\ell\bigr)\,d\mu(t) \\
&= \int_I r(t)\,d\mu(t) \\
&= \mu\{t \in I : t > p/q\} + \sigma(\ell)\mu\left\{\frac{p}{q}\right\}.
\end{aligned}$$

If we choose ℓ very large and negative, then by the properties of a sigmoidal function, $\sigma(\ell)$ is very small. This forces

$$\mu\{t \in I : t > p/q\} = 0 \quad \text{whenever } p/q \in I,$$

and then we conclude that $\mu\{p/q\} = 0$ also, whenever $p/q \in I$. This forces μ to be the trivial measure and so E is dense in $C(I)$ as required. ∎

Theorem 3.2. *Let σ be a sigmoidal function in $C(\mathbb{R})$. Then the set of functions*

$$\{t \longmapsto \sigma(zt + k) : z \in \mathbb{Z}^n,\, k \in \mathbb{Z}\} \qquad (t \in \mathbb{R}^n)$$

is fundamental in $C(\mathbb{R}^n)$.

Proof: We intend to apply **2.2**. Firstly, **3.1** shows that the set

$$G = \{t \longmapsto \sigma(kt + \ell) : k, \ell \in \mathbb{Z}\} \qquad (t \in \mathbb{R})$$

is fundamental in $C(\mathbb{R})$. Secondly, identifying the dual of $\mathbb{R}^n$ with $\mathbb{R}^n$, it follows easily that if $\Phi = \mathbb{Z}^n$, then the set $\Phi_0 = \{\phi/\|\phi\| : \phi \in \Phi, \phi \neq 0\}$ is dense in $S(\mathbb{R}^n)$. Thus from **2.2**, the set

$$A(G, \Phi) = \{g \circ \phi : g \in G, \phi \in \Phi\}$$

is fundamental in $C(\mathbb{R}^n)$. An element of $A(G, \Phi)$ has the form

$$(g \circ \phi)(t) = g(zt) = \sigma(kzt + \ell) = \sigma(z't + \ell)$$

where $\phi(t) = zt$, $t \in \mathbb{R}^n$ and $z' \in \mathbb{Z}^n$. ∎

We now move on to consider constructive proofs of this type of result. We begin with the analysis of Xu, Light and Cheney as found in [44] and [45]. As a general principle, we shall try to use the idea of convolution, constructing a convolution kernel from a sigmoidal function. As we have already seen, it is the ridge function nature of functions in $C(\mathbb{R}^n)$ of the form $t \longmapsto \sigma(zt+k)$, $z \in \mathbb{Z}^n$, $k \in \mathbb{Z}$, $t \in \mathbb{R}^n$, which makes the convolution approach difficult. Thus, we may as well start with a general $\phi \in C(\mathbb{R})$ and $x \in \mathbb{R}^n$ and consider the ridge function

$$f(t) = \phi(xt), \quad t \in \mathbb{R}^n .$$

From this function we want to construct a second function which is in $L^1(\mathbb{R}^n)$ and whose integral is non-zero. With any function whose behaviour for large values of x is uneven, in the sense that as x moves around on the surface of a sphere of large radius, the function varies considerably, it is a sound principle to average the function over that sphere. To this end, let S^{n-1} denote the unit sphere in $\mathbb{R}^n$ and w_{n-1} the surface area of S^{n-1}. We set

$$g(x) = \frac{1}{\omega_{n-1}} \int_{S^{n-1}} \phi(xu) \, dS^{n-1}(u) , \quad x \in \mathbb{R}^n .$$

The following result can be found in Madych [27], although it is certainly older than that, appearing in [23].

Lemma 3.3. *Let $\phi \in C(\mathbb{R})$, $n \in \mathbb{N}$ and $n \geq 2$. Define $g : \mathbb{R}^n \to \mathbb{R}$ by the equation prior to the Lemma. Then g is a radial function, and if $g(x) = g_0(\|x\|)$ we have*

$$g_0(r) = \omega_{n-2} \, \omega_{n-1}^{-1} \int_{-1}^{1} \phi(rs)(1-s^2)^{(n-3)/2} \, ds .$$

Proof: Suppose $x \neq 0$, and select the pole of the coordinate system to be $\rho = x/r$ where $r = \|x\|$. Any point $u \in S^{n-1}$ can be expressed in the form $u = (\cos\theta, v \sin\theta)$ where $\cos\theta = u\rho$ and $v \in S^{n-2}$. Then

$$dS^{n-1}(u) = \sin^{n-2}\theta \, d\theta \, dS^{n-2}(v) .$$

Now $xu = r\rho u = r\cos\theta$ and so

$$\begin{aligned}
g(x) &= \omega_{n-1}^{-1} \int_{S^{n-1}} \phi(xu) \, dS^{n-1}(u) \\
&= \omega_{n-1}^{-1} \int_{S^{n-2}} \int_0^{\pi} \phi(r\cos\theta) \sin^{n-2}\theta \, d\theta \, dS^{n-2}(v) \\
&= \omega_{n-2} \omega_{n-1}^{-1} \int_0^{\pi} \phi(r\cos\theta) \sin^{n-3}\theta \sin\theta \, d\theta \\
&= \omega_{n-2} \omega_{n-1}^{-1} \int_{-1}^{1} \phi(rs)(1-s^2)^{(n-3)/2} \, ds .
\end{aligned}$$
∎

We want to be able to tell not only when a ridge function is $L^1(\mathbb{R}^n)$ but also when such a function can be integrated against $\|x\|^m$. The next lemma helps in this process.

Lemma 3.4. *Let $\phi \in C(\mathbb{R})$ and let n be a natural number with $n \geq 2$. Define g as in 3.3. Then, if $\|\cdot\|^m g \in L^1(\mathbb{R}^n)$,*

$$\int_{\mathbb{R}^n} \|x\|^m |g(x)|\, dx = \omega_{n-2} \int_0^\infty r^{m+1} \left|\int_{-r}^r \phi(t)(r^2 - t^2)^{(n-3)/2}\, dt\right| dr\ .$$

Proof: From 3.3, we have

$$\int_{\mathbb{R}^n} \|x\|^m |g(x)|\, dx = \int_0^\infty \int_{S^{n-1}} r^{n+m-1} |g_0(r)|\, dS^{n-1}\, dr$$

$$= \omega_{n-1} \int_0^\infty r^{n+m-1} |g_0(r)|\, dr$$

$$= \omega_{n-2} \int_0^\infty r^{n+m-1} \left|\int_{-1}^1 \phi(rs)(1 - s^2)^{(n-3)/2}\, ds\right| dr$$

$$= \omega_{n-2} \int_0^\infty r^{m+1} \left|\int_{-r}^r \phi(t)(r^2 - t^2)^{(n-3)/2}\, ds\right| dr\ . \blacksquare$$

We see from **3.4** that $\int_{\mathbb{R}^n} \|x\|^m |g(x)|\, dx = 0$ if ϕ is an odd function, and the effect of the integral on the right-hand side of the equation in **3.4** is to "remove" the odd part of the function ϕ. Henceforward, we can and do assume ϕ is even. We now tackle the integrability of g.

Theorem 3.5. *Let $\phi \in C(\mathbb{R})$ and define g by*

$$g(x) = \omega_{n-1}^{-1} \int_{S^{n-1}} \phi(xu)\, dS^{n-1}(u)\ , \qquad x \in \mathbb{R}^n\ .$$

Let n be odd and let ϕ satisfy
(i) $\int_{-\infty}^\infty \phi(t) t^{2j}\, dt = 0$, $0 \leq j \leq (n-3)/2$
(ii) $\int_{-\infty}^\infty |\phi(t) t^{n+m-1}|\, dt < \infty$ for some $m \geq 0$.
Then the function $x \mapsto \|x\|^m g(x)$ is integrable on $\mathbb{R}^n$.

Proof: Set $\lambda = (n-3)/2$. Using **3.4**, and (i),

$$\int_{\mathbb{R}^n} \|x\|^m |g(x)|\, dx = 2\omega_{n-2} \int_0^\infty r^{m+1} \left|\int_0^r \phi(t)(r^2 - t^2)^\lambda\, dt\right| dr$$

$$= 2\omega_{n-2} \int_0^\infty r^{m+1} \left|\int_r^\infty \phi(t)(r^2 - t^2)^\lambda\, dt\right| dr$$

$$\leq 2\omega_{n-2} \int_0^\infty \int_r^\infty |\phi(t)|(r^2 - t^2)^\lambda r^{m+1}\, dt\, dr\ .$$

Now we apply Fubini's theorem to obtain

$$\int_{\mathbb{R}^n} \|x\|^m |g(x)|\, dx \leq 2\omega_{n-2} \int_0^\infty |\phi(t)| \int_0^t r^{m+1}(r^2 - t^2)^\lambda\, dr\, dt\ .$$

It will help to define, for $m, n, \in \mathbb{N}$, the constants c_{mn} by

$$c_{mn}t^{n+m-1} = \int_0^t r^{m+1}(t^2 - r^2)^\lambda dr . \tag{3}$$

Then

$$\int_{\mathbb{R}^n} \|x\|^m |g(x)| \, dx \leq 2\omega_{n-2} c_{mn} \int_0^\infty |\phi(t)| t^{n+m-1} \, dt < \infty. \qquad \blacksquare$$

Of course, it may turn out that $g \in L^1(\mathbb{R}^n)$ but that $\int_{\mathbb{R}^n} g = 0$, which will preclude our using g to construct a convolution kernel. The next two results address this matter.

Lemma 3.6. Let ϕ and g be as in 3.5 and let n be odd. In order that $\int_{\mathbb{R}^n} g(t) \, dt \neq 0$, it is necessary and sufficient that

$$\int_{-\infty}^\infty \phi(t) t^{n-1} \, dt \neq 0 .$$

Proof: Following the proof of 3.5, we see that

$$\int_{\mathbb{R}^n} g(x) \, dx = -2\omega_{n-2} c_{0n} \int_{-\infty}^\infty \phi(t) t^{n-1} \, dt .$$

By definition $\omega_{n-2} \neq 0$ and

$$c_{0n} t^{n-1} = \int_0^t r(t^2 - r^2)^\lambda \, dr .$$

Putting $t = 1$, we see that $c_{0n} > 0$. $\blacksquare$

The situation when n is even is very different! We investigate first conditions for the function g to belong to $L^1(\mathbb{R}^n)$.

Lemma 3.7. Let ϕ and g be as in 3.5 and let n be even. Let
(i) $\int_{-\infty}^\infty \phi(t) t^{2j} \, dt = 0$, $0 \leq j \leq (n-2)/2$
(ii) $\int_{-\infty}^\infty |\phi(t) t^{n-1}| \, dt < \infty$.
Then $g \in L^1(\mathbb{R}^n)$.

Proof: Again set $\lambda = (n-3)/2$, but now set $\mu = (n-2)/2$. Fix $n \geq 4$ and even. For $0 < t \leq 1$, define

$$v(t) = \left\{(1-t)^\lambda - \sum_{j=0}^\mu \binom{\lambda}{j}(-1)^j t^j\right\} t^{-\mu-1} ,$$

and set $v(0) = \lim_{t\to 0} v(t)$. Taylor's Theorem assures us that this limit exists and that $v \in C[0,1]$. By **3.3**, assuming ϕ is even,

$$|rg(x)| = |rg_0(r)|$$

$$= 2\omega_{n-2}\omega_{n-1}^{-1}\left|\int_0^r \phi(t)\left(1 - \frac{t^2}{r^2}\right)^\lambda dt\right|$$

$$= 2\frac{\omega_{n-2}}{\omega_{n-1}}\left|\int_0^r \phi(t)\left\{v(t^2r^{-2})(t^2r^{-2})^{\mu+1} + \sum_{j=0}^\mu \binom{\lambda}{j}(-1)^j(t^2r^{-2})^j\right\}dt\right|$$

$$\leq 2\frac{\omega_{n-2}}{\omega_{n-1}}\left\{\int_0^r |\phi(t)|\,\|v\|_\infty(t^2r^{-2})^{\mu+1}\,dt \right.$$

$$\left. + \sum_{j=0}^\mu \left|\binom{\lambda}{j}\right|\left|\int_0^r \phi(t)(t^2r^{-2})^j\,dt\right|\right\} =: I_1 + I_2 \,.$$

It is elementary that

$$I_1 \leq r^{-n}\|v\|_\infty \int_0^r |\phi(t)t^n|\,dt \,.$$

Also, for $0 \leq 2j \leq n-2$,

$$\left|\int_0^r \phi(t)t^{2j}\,dt\right| = \left|\int_r^\infty \phi(t)t^{2j}\,dt\right|$$

$$\leq \int_r^\infty t^{2j-n+2}|\phi(t)|t^{n-2}\,dt$$

$$\leq r^{2j-n+2}\int_r^\infty |\phi(t)|t^{n-2}\,dt \,.$$

Now, substituting all this computation at the correct point gives

$$\int_{\mathbb{R}^n} |g(x)|\,dx = \omega_{n-1}\int_0^\infty r^{n-1}|g_0(r)|\,dr$$

$$\leq 2\omega_{n-2}\int_0^\infty r^{n-2}\left\{r^{-n}\|v\|_\infty \int_0^r |\phi(t)t^n|\,dt \right.$$

$$\left. + r^{-n+2}\sum_{j=0}^\mu \left|\binom{\lambda}{j}\right|\int_r^\infty |\phi(t)|t^{n-2}\,dt\right\}dr$$

$$= 2\omega_{n-2}\left\{\|v\|_\infty \int_0^\infty \int_0^r r^{-2}t^n|\phi(t)|\,dt\,dr \right.$$

$$\left. + A\int_0^\infty \int_r^\infty |\phi(t)|t^{n-2}\,dt\,dr\right\},$$

where $A = \sum_{j=0}^{\mu} |\binom{\lambda}{j}|$. Then, using Fubini's Theorem gives

$$\int_{\mathbb{R}^n} |g(x)|\, dx \leq 2\omega_{n-2} A \|v\|_\infty \left\{ \int_0^\infty \int_t^\infty r^{-2} t^n |\phi(t)|\, dr\, dt \right.$$
$$\left. + \int_0^\infty \int_0^t |\phi(t)| t^{n-2}\, dr\, dt \right\}$$
$$= 4\omega_{n-2} A \|v\|_\infty \int_0^\infty |\phi(t)| t^{n-1}\, dt < \infty \ .$$

The case $n = 2$ needs special treatment. The reader may refer to [45] for details. ∎

The next Lemma has a strong parallel in the work of Jackson [22].

Lemma 3.8. Let ϕ and g be as in **3.5**, and let n be even. Then we have $\int_{\mathbb{R}^n} g(x)\, dx = 0$.

Proof: Using arguments similar to **3.4** and setting $\lambda = (n-3)/2$, we have

$$\int_{\mathbb{R}^n} g(x)\, dx = 2\omega_{n-2} \int_0^\infty \int_0^r \phi(t) r \left(r^2 - t^2 \right)^\lambda dt\, dr \ .$$

We now set
$$Q(r,t) = r(r^2 - t^2)^\lambda$$
and, with $\mu = (n-2)/2$,
$$P(r,t) = r \sum_{j=0}^{\mu} \binom{\lambda}{j} (r^2)^{\lambda-j} (-t^2)^j \ .$$

Thus $P(r, \cdot)$ is the Taylor polynomial associated with $Q(r, \cdot)$. Then,

$$\int_0^\infty \int_0^r \phi(t) Q(r,t)\, dt\, dr = \int_0^\infty \left[\int_0^r \phi(t) Q(r,t)\, dt - \int_0^\infty \phi(t) P(r,t)\, dt \right] dr$$
$$= \int_0^\infty \int_0^r \phi(t) \{Q(r,t) - P(r,t)\}\, dt\, dr$$
$$- \int_0^\infty \int_r^\infty \phi(t) P(r,t)\, dt\, dr =: I_1 - I_2 \ .$$

Now, using Fubini's theorem,

$$I_2 = \int_0^\infty \phi(t) \int_0^t P(r,t)\, dr\, dt = \int_0^\infty \phi(t) V(t)\, dt \ ,$$

where

$$V(t) = \int_0^t P(r,t)\, dr = \sum_{j=0}^{\mu} \binom{\lambda}{j} (-1)^j \frac{t^{n-1}}{n - 2j - 1} \ .$$

Also,

$$\int_t^s \{Q(r,t) - P(r,t)\}\, dr = \int_t^s \left\{ r(r^2-t^2)^\lambda - \sum_{j=0}^\mu \binom{\lambda}{j} t^{2j} r^{2\lambda-2j+1} \right\} dr$$

$$= \frac{(s^2-t^2)^{\lambda+1}}{2\lambda+2} - \sum_{j=0}^\mu \binom{\lambda}{j}(-1)^j \frac{t^{2j}}{2\lambda-2j+2}\left(s^{2\lambda-2j+2} - t^{2\lambda-2j+2}\right)$$

$$= \sum_{j=0}^\infty \binom{\lambda+1}{j}(-1)^j \frac{s^{2\lambda+2-2j}t^{2j}}{2\lambda+2} - \sum_{j=0}^\mu \binom{\lambda}{j}(-1)^j \frac{s^{n-1-2j}t^{2j} - t^{n-1}}{n-1-2j}$$

$$= \sum_{j=0}^\infty \binom{\lambda}{j}(-1)^j \frac{s^{n-1-2j}t^{2j}}{n-1-2j} - \sum_{j=0}^\mu \binom{\lambda}{j}(-1)^j \frac{s^{n-1-2j}t^{2j} - t^{n-1}}{n-1-2j}$$

$$= \sum_{j=\mu+1}^\infty \binom{\lambda}{j}(-1)^j \frac{s^{n-1-2j}t^{2j}}{n-1-2j} + V(t).$$

Now $n-1-2j < 0$ for $j > \mu+1$ and so

$$I_1 = \int_0^\infty \phi(t) \int_t^\infty \{Q(r,t) - P(r,t)\}\, dr\, dt$$

$$= \int_0^\infty \phi(t) \lim_{s\to\infty} \int_t^s \{Q(r,t) - P(r,t)\}\, dr\, dt$$

$$= \int_0^\infty \phi(t) \lim_{s\to\infty} \left(\sum_{j=\mu+1}^\infty \binom{\lambda}{j}(-1)^j \frac{s^{n-1-2j}t^{2j}}{n-1-2j} + V(t) \right) dr\, dt$$

$$= \int_0^\infty \phi(t) V(t)\, dt.$$

Thus $I_1 = I_2$ and so $\int_{\mathbb{R}^n} g(x)\, dx = 0$. ∎

It is worth noting before we proceed that the moment conditions on ϕ are not only sufficient for $g \in L^1(\mathbb{R}^n)$, but also in a sense necessary – again see [45] for details. The problem caused by **3.8** can be avoided by "averaging" the kernel in one dimension higher. This result is given with a sketch of proof.

Lemma 3.9. *Let n be an even integer, $n \geq 2$. Let $\phi \in C(\mathbb{R})$ satisfy*
1. $\int_{-\infty}^\infty \phi(t) t^{2j}\, dt = 0$, $0 \leq j \leq (n-3)/2$
2. $\int_{-\infty}^\infty |\phi(t) t^{n+m-1}|\, dt < \infty$ *for some $m \geq 0$.*

For $x = (x_1, \ldots, x_n) \in \mathbb{R}^n$ define $\bar{x} = (x_1, \ldots, x_n, 0) \in \mathbb{R}^{n+1}$. Put

$$h(x) = \frac{1}{\omega_n} \int_{S^n} \phi(u\bar{x})\, dS^n(u).$$

Then $\int_{\mathbb{R}^n} \|x\|^m |h(x)|\, dx < \infty$. The condition $\int_{-\infty}^\infty \phi(t)|t^{n-1}|\, dt \neq 0$ is necessary and sufficient for $\int_{\mathbb{R}^n} h(x)\, dx \neq 0$.

Proof: Again, we are assuming ϕ is even. Put

$$h_0(r) = \omega_{n-1}\omega_n^{-1}r^{-1}\int_{-r}^{r}\phi(t)(1-t^2r^{-2})^{(n-2)/2}\,dt.$$

By 3.4, with $r = \|x\|$,

$$h_0(x) = h_0(\|\overline{x}\|) = h_0(\|x\|) = h_0(r).$$

Hence,

$$\int_{\mathbb{R}^n} h(x)\,dx = \int_0^\infty h_0(r)r^{n-1}\omega_{n-1}\,dr$$

$$= 2\omega_n^{-1}\omega_{n-1}^2 \int_0^\infty\int_0^r \phi(t)r^{n-2}(1-t^2r^{-2})^{(n-2)/2}\,dt\,dr$$

$$= 2\omega_n^{-1}\omega_{n-1}^2 \int_0^\infty \left(\int_0^\infty - \int_r^\infty\right)\phi(t)r^{n-2}(1-t^2r^{-2})^{(n-2)/2}\,dt\,dr.$$

$$= 2\omega_n^{-1}\omega_{n-2}^2 \int_0^\infty \phi(t)\left[\int_0^\infty (r^2-t^2)^{(n-2)/2}\,dr - \int_0^t (r^2-t^2)^{(n-2)/2}\,dr\right] dt$$

$$= -2\omega_n^{-1}\omega_{n-2}^2 \int_0^\infty \phi(t)\int_0^t (r^2-t^2)^{(n-2)/2}\,dr\,dt,$$

where we have used condition (i). Now

$$\int_0^t (r^2-t^2)^{(n-2)/2}\,dr = \gamma_n t^{n-1}$$

for some constant γ_n which is non-zero. (Put $t=1$ in the above equation to see this.) The proof now follows that of **3.6**. ∎

We now establish the major results towards which we have been aiming.

Theorem 3.10. *Let n be odd, $n \geq 3$. Let ϕ be a uniformly continuous function on $\mathbb{R}$ satisfying*

1. $\int_0^\infty |\phi(t)t^{n-1}|\,dt < \infty$
2. $\int_{-\infty}^\infty \phi(t)t^{2j}\,dt = 0$, $0 \leq j \leq (n-3)/2$
3. $\int_{-\infty}^\infty \phi(t)t^{n-1}\,dt \neq 0$.

Then the set of functions

$$\{x \mapsto \phi(xv+\alpha) : v \in \mathbb{R}^n,\ \alpha \in \mathbb{R}\}$$

is fundamental in $C(\mathbb{R}^n)$.

Proof: Set

$$g(x) = \frac{1}{\omega_{n-1}}\int_{S^{n-1}} \phi(xu)\,dS^{n-1}(u), \qquad x \in \mathbb{R}^n.$$

By 3.5, $g \in L^1(\mathbb{R}^n)$ and by 3.6 we can assume $\int_{\mathbb{R}^n} g(x)\,dx = 1$. Now fix $f \in C(\mathbb{R}^n)$, K compact in $\mathbb{R}^n$ and $\epsilon > 0$. We can assume that $K = [-\nu, \nu]^n$, $\nu \in \mathbb{N}$, and that $f(x) = 0$ for $x \in \mathbb{R}^n \setminus 2K$. Define $g_m(x) = m^n g(mx)$ for $m \in \mathbb{N}$. Then by standard theory (see, for example, Stein and Weiss [39]) there is a constant C such that

$$\|g_m * f - f\|_K \leq C\omega(f, m^{-1}),$$

where $\|\cdot\|_K$ is the usual supremum norm on K and ω the modulus of continuity. Choose m so that $C\omega(f, m^{-1}) < \epsilon/3$. Write

$$(g_m * f)(x) = \int_{2K} g_m(x-y) f(y)\, dy = \int_{2mK} g(mx - z) f(z/m)\, dz.$$

Now let P be a partition of the set $2mK$ into a finite disjoint family of Borel sets, each of diameter at most δ. For each $A \in P$, let $z_A \in A$ and define

$$b_A = \int_A f(z/m)\, dz.$$

Now,

$$\left|\int_{2mK} g(mx-z) f(z/m)\, dz - \sum_{A \in P} b_A g(mx - z_A)\right|$$

$$\leq \sum_{A \in P} \int_A |g(mx-z) - g(mx - z_A)||f(z/m)|\, dz$$

$$\leq \omega(g;\delta) \sum_{A \in P} \int_A |f(z/m)|\, dz$$

$$= \omega(g;\delta) \int_K m^n |f(y)|\, dy.$$

It follows easily from the definition of g that $\omega(g;\delta) \leq \omega(\phi;\delta)$ (because g is derived from ϕ by a normalized integral). Since ϕ is uniformly continuous by hypothesis we may choose δ and P so that the final term above is bounded by $\epsilon/3$.

Now we apply a quadrature to g. Let Q be a partition of S^{n-1} into a finite, disjoint collection of Borel sets, each of diameter at most θ. By analogy with the above, set

$$e_B = \omega_{n-1}^{-1} \int_B dS^{n-1}(u),$$

and let $u_B \in B$, $B \in Q$. Then again, setting $\psi(u) = \phi((mx - z_A)u)$, $u \in S^{n-1}$,

$$\left|g(mx - z_A) - \sum_{B \in Q} e_B \phi((mx - z_A) u_B)\right|$$

$$= \left|\omega_{n-1}^{-1} \int_{S^{n-1}} \phi((mx - z_A)u)\, du - \sum_{B \in Q} e_B \phi((mx - z_A) u_B)\right|$$

$$\leq \omega(\psi;\theta) \leq \omega(\phi : \|mx - z_A\|\theta) \leq 3m\nu\omega(\phi;\theta).$$

Consequently,

$$\left|\sum_{A\in P} b_A g(mx - z_A) - \sum_{A\in P} b_A \sum_{B\in Q} e_B \phi((mx - z_A, u_B))\right|$$

$$\leq \sum_{A\in P} |b_A| 3m\nu\omega(\phi;\theta) \leq 3m\nu\omega(\phi:\theta)\|f\|_1 \ .$$

Select θ and the partition Q so that this last bound is at most $\epsilon/3$. Then, for $x \in K$,

$$\left|f(x) - \sum_{A\in P}\sum_{B\in Q} c_{AeB} \phi(mxu_B - z_A u_B)\right|$$

$$\leq \left|f(x) - \int_{2mK} g(mx - z) f(z/m)\, dz\right|$$

$$+ \left|\int_{2mK} g(mx - z) f(z/m)\, dz - \sum_{A\in P} b_A g(mx - z_A)\right|$$

$$+ \left|\sum_{A\in P} b_A g(mx - z_A) - \sum_{A\in P} b_A \sum_{B\in Q} e_B \phi(mxu_B - z_A u_B)\right| < \epsilon \ . \ \blacksquare$$

A similar theorem can be established for n even. Again the reader is referred to [45] for details. The problem now is how to construct functions ϕ satisfying the appropriate moment conditions. Some examples of such functions have been given by Madych [27] in the context of Radon transforms. We provide one such example. (Madych provides a technique for generating examples via a formula akin to the Rodrigues' formula for orthogonal polynomials.)

Example (Madych [27]). Take

$$\phi(t) = (1+t^2)^{-n/2} T_n[(1+t^2)^{-1/2}] \ ,$$

where T_n is the usual Chebyshev polynomial of degree n. It is shown in [27] that for $n \geq 3$,

$$\int_{-\infty}^{\infty} \phi(t) t^j\, dt = 0 \ , \qquad 0 \leq j \leq n-2 \ .$$

Also, if we want to verify the remaining condition of **3.7**, then

$$\int_{-\infty}^{\infty} |\phi(t) t^{n-1}|\, dt = 2 \int_0^{\infty} |\phi(t) t^{n-1}|\, dt$$

$$= 2 \int_0^{\pi/2} \left|(\sec^2\theta)^{-n/2} \cos n\theta (\tan\theta)^{n-1} \sec^2\theta\right| d\theta$$

$$= 2 \int_0^{\pi/2} \left|\frac{\cos n\theta}{\cos\theta}(\sin\theta)^{n-1}\right| d\theta \ .$$

Now since n is odd, $(\cos n\theta)/(\cos \theta)$ is a linear combination of a finite number of powers of $\cos \theta$ and so this last integral is finite.

Of course, it is also possible to start with a much weaker assumption on ϕ and then proceed via some differencing to the required result. We illustrate this for the case of sigmoidal functions. Suppose $\sigma \in C(\mathbb{R})$ is a sigmoidal function. Set $\phi(t) = \sigma(t+1) - \sigma(t)$, $t \in \mathbb{R}$. Then $\phi(t) \to 0$ as $t \to \pm\infty$. The rate at which $\phi(t) \to 0$ as $t \to \pm\infty$ depends on the rate at which $\sigma(t) \to 0$ as $t \to -\infty$ and $\sigma(t) \to 1$ as $t \to +\infty$. However, with quite mild assumptions on σ, such rates may be improved by differencing.

Let Δ be the forward difference operator $(\Delta\phi)(t) = \phi(t+1) - \phi(t)$. Suppose ϕ is even and has the power series expansion at ∞,

$$\phi(t) = \sum_{k=1}^{\infty} a_k t^{-k-\alpha} \quad \text{for } 0 < \alpha \le 1, t > T.$$

Then

$$(\Delta^{n-1}\phi)(t) = \sum_{j=0}^{n-1} (-1)^{n-1-j} \binom{n-1}{j} \phi(t+j)$$

and so for $t > T$,

$$(\Delta^{n-1}\phi)(t) = (n-1)! \sum_{k=1}^{\infty} a_k \xi_t^{-k-n+1-\alpha},$$

for some $t < \xi_t < t+n-1$. It follows that $(\Delta^{n-1}\phi)(t) = O(t^{-n-\alpha})$ as $t \to \infty$. Consequently, the function $t \mapsto t^{n-1}(\Delta^{n-1}\phi)(t)$ is in $L^1(\mathbb{R})$. Put $\psi = \Delta^{n-1}\phi$. Then for $0 \le k \le n-1$,

$$\int_{-\infty}^{\infty} \psi(t) t^k \, dt = \int_{-\infty}^{\infty} (\Delta^{n-1}\phi)(t) t^k \, dt$$

$$= (-1)^{n-1} k! \int_{-\infty}^{\infty} \phi(t+n-1)(\Delta^{n-1}V_k)(t) \, dt$$

$$= \begin{cases} 0 & \text{if } 0 \le k \le n-2 \\ (-1)^{n-1} k! \int_{-\infty}^{\infty} \phi(t) \, dt & \text{if } k = n-1. \end{cases}$$

Here we have used the notation V_k for the normalized monomial $V_k(t) = t^k/k!$. This leads to the following result.

Theorem 3.11. *Let $\phi \in C(\mathbb{R})$ be an even function and let n be odd. Suppose*
(a) $\int_{-\infty}^{\infty} \phi(t) \, dt \ne 0$
(b) *ϕ has a descending power series expansion at ∞ of the following form:*

$$\phi(t) = \sum_{k=1}^{\infty} a_k t^{-k-\alpha} \quad \text{for } 0 < \alpha \le 1, t > T.$$

Then the set of functions

$$\{x \mapsto \phi(xv + \alpha) : v \in \mathbb{R}^n, \alpha \in \mathbb{R}\}$$

is fundamental in $C(\mathbb{R}^n)$.

Example. Consider the sigmoidal function $\sigma \in C(\mathbb{R})$ defined by

$$\sigma(t) = \frac{e^t}{1+e^t}, \qquad t \in \mathbb{R}.$$

Then ϕ defined by

$$\phi(t) = \sigma(t+1) - \sigma(t), \qquad t \in \mathbb{R}$$

has the desired behaviour at infinity. (This function decays to zero like e^{-t} as $t \to \pm\infty$, and so the differencing is only needed to obtain the zero moments.) Consequently, **3.11** may be applied to obtain a constructive proof of the fundamentality of the set of functions

$$\{x \longmapsto \sigma(xv + \alpha) : v \in \mathbb{R}^n, \alpha \in \mathbb{R}\}$$

using the convolution techniques of **3.10**.

Indeed, given $f \in C(\mathbb{R}^n)$, K compact in $\mathbb{R}^n$ and $\epsilon > 0$ we can construct $\{v_i\}_{i=1}^N$ in $\mathbb{R}^n$ and $\{c_{ij}\}_{i,j=1}^N$, $\{\alpha_i\}_{i=1}^N$ in $\mathbb{R}$ such that

$$\sup_{x \in K} \left| f(x) - \sum_{i,j=1}^N c_{ij} \psi(xv_i + \alpha_j) \right| < \epsilon.$$

Unscrambling all this gives the approximation

$$\sum_{i,j=1}^N c_{ij} \psi(xv_i + \alpha_j)$$

$$= \sum_{i,j=1}^N c_{ij} \sum_{\ell=1}^n e_\ell [\sigma(xv_i + \alpha_j + \ell + 1) - \sigma(xv_i + \alpha_j + \ell)]$$

$$= \sum_{i,j=1}^N c_{i,j} \left\{ \sum_{\ell=0}^{n+1} e'_\ell \sigma(xv_i + \alpha_j + \ell) + e''_\ell \sigma(xv_i + \alpha_j + \ell) \right\}$$

$$= \sum_{i,j=1}^N c_{ij} \sum_{\ell=0}^{n+1} (e'_\ell + e''_\ell) \sigma(xv_i + \alpha_j + \ell).$$

This is the required sigmoidal approximation. Note that even if the decay properties of σ at $t = \pm\infty$ are not very strong, it can be expected that the differencing applied to generate the function ψ will greatly improve this decay.

There are alternative approaches to these ideas of generating in a "practical" way an appropriate ridge function approximation to a given continuous function. We consider now the method proposed by Chui and Li in their second paper [8]. The first step they take has independent interest. A very simple example of a ridge function is the function $x \longmapsto (vx)^k$, $x \in \mathbb{R}^n$, where

v is a fixed vector in $\mathbb{R}^n$ and $k \in \mathbb{Z}_+$. Such a function might be called a "ridge monomial" function. A linear combination of such functions is called a ridge polynomial and takes the form

$$R(x) = \sum_{k=0}^{\ell} \sum_{i=1}^{m} c_{ki}(v_i x)^k, \qquad x \in \mathbb{R}^n.$$

Of course, having introduced this special class of polynomials, one can immediately observe two things. Firstly, each ridge polynomial is a member of $\Pi_\ell(\mathbb{R}^n)$ for the appropriate value of ℓ. Secondly, $\Pi_\ell(\mathbb{R}^n)$ can be generated from ridge polynomials simply by observing that a special case of the argument provided in **2.6** does the job. We reiterate this argument next. The reader should also note the work of [27] which discusses related matters.

Lemma 3.12. *Fixing k, we define (for $x \in \mathbb{R}^n$):*

$$H = \text{span } \{x \mapsto x^\alpha : |\alpha| = k\}$$
$$M = \text{span } \{x \mapsto (vx)^k : v \in \mathbb{R}^n, \|v\|_\infty \leq 1\}$$

Then $H = M$.

Proof: By the multinomial theorem,

$$(vx)^k = \sum_{|\alpha|=k} \frac{k!}{\alpha!} v^\alpha x^\alpha.$$

This shows that $M \subset H$. For the reverse inclusion, let

$$m = \dim H = \#\{\alpha : |\alpha| = k\}.$$

Select points $v_1, \ldots, v_m$ so that the matrix (v_i^β), $1 \leq i \leq m$, $|\beta| = k$, is nonsingular. We may assume that $\|v_i\|_\infty \leq 1$. Then for any α satisfying $|\alpha| = k$ the following equation can be solved for $c_1, \ldots, c_m$:

$$x^\alpha = \sum_{i=1}^{m} c_i (v_i x)^k = \sum_{i=1}^{m} c_i \sum_{|\beta|=k} \frac{k!}{\beta!} v_i^\beta x^\beta. \qquad \blacksquare$$

Of course, if one is interested in the details of algorithms, one can easily do more than **3.12**. One can compute how various polynomials are represented by ridge polynomials. This is done in [8], but we content ourselves with a broad outline of the algorithm.

Initial Data: Let $\sigma \in C(\mathbb{R})$ be a sigmoidal function and let f be given in $C(\mathbb{R}^n)$. We wish to approximate f to accuracy ϵ using "sigmoidal ridge functions"; that is, we want to write f in the form

$$f(x) \approx \sum_{i=1}^{N} c_i \sigma(xy_i - r_i), \qquad x \in \mathbb{R}^n,$$

where $y_i \in \mathbb{R}^n$, c_i and r_i are real numbers, $1 \le i \le N$. This approximation is to be effected over a compact set $K \subset \mathbb{R}^n$, which we take to be $[0,1]^n$.

Step 1: Approximate f by a polynomial. In [8], this is done by using the Bernstein operator

$$(B_m f)(x) = \sum_{\substack{|\alpha| \le m \\ \alpha \in \mathbb{Z}_+^n}} f\left(\frac{\alpha}{m}\right) \binom{m}{\alpha} x^\alpha (1 - xe)^{m-|\alpha|}, \qquad x \in \mathbb{R}^n,$$

where $e \in \mathbb{R}^n$ is the vector all of whose entries are unity. It is well-known that $B_m f \to f$ as $m \to \infty$ for all $f \in C(K)$. One therefore chooses m sufficiently large so that $\|f - B_m f\|_{K,\infty} < \epsilon/2$.

Step 2: Rewrite $B_m f$ as

$$(B_m f)(x) = \sum_{k=0}^{m} \sum_{j=1}^{N} d_{jk} (x y_{mj})^k$$

where $\|y_{mj}\|_\infty \le 1$ for $1 \le j \le N$. This is an application of 3.12.

Step 3: Approximate each term $(x y_{mj})^k$ by a linear combination of sigmoidal ridge functions, to within accuracy $\epsilon/2$.

All of the above is straightforward except for Step 3 which needs some explanation. The technique is again to use a convolution operator. We take $\sigma \in C(\mathbb{R})$ and define $\sigma_\ell \in C(\mathbb{R})$, $\ell = 1, 2, \ldots$, by

$$\sigma_\ell(x) = \ell \{\sigma(\ell x + 1) - \sigma(\ell x)\}, \qquad x \in \mathbb{R}.$$

Note that our approach is essentially the same as that of [8] except that we do *not* demand that σ be of bounded variation at this stage. Our approach follows unpublished results of X. Sun and relies on the second mean value theorem for integrals to avoid imposing the condition of bounded variation. The following is an elementary result.

Lemma 3.13. *Let σ be a continuous sigmoidal function. Then*

(i) $\left|\int_a^b \sigma_\ell(t)\, dt\right| \le 2\|\sigma\|_\infty$

(ii) *if $b > 0 > a$ then $\lim_{\ell \to \infty} \int_a^b \sigma_\ell(t)\, dt = 1$*

(iii) *if $b \ge a > 0$ or $0 > b \ge a$ then $\lim_{\ell \to \infty} \int_a^b \sigma_\ell(t)\, dt = 0$.*

Lemma 3.14. *Let σ be a continuous sigmoidal function. If $b > a > 0$, then*

$$\lim_{\ell \to \infty} \int_{-b}^{b} \sigma_\ell(t - s)\, ds = 1$$

uniformly for t in $[-a, a]$.

Proof: By the definition of σ_ℓ,

$$\int_{-b}^{b} \sigma_\ell(t-s)\, ds = \int_{t-b}^{t+b} \sigma_\ell(s)\, ds = \int_{\ell(t+b)}^{\ell(t+b)+1} \sigma(s)\, ds - \int_{\ell(t-b)}^{\ell(t-b)+1} \sigma(s)\, ds.$$

Given $\epsilon > 0$, select M so that

$$|\sigma(s)| \leq \epsilon/2 \quad \text{for } s \leq -M(b-a)+1$$

and

$$|\sigma(s) - 1| \leq \epsilon/2 \quad \text{for } s \geq M(b-a).$$

Now if $\ell \geq M$ and $|t| \leq a$, then

$$\left| \int_{-b}^{b} \sigma_\ell(t-s)\, ds - 1 \right| = \left| \int_{\ell(t+b)}^{\ell(t+b)+1} [\sigma(s) - 1]\, ds - \int_{\ell(t-b)}^{\ell(t-b)+1} \sigma(s)\, ds \right|.$$

In the first integral on the right-hand side of the above inequality we have

$$s \geq \ell(t+b) \geq \ell(-a+b) \geq M(b-a)$$

and so $|\sigma(s) - 1| \leq \epsilon/2$. In the second integral

$$s \leq \ell(t-b) + 1 \leq \ell(a-b) + 1 \leq -M(b-a) + 1,$$

and so $|\sigma(s)| < \epsilon/2$. This completes the proof. ∎

Lemma 3.15. *Let σ be a continuous sigmoidal function and ϵ, δ positive real numbers. Then there exists L such that for $\ell \geq L$,*

$$\left| \int_u^v \sigma_\ell(s)\, ds \right| < \epsilon$$

for either $v \geq u \geq \delta$ or $-\delta \geq v \geq u$.

Proof: Select L so that for $t \geq L\delta$ we have $|\sigma(t) - 1| < \epsilon/2$, and for $t \leq -L\delta + 1$ we have $|\sigma(t)| < \epsilon/2$. Assume that $v \geq u \geq \delta$ and $\ell \geq L$. Then

$$\left| \int_u^v \sigma_\ell(s)\, ds \right| = \left| \int_{\ell v}^{\ell v + 1} \sigma(s)\, ds - \int_{\ell u}^{\ell u + 1} \sigma(s)\, ds \right|$$

$$\leq \int_{\ell v}^{\ell v + 1} |\sigma(s) - 1|\, ds + \int_{\ell u}^{\ell u + 1} |\sigma(s) - 1|\, ds < \epsilon. \quad \blacksquare$$

The next theorem introduces the required convolution.

Theorem 3.16. *Let $\sigma \in C(\mathbb{R})$ be a sigmoidal function. If f is continuous and of bounded variation on $[-b, b]$ then the sequence of functions $\{S_\ell f\}$ defined by*

$$(S_\ell f)(t) = \int_{-b}^{b} \sigma_\ell(t-s) f(s)\, ds, \quad |t| \leq a < b$$

converges uniformly to f on $[-a,a]$.

Proof: Observe that S_ℓ is linear and f can be written as the difference of two monotone functions. Hence we can assume that f is non-increasing on $[-b,b]$. Define u by $u(t) = 1$, $t \in [-b,b]$. Then

$$S_\ell f - f = (S_\ell f - fS_\ell u) + (fS_\ell u - f) = (S_\ell f - fS_\ell u) + f(S_\ell u - u) \, .$$

Now **3.14** shows that $f(S_\ell u - u) \to 0$ uniformly on $[-a,a]$. Take $\epsilon > 0$ and select $\delta \in (0, b-a)$ so that $|f(t-s) - f(t)| \le \epsilon$ whenever $|t| \le a$ and $|s| \le \delta$. Then for $|t| \le a$,

$$(S_\ell f)(t) - f(t)(S_\ell u)(t) = \int_{-b}^{b} \sigma_\ell(t-s)\{f(s) - f(t)\}\,ds$$

$$= \int_{t-b}^{t+b} \sigma_\ell(s)\{f(t-s) - f(t)\}\,ds$$

$$= \left\{ \int_{t-b}^{-\delta} + \int_{-\delta}^{\delta} + \int_{\delta}^{t+b} \right\} \sigma_\ell(s)\{f(t-s) - f(t)\}\,ds$$

$$=: I_\ell + J_\ell + K_\ell \, .$$

Now $f(t-s) - f(t)$ is a non-decreasing as a function of s, and so in each of the three integrals, the second mean value theorem applies. For example, there is a point ξ in $[-\delta, \delta]$ so that

$$|J_\ell| \le \left| \{f(t+\delta) - f(t)\} \int_{-\delta}^{\xi} \sigma_\ell(s)\,ds + \{f(t-\delta) - f(t)\} \int_{\xi}^{\delta} \sigma_\ell(s)\,ds \right|$$

$$\le \epsilon \left| \int_{-\delta}^{\xi} \sigma_\ell(s)\,ds \right| + \epsilon \left| \int_{\xi}^{\delta} \sigma_\ell(s)\,ds \right| \le 4\epsilon \|\sigma\|_\infty \, .$$

Similarly, there is a $\theta \in [t-b, -\delta]$ so that

$$|I_\ell| = \left| \{f(b) - f(t)\} \int_{t-b}^{\theta} \sigma_\ell(s)\,ds + \{f(t+\delta) - f(t)\} \int_{\theta}^{-\delta} \sigma_\ell(s)\,ds \right|$$

$$\le 2\|f\|_\infty \left\{ \left| \int_{t-b}^{\theta} \sigma_\ell(s)\,ds \right| + \left| \int_{\theta}^{-\delta} \sigma_\ell(s)\,ds \right| \right\} \, .$$

Now $t - b \le \theta \le \delta$, so **3.15** may be used to infer the existence of L such that

$$|I_\ell| \le 2\|f\|_\infty (\epsilon + \epsilon) = 4\epsilon \|f\|_\infty \, ,$$

for all $\ell \ge L$. The same is true for K_ℓ and so the result follows. ∎

Step 3 of the algorithm may now be elaborated, as follows:

Step 3: Approximate each ridge polynomial $(xy)^k$ using the convolution operator S_ℓ. This is done by first choosing a such that $K \subset [-a,a]^n$. Then, since $\|x\|_\infty \leq a$ and $\|y\|_\infty \leq 1$ it follows that $|xy| \leq na$ for all $x \in K$ and $1 \leq j \leq N$. Now take $b > na$ and choose ℓ so that

$$\sup_{t \in [-na, na]} |(S_\ell p_i - p_i)(t)| < \epsilon/2, \quad i = 0, 1, \ldots, m$$

where $p_i(t) = t^i$, $t \in [-b, b]$. This involves an application of **3.16**. It then follows that

$$\sup_{x \in K} \left| (xy_{mj})^k - \int_{-b}^{b} \left[\sigma\big(\ell\{xy_{mj} - s + 1\}\big) - \sigma\big(\ell\{xy_{mj} - s\}\big) \right] \ell s^k \, ds \right| < \epsilon/2 .$$

The final stage is to apply some univariate quadrature rule to the convolution integral so as to generate a finite sum of ridge functions. Of course, in manufacturing an actual sum of ridge functions, one must keep a more precise account of the above analysis, but steps 1-3 embody the essential mathematical details from [8].

We now move on to address some work which appeared first in Chen, Chen and Liu [5]. Ideas similar to those in [5], but in greater generality, form part of the paper by Mhaskar and Micchelli [29]. We will not discuss the full generality of kth degree sigmoidal functions initially, but will continue to assume $\sigma \in C(\mathbb{R})$ and $\lim_{t \to \infty} \sigma(t) = 1$, $\lim_{t \to -\infty} \sigma(t) = 0$. The general idea of [5] and [29] is to construct a suitable operator which is in general polynomial preserving, although in the case of a zeroth degree sigmoidal function one only gets preservation of constants.

Defininition 3.17. *Given a sigmoidal function $\sigma \in C(\mathbb{R})$ and $n \in \mathbb{N}$, we define A_n to be the smallest positive integer such that*
(a) $|\sigma(x)| \leq n^{-1}$ for $x \leq -A_n$
(b) $1 - n^{-1} \leq \sigma(x) \leq 1 + n^{-1}$ for $x \geq A_n$.
The operator G_n is then defined by

$$(G_n f)(x) = f(0) + \sum_{\nu=1}^{n} \left\{ f\left(\frac{\nu}{n}\right) - f\left(\frac{\nu-1}{n}\right) \right\} \sigma(A_n(nx - \nu)), \quad x \in \mathbb{R} .$$

Theorem 3.18. *There exists a constant $c > 0$ such that for $f \in C(\mathbb{R})$,*

$$\sup_{x \in [0,1]} |f(x) - (G_n f)(x)| \leq c \omega(f, 1/n) .$$

Proof: Fix $x \in [0, 1]$ and let μ denote the largest integer which does not exceed nx. Then

$$|f(x) - (G_n f)(x)|$$

$$= \left| f(x) - f(0) - \sum_{\nu=1}^{n} \left\{ f\left(\frac{\nu}{n}\right) - f\left(\frac{\nu-1}{n}\right) \right\} \sigma(A_n(nx-\nu)) \right|$$

$$\leq \left| f(x) - f(0) - \sum_{\nu=1}^{\mu} \left\{ f\left(\frac{\nu}{n}\right) - f\left(\frac{\nu-1}{n}\right) \right\} \sigma(A_n(nx-\nu)) \right|$$

$$+ \sum_{\nu=\mu+1}^{n} \left| f\left(\frac{\nu}{n}\right) - f\left(\frac{\nu-1}{n}\right) \right| |\sigma(A_n(nx-\nu))|$$

$$\leq \left| f(x) - f(0) - \sum_{\nu=1}^{\mu} \left\{ f\left(\frac{\nu}{n}\right) - f\left(\frac{\nu-1}{n}\right) \right\} \right|$$

$$+ \sum_{\nu=1}^{\mu} \left| f\left(\frac{\nu}{n}\right) - f\left(\frac{\nu-1}{n}\right) \right| |1 - \sigma(A_n(nx-\nu))|$$

$$+ \sum_{\nu=\mu+1}^{n} \left| f\left(\frac{\nu}{n}\right) - f\left(\frac{\nu-1}{n}\right) \right| |\sigma(A_n(nx-\nu))|$$

$$\leq \left| f(x) - f\left(\frac{\mu}{n}\right) \right| + \omega(f; 1/n) \left[\sum_{\nu=1}^{\mu} |1 - \sigma(A_n(nx-\nu))| \right.$$

$$\left. + \sum_{\nu=\mu+1}^{n} |\sigma(A_n(nx-\nu))| \right]$$

$$\leq \omega(f; 1/n) \left[2\|\sigma\|_\infty + 2 + \sum_{\nu=1}^{\mu-1} |1 - \sigma(A_n(nx-\nu))| \right.$$

$$\left. + \sum_{\nu=\mu+2}^{n} |\sigma(A_n(nx-\nu))| \right].$$

Now observe that $\nu \leq \mu-1$ implies $nx-\nu \geq 1$ and so $|1-\sigma(A_n(nx-\nu))| \leq n^{-1}$ by the definition of A_n. Also $\nu \geq \mu+2$ implies $nx-\nu \leq -1$, so that $|\sigma(A_n(nx-\nu))| \leq n^{-1}$. This gives

$$|f(x) - (G_n f)(x)| \leq \omega(f; 1/n) \left[2\|\sigma\|_\infty + 2 + \sum_{\nu=1}^{n} n^{-1} \right]$$

$$= (2\|\sigma\|_\infty + 3)\omega(f; 1/n) . \qquad \blacksquare$$

Before turning to our next topic, let us briefly outline the use to which Mhaskar and Micchelli put the concept of a kth degree sigmoidal function. The fundamental idea is that a kth degree sigmoidal function can model a truncated power very closely over a large part of the real line.

Lemma 3.19. Let $\sigma \in C(\mathbb{R})$ be a kth degree sigmoidal function. For $n \in \mathbb{N}$ let A_n be the smallest positive integer such that

(i) $|\sigma(x)| \leq n^{-k-1}|x|^k$ if $x \leq -A_n$

(ii) $|\sigma(x) - x^k| \leq n^{-k-1}x^k$ if $x \geq A_n$.

Suppose also that $K > 0$ is chosen so that $|\sigma(x)| \leq K(1+|x|)^k$ for $x \in \mathbb{R}$. Let y_+^k denote the usual truncated power function, given by

$$y_+^k = \begin{cases} 0 & y \leq 0 \\ y^k & y > 0 \end{cases}.$$

Then

$$|y_+^k - A_n^{-k}\sigma(A_n y)| \leq \begin{cases} n^{-k-1}|y|^k & \text{for } |y| \geq 1 \\ K 2^{k+1} & \text{for } |y| < 1 \end{cases}.$$

Proof: From (ii),

$$|\sigma(A_n y) - A_n^k y^k| \leq n^{-k-1} A_n^k y^k \quad \text{for } y \geq 1,$$

so that

$$|A_n^{-k}\sigma(A_n y) - y^k| \leq n^{-k-1} y^k \quad \text{for } y \geq 1.$$

From (i),

$$|A_n^{-k}\sigma(A_n y)| \leq n^{-k-1}|y^k| \quad \text{for } y \leq -1.$$

Thus,

$$|y_+^k - A_n^{-k}\sigma(A_n y)| \leq n^{-k-1}|y^k| \quad \text{for } |y| \geq 1.$$

For $|y| < 1$,

$$\begin{aligned}|y_+^k - A_n^{-k}\sigma(A_n y)| &\leq |y_+^k| + A_n^{-k}|\sigma(A_n y)| \\ &\leq 1 + A_n^{-k} K(1 + |A_n y|)^k \\ &\leq K 2^{k+1}.\end{aligned} \qquad \blacksquare$$

With **3.19** in place, the way to a sigmoidal approximation is now straightforward. We first develop a B-spline approximation to $f \in C(\mathbb{R})$. This has the form

$$(s_n f)(x) = \sum_{\nu \in \mathbb{Z}} \phi_{\nu,n}(f) M(nx - \nu),$$

where each $\phi_{\nu,n}$ is a simple linear functional (in fact, a linear combination of point evaluation functionals) and M is the cardinal B-spline of order k. It is well-known (see Schumaker [38], for example) that s_n can be written in terms of truncated powers. In fact, for $x \in [0,1]$, we can write

$$(s_n f)(x) = P_f(x) + \sum_{\nu=0}^{n} \psi_{\nu,n}(f)(nx - \nu)_+^k,$$

where again each $\psi_{\nu,n}$ is a linear combination of point-evaluation functionals and P_f is a polynomial of degree k. The convergence properties of s_n are also well-understood. For example, there is a constant $B > 0$ such that if $f \in C^{k+1}[-1, 2]$, then for $n > k$,

$$|(s_n f)(x) - f(x)| \leq Bn^{-k-1} \sup_{u \in [-1,2]} |f^{k+1}(u)| \quad \text{for } x \in [0, 1] \ .$$

Now Mhaskar and Micchelli propose to obtain a sigmoidal approximation simply by replacing the truncated powers in s_n by sigmoidal functions. This gives a new operator :

$$(T_n f)(x) = P_f(x) + \sum_{\nu=0}^{n} \psi_{\nu,n}(f) A_n^{-k} \sigma\big(A_n(nx - \nu)\big) \ , \qquad x \in [0, 1] \ .$$

To establish the convergence properties of T_n we need one more fact from spline theory. Again, there is a constant $E > 0$ such that if $f \in C^{k+1}[-1, 2]$, then for $n > k$,

$$|\psi_{\nu,n}(f)| \leq En^{-k-1} \sup_{u \in [-1,2]} |f^{k+1}(u)| \ , \qquad 0 \leq \nu \leq n \ .$$

Theorem 3.20. *There is a constant $\Gamma > 0$ such that if $f \in C^{k+1}[-1, 2]$ then for $n > k$,*

$$|(T_n f)(x) - f(x)| \leq \Gamma n^{-k-1} \sup_{u \in [-1,2]} |f^{k+1}(u)| \quad \text{for } x \in [0, 1] \ .$$

Proof: One simply observes, with $J = \max(B, E)$

$$|(T_n f)(x) - f(x)| \leq |(T_n f)(x) - (s_n f)(x)| + |(s_n f)(x) - f(x)|$$

$$\leq \sum_{\nu=0}^{n} |\psi_{\nu,n}(f)| \left|(nx - \nu)_+^k - A_n^{-k} \sigma\big(A_n(nx - \nu)\big)\right| + Bn^{-k-1} \sup_{u \in [-1,2]} |f^{k+1}(u)|$$

$$\leq Jn^{-k-1} \sup_{u \in [-1,2]} |f^{k+1}(u)| \left(1 + \sum_{\nu=0}^{n} \left|(nx - \nu)_+^k - A_n^{-k} \sigma(A_n(nx - \nu))\right|\right) \ .$$

To establish the claim of the theorem we now need only show the term in the parentheses above is bounded. In a manner reminiscent of **3.18** we write, for $x \in [0, 1]$, using **3.19**,

$$\sum_{\nu=0}^{n} \left|(nx - \nu)_+^k - A_n^{-k} \sigma\big(A_n(nx - \nu)\big)\right|$$

$$= \left(\sum_{\substack{0 \leq \nu \leq n \\ |nx - \nu| \geq 1}} + \sum_{\substack{0 \leq \nu \leq n \\ |nx - \nu| < 1}} \right) \left|(nx - \nu)_+^k - A_n^{-k} \sigma\big(A_n(nx - \nu)\big)\right|$$

$$\leq \sum_{\substack{0 \leq \nu \leq n \\ |nx - \nu| \geq 1}} n^{-k-1} |nx - \nu|^k + \sum_{\substack{0 \leq \nu \leq n \\ |nx - \nu| < 1}} K 2^{k+1}$$

$$\leq \sum_{\nu=0}^{n} n^{-k-1} n^k + 2.K 2^{k+1} \leq 1 + K 2^{k+2} \ . \qquad \blacksquare$$

Of course, the operator T_n does not generate a pure sigmoidal approximation, since the sigmoidal part is augmented by a polynomial. This process of augmenting an approximating family by (low degree) polynomials is a common one in approximation theory. However, in the context of neural networks (see Section 5) the presence of a polynomial term is unhelpful. The polynomial part of T_n can be approximated by sigmoidal functions using Step 3 of the algorithm of Chui and Li as outlined previously. This may indeed be an efficient procedure, since T_n is as cheap to compute as a spline approximation, while if the polynomial part of T_n has low degree, then it can also be rewritten inexpensively in terms of sigmoidal functions. However, we do not believe that the concept of a kth degree sigmoidal function will necessarily find favour with the practitioners in neural networks.

We mention that Chen, Chen and Liu [5] and Mhaskar and Micchelli deal in a similar fashion with the multivariate case. There the idea is first to approximate the given function by a suitable average of Fourier series. The average is taken so as to generate an approximation with a sufficiently high degree of accuracy. This approximation involves linear combinations of functions of the form $x \longmapsto e^{izx}$, $x \in \mathbb{R}^n$, $z \in \mathbb{Z}^n$. Now the univariate result is essentially what is used to generate a sigmoidal approximation to each of these functions.

Section 5 indicates another approach adopted by Lenze [24] who kindly made available to us a pre-preprint of his interesting work.

§4. The Interpolation Problem

We finally turn our attention to problems about interpolation with ridge functions. We study the following problem. Let $a_1, a_2, \ldots, a_k \in \mathbb{R}^n$ be distinct non-zero vectors. Then we wish to characterize the sets of distinct points $\{x_1, \ldots, x_m\} \subset \mathbb{R}^n$ such that the following interpolation problem is solvable: given $\alpha_1, \ldots, \alpha_m \in \mathbb{R}$ we seek functions $f_i : \mathbb{R} \to \mathbb{R}$, $i = 1, \ldots, m$, such that

$$f_1(a_1 x_i) + \ldots + f_k(a_k x_i) = \alpha_i, \qquad i = 1, 2, \ldots, m .$$

If we set

$$F(x) = \sum_{j=1}^{k} f_j(a_j x) ,$$

then we require $F(x_i) = \alpha_i$, $i = 1, 2, \ldots, m$. Observe that since the functions $f_1, \ldots, f_k$ are at our disposal, there is no loss of generality in assuming that $a_1, \ldots, a_k \in B(\mathbb{R}^n)$ where B is some unit ball centered at the origin.

We begin by discussing the case $n = k = 2$. In this case, one can assume that $a_1 = (1,0)$ and $a_2 = (0,1)$. Then

$$F(x) = f_1(u) + f_2(v) \text{ where } x = (u,v) \in \mathbb{R}^2 .$$

The interpolation problem in this case was first solved by Dyn, Light and Cheney [15], and relies on a very simple geometrical notion.

Definition 4.1. *A path in $\mathbb{R}^2$ is a finite ordered set $\{x_1,\ldots,x_m\}$ of points in $\mathbb{R}^2$ such that the line segments joining consecutive points are of positive length and are alternately horizontal and vertical. A path is said to be closed if m is even, if $x_m \neq x_1$ and if the line segment joining x_1 to x_2 is perpendicular to that joining x_m and x_1.*

It is perhaps surprising that interpolation is possible if and only if the set of interpolation points $\{x_1,\ldots,x_k\}$ does not contain a closed path. One half of this result is easy.

Theorem 4.2. *Suppose $\{x_1,\ldots,x_m\} \subset \mathbb{R}^2$. If the set $\{x_1,\ldots,x_m\}$ contains a closed path then one cannot interpolate arbitrary data on this set of points by functions of the form $F(x) = f(u) + g(v)$, where $x = (u,v) \in \mathbb{R}^2$, and $f, g : \mathbb{R} \to \mathbb{R}$.*

Proof: Suppose, by reordering the points if necessary, that the closed path in $\{x_1,\ldots,x_m\}$ is $\{x_1,\ldots,x_\ell\}$. Then, if $x_i = (u_i, v_i)$, it is easily seen that for any functions $f, g : \mathbb{R} \to \mathbb{R}$,

$$\sum_{i=1}^{\ell}(-1)^i\{f(u_i) + g(v_i)\} = 0 .$$

Thus one can interpolate only data $\alpha_1,\ldots,\alpha_m$ that satisfies $\sum_{i=1}^{\ell}(-1)^i \alpha_i = 0$. ∎

For the reverse implication, we prefer to provide only an outline of the proof. We begin with a special case of an old result due to Schoenberg [36].

Lemma 4.3. *Let $u_1,\ldots,u_m$ be distinct real numbers. If $c \in \mathbb{R}^m \setminus 0$ and satisfies $\sum_{j=1}^{m} c_j = 0$, and if B is the $m \times m$ matrix with ij^{th} element $|u_i - u_j|$, then $c^T B c < 0$. Furthermore, B is non-singular.*

Now take interpolation points $x_i = (u_i, v_i)$, $i = 1, 2, \ldots, m$, where each $x_i \in \mathbb{R}^2$. The two functions we seek for interpolation may be taken as $f, g : \mathbb{R} \to \mathbb{R}$ where

$$f(u) = \sum_{i=1}^{m} \lambda_i |u - u_i|,\; u \in \mathbb{R} \quad \text{and} \quad g(v) = \sum_{i=1}^{m} \mu_i |v - v_i|, \quad v \in \mathbb{R} .$$

Here $\lambda_1,\ldots,\lambda_m$ and $\mu_1,\ldots,\mu_m$ will be uniquely determined by the data values $\alpha_1,\ldots,\alpha_m$ prescribed at $x_1,\ldots,x_m$, providing $\{x_1,\ldots,x_m\}$ does not contain a closed path. The interpolation equations may be written

$$\sum_{i=1}^{m} \lambda_i |u_j - u_i| + \sum_{i=1}^{m} \mu_i |v_j - v_i| = \alpha_j, \quad j = 1,\ldots,m .$$

Alternatively, we may write these equations in matrix form as

$$B\lambda + C\mu = \alpha$$

where
$$B = (|u_j - u_i|), \quad C = (|v_j - v_i|), \quad \lambda = (\lambda_1, \ldots, \lambda_m),$$
$\mu = (\mu_1, \ldots, \mu_m)$ and $\alpha = (\alpha_1, \ldots, \alpha_m)$. It turns out that one can be even more specific here by demanding that $\lambda = \mu$. Then we have to solve $(B + C)\lambda = \alpha$. As usual, this is equivalent to showing that $B + C$ is invertible.

Theorem 4.4. *Suppose $\{x_1, \ldots, x_m\}$ is a set in $\mathbb{R}^2$ which does not contain a closed path. Then one can interpolate arbitrary data on this set of points by functions of the form $F(x) = f(u) + g(v)$, where $x = (u, v) \in \mathbb{R}^2$ and $f, g : \mathbb{R} \to \mathbb{R}$.*

Proof: As indicated prior to the theorem we choose
$$F(x) = \sum_{i=1}^{m} \lambda_i (|u - u_i| + |v - v_i|)$$
where $x = (u, v)$ and $x_i = (u_i, v_i)$, $i = 1, 2, \ldots, m$. Again, in the terminology established previously, the interpolation question has an affirmative answer if the matrix $B + C$ is non-singular.

Suppose $B+C$ is singular. We will sketch the reasoning which establishes that $\{x_1, \ldots, x_m\}$ contains a closed path. The details may be found in [15]. Firstly, all entries of $B + C$ are non-negative, and only the diagonal entries are zero. It follows (for example from the Courant-Fischer theorem, or the Frobenius theorem) that $B + C$ has a positive eigenvalue. From **4.3**, we see that if $\sum_{j=1}^{m} \theta_j = 0$, then $\theta^T(B + C)\theta = \theta^T B \theta + \theta^T C \theta \leq 0$. The weak inequality arises here because we can no longer assume that $B = (|u_i - u_j|)$ with $u_1, \ldots, u_m$ distinct. From the Courant-Fischer theorem again, it follows that the second largest eigenvalue of $B + C$ is non-positive. Since $B + C$ is singular, it follows that this eigenvalue is zero. Hence, there is a vector $\theta \in \mathbb{R}^m$ with $\sum_{j=1}^{m} \theta_j = 0$ and $\theta^T B \theta = \theta^T C \theta = 0$. Here comes the part which needs a bit of cleaning up! Firstly $\theta^T B \theta = 0$ only if two rows of B are identical because $u_i = u_k$, for some pair i, k. Then $\theta_i = -\theta_k$. Now $\theta^T C \theta = 0$ only if two columns of C are identical because $v_k = v_\ell$. Hence $\theta_i = -\theta_k = \theta_\ell$. Eventually this process must terminate with $\theta_i = -\theta_k = \theta_\ell = \ldots = \pm \theta_j$ where the index j occurs twice in the list – indicating the presence of a closed path. ∎

Braess and Pinkus [2] consider the next most complicated case in $\mathbb{R}^2$, that of three vectors a_1, a_2 and a_3. In this case, we may assume $a_1 = (1, 0)$, $a_2 = (0, 1)$ and $a_3 = (1, -1)$. Thus we seek to interpolate by a function
$$F(x) = f(u) + g(v) + h(u - v)$$
where f, g and h are at our disposal. It will be convenient to adopt the following terminology. A set $\{x_1, \ldots, x_m\} \subset \mathbb{R}^2$ will be classed an NI-set if the related interpolation problem is not always solvable on this set.

If we return to the case of two vectors $a_1 = (1,0)$ and $a_2 = (0,1)$ then a set $\{x_1, \ldots, x_m\}$ is an NI-set if and only if it contains a closed path. Figure A illustrates a closed path and shows how this particular path is a "sum of rectangles". This is a feature of every closed path. In addition, **4.2** and **4.4** show that the important feature of a closed path is the fact that it determines a linear functional. This functional is obtained very simply by taking the point evaluation functionals for points in the path with weights alternatively plus and minus one. Figure A also illustrates how each rectangle's weights have to be chosen in order to achieve a total weight of zero at points off the closed path and +1 and −1 alternately on the closed path. One can now rephrase **4.2** and **4.4**.

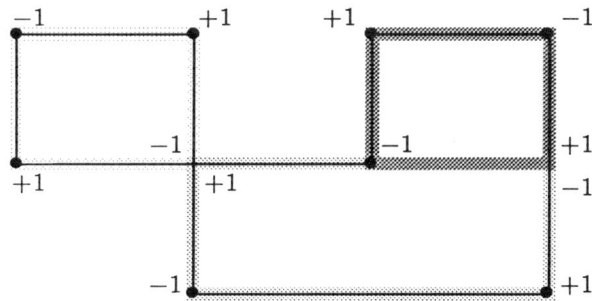

Figure A. A closed path decomposed into rectangles.

Theorem 4.5. *A set $\{x_1, \ldots, x_m\} \subset \mathbb{R}^2$ is an NI-set if and only if it contains a subset which is an algebraic sum of rectangles.*

Now let us return to the case $a_1 = (1,0)$, $a_2 = (0,1)$ and $a_3 = (1,-1)$. Figure B illustrates a simple NI-set in this case. Braess and Pinkus call this a brick. One may conjecture that every NI-set is an algebraic sum of bricks. This is not the case. Figure C shows two configurations that are not bricks, but are NI-sets — the "Escher" brick and the "exceptional hexagon."

It turns out that the exceptional hexagon is the basic building block for NI-sets in the case of the three vectors a_1, a_2 and a_3. Figure D shows how a regular brick is the sum of two Escher bricks, and how each Escher brick is the algebraic sum of two hexagons. (Note that henceforward, we drop the adjective exceptional.) The result of Braess and Pinkus [2] is in Theorem 4.6, below.

Theorem 4.6. *Let X be a finite subset of $\mathbb{R}^2$. One can interpolate arbitrary data given at the points in X by functions of the form $F(x) = f(u) + g(v) + h(u-v)$, where $f, g, h : \mathbb{R} \to \mathbb{R}$ are at our disposal and $x = (u,v)$, if and only if X does not contain an algebraic sum of hexagons.*

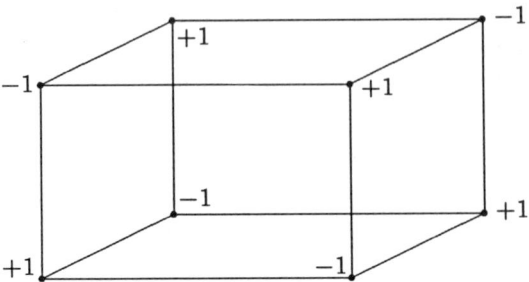

Figure B. A simple NI-set when $a_1 = (1,0)$, $a_2 = (0,1)$, $a_3 = (1,-1)$.

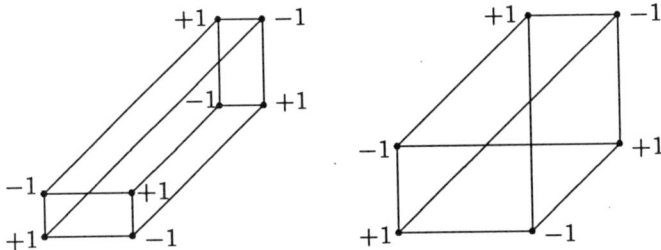

Figure C. The Escher brick (left) and the exceptional hexagon (right).

Proof: The proof is very complicated. Basically, one prunes away hexagons or bricks from an NI-set until one sees the required algebraic sum. The reader should consult [2] for the details. ∎

It seems unlikely that the pleasing geometrical ideas of [15] and [2] will carry over in an intuitively obvious way even to the case of vectors $a_1, a_2, \ldots, a_k$ in $\mathbb{R}^2$, and certainly not to the case of large numbers of vectors in $\mathbb{R}^n$. These considerations have motivated the work of Sun [42], which we now describe. Let $A = \{a_1, \ldots, a_k\}$ be a set of directions in $\mathbb{R}^n$ and $X = \{x_1, \ldots, x_m\}$ be a set of points in $\mathbb{R}^n$ at which we are to interpolate. Sun introduces the concept of an incidence matrix $In(A, X)$. The ij^{th} element of $In(A, X)$ is the cardinality of the set $\{a_\nu : a_\nu x_i = a_\nu x_j, 1 \leq \nu \leq k\}$. Another way of viewing $In(A, X)$ is to introduce the notation

$$\delta_{ij\nu} = \begin{cases} 1 & a_\nu x_i = a_\nu x_j \\ 0 & \text{otherwise} \end{cases}, \quad 1 \leq i,j \leq m, \ 1 \leq \nu \leq k.$$

Then the ij^{th} element of $In(A, X)$ is $\sum_{\nu=1}^{k} \delta_{ij\nu}$. We will also define $\tau_\nu(i)$ to

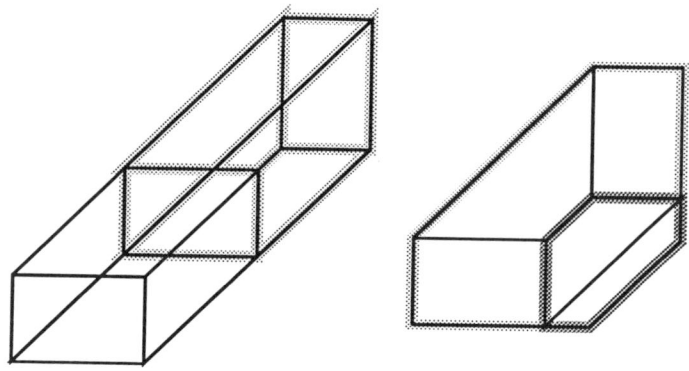

Regular brick =
sum of two Escher bricks

Escher brick =
sum of two hexagons

Figure D.

be the smallest integer such that $a_\nu x_{\tau_\nu(i)} = a_\nu x_i$, $1 \leq i \leq m$, $1 \leq \nu \leq k$. The mapping τ_ν from $\{1, 2, \ldots, m\}$ to itself is well-defined, and $\tau_\nu(i) \leq i$.

Lemma 4.7. *The matrix $In(A, X)$ defined above is non-negative definite.*

Proof: Let $c = (c_1, \ldots, c_m)$ be a vector in $\mathbb{R}^m$. Then

$$c^T In(A, X) c = \sum_{i,j=1}^m c_i c_j \sum_{\nu=1}^k \delta_{ij\nu} = \sum_{\nu=1}^k \sum_{i,j=1}^m c_i c_j \delta_{\tau_\nu(i)\tau_\nu(j)\nu}$$

Let $\mathcal{R}(\tau_\nu)$ denote the range of τ_ν. Then by observing that

$$\delta_{\tau_\nu(i)\tau_\nu(j)\nu} = \begin{cases} 1 & \tau_\nu(i) = \tau_\nu(j) \\ 0 & \text{otherwise} \end{cases},$$

we obtain

$$c^T In(A, X) c = \sum_{\nu=1}^k \sum_{\beta \in \mathcal{R}(\tau_\nu)} \left(\sum \{c_i : \tau_\nu(i) = \beta\} \right)^2 \geq 0. \qquad \blacksquare$$

We now turn to the question of the possibility of interpolation. A vague principle from **4.4** is that when interpolation is possible, it is possible with a wide variety of univariate functions $f_1, \ldots, f_m$. We exploit this fact in our next result.

The following notation will be useful in the next theorem. Let $\{a_1, \ldots, a_k\}$ and $\{x_1, \ldots, x_m\}$ be finite subsets of $\mathbb{R}^n$. Define $g \in C(\mathbb{R})$ by $g(u) = e^{-u^2}$. Given $\lambda = (\lambda_1, \ldots, \lambda_m)$ in $\mathbb{R}^m$, define

$$f_i(t) = \sum_{j=1}^m \lambda_j g(t - a_i x_j), \quad t \in \mathbb{R}, \ i = 1, 2, \ldots, k.$$

We now seek to determine λ so that, if data $\beta_1, \ldots, \beta_m$ are given at $x_1, \ldots, x_m$, then
$$\sum_{\nu=1}^{k} f_\nu(a_\nu x_\ell) = \beta_\ell, \qquad \ell = 1, \ldots, m.$$
That is
$$\sum_{j=1}^{m} \lambda_j \sum_{\nu=1}^{k} g(a_\nu(x_\ell - x_j)) = \beta_\ell, \qquad \ell = 1, \ldots, m. \tag{4}$$
We will need the following elementary result concerning Fourier transforms. It is an old result that can be found in [33, p.170], for example.

Lemma 4.8. *For all $x \in \mathbb{R}$,*
$$e^{-x^2} = \pi^{-1/2} \int_{\mathbb{R}} e^{2iux} e^{-u^2} \, du.$$

Theorem 4.9. *Let $A = \{a_1, \ldots, a_k\}$ and $X = \{x_1, \ldots, x_m\}$ be sets in $\mathbb{R}^n$. If the incidence matrix $In(A, X)$ is positive definite, then the interpolation problem in (4) is solvable for arbitrary data $\beta_1, \ldots, \beta_m$.*

Proof: Suppose $In(A, X)$ is positive definite. We will show that the matrix with element in the (ℓ, j) position given by
$$\sum_{\nu=1}^{k} g(a_\nu(x_\ell - x_j))$$
is positive definite. To establish this, take $c_1, \ldots, c_m \in \mathbb{R}$, not all 0, and consider
$$\sum_{\ell,j=1}^{m} c_\ell c_j \sum_{\nu=1}^{k} g(a_\nu(x_\ell - x_j)) =$$
$$= \sum_{\nu=1}^{k} \pi^{-1/2} \int_{\mathbb{R}} e^{-u^2} \sum_{\ell,j=1}^{m} c_\ell c_j e^{2iua_\nu(x_\ell - x_j)} \, du$$
$$= \pi^{-1/2} \sum_{\nu=1}^{k} \int_{\mathbb{R}} e^{-u^2} \left| \sum_{\ell=1}^{m} c_\ell e^{2iua_\nu x_\ell} \right|^2 du.$$

Consider the functions
$$\psi_\nu(u) = \sum_{\ell=1}^{m} c_\ell e^{2iua_\nu x_\ell} = \sum_{\beta \in \mathcal{R}(\tau_\nu)} \left(\sum \{c_l : \tau_\nu(l) = \beta\} \right) e^{2iua_\nu x_\beta}.$$

Since $In(A, X)$ is positive definite, the proof of **4.7** shows that there is at least one value $\nu_0 \in \{1, \ldots, k\}$ such that ψ_{ν_0} is a non-trivial linear combination of exponentials. Also we have
$$\sum_{\ell,j=1}^{m} c_\ell c_j \sum_{\nu=1}^{k} g(a_\nu(x_\ell - x_j)) \geq \pi^{-1/2} \int_{\mathbb{R}} e^{-u^2} \left| \sum_{\ell=1}^{m} c_\ell e^{2iua_{\nu_0} x_\ell} \right|^2 du.$$

By the linear independence of the exponentials, we finally obtain

$$\sum_{\ell,j=1}^{m} c_\ell c_j \sum_{\nu=1}^{k} g(a_\nu(x_\ell - x_j)) > 0 . \qquad \blacksquare$$

Sun's concept of an incidence matrix (called a characteristic matrix in [42]) actually provides a characterization of NI-sets, as the next result shows.

Theorem 4.10. Let $A = \{a_1, \ldots, a_k\}$ and $X = \{x_1, \ldots, x_m\}$ be subsets of $\mathbb{R}^n$. If the incidence matrix $In(A, X)$ is not positive definite, then X is an NI-set.

Proof: If $In(A, X)$ is not positive definite, then there exists a non-trivial vector $c \in \mathbb{R}^m$ such that $c^T In(A, X) c = 0$. The proof of 4.7 shows that for any $1 \leq \nu \leq k$,

$$\sum_{\beta \in \mathcal{R}(\tau_\nu)} \left(\sum \{c_i : \tau_\nu(i) = \beta\} \right)^2 = 0 .$$

Now consider the functional $\sum_{i=1}^{m} c_i \widehat{x}_i$ where $\widehat{x}_i$ represents the point evaluation at x_i, $1 \leq i \leq m$. Then for any ridge function

$$F(x) = \sum_{\nu=1}^{k} f_\nu(a_\nu x) , \qquad x \in \mathbb{R}^n ,$$

we have

$$\left(\sum_{i=1}^{m} c_i \widehat{x}_i \right)(F) = \sum_{i=1}^{m} c_i \sum_{\nu=1}^{k} f_\nu(a_\nu x_i) = \sum_{\nu=1}^{k} \sum_{i=1}^{m} c_i f_\nu(a_\nu x_i) .$$

Fix $1 \leq \nu \leq k$ and write

$$\sum_{i=1}^{m} c_i f_\nu(a_\nu x_i) = \sum_{\beta \in \mathcal{R}(\tau_\nu)} \left(\sum \{c_i : \tau_\nu(i) = \beta\} \right) f_\nu(a_\nu x_\beta) = 0 .$$

This effectively demonstrates that X is an NI-set, since if data $\alpha_1, \ldots, \alpha_m$ are prescribed at $x_1, \ldots, x_m$, then there must be the following dependence amongst the α_i, $1 \leq i \leq m$:

$$\sum_{i=1}^{m} c_i \alpha_i = \sum_{i=1}^{m} c_i F(x_i) = \sum_{i=1}^{m} c_i \sum_{\nu=1}^{k} f_\nu(a_\nu x_i) = \sum_{\nu=1}^{k} \sum_{i=1}^{m} c_i f_\nu(a_\nu x_i) = 0 . \qquad \blacksquare$$

§5. Appendix: Neural Networks

In this appendix we provide a brief introduction to the field of neural networks. Our main objective is to identify the link between the foregoing approximation problems and this rapidly expanding field.

For many years, computers have been able to perform symbolic manipulation faster and more reliably than humans. Yet the human capacity to learn and to organize data easily outstrips that of any machine. Of course, the human brain has a radically different architecture from most computers. Even the most advanced computers have a limited number of sophisticated processors linked together in a fairly naive fashion. In contrast, the brain may be described as a "massively parallel" distributed architecture system in which each individual processor (a neuron) has very limited computational power. This provides the motivation for the study and construction of neural networks. Basically, a neural network consists of a large number of simple processors (hereafter referred to as "units") connected in a complex fashion. This idea of using the human brain as a model for computer architecture has its roots in such papers as McCullogh and Pitts [28]. In the late 1960's there was a flurry of activity in this area, which was effectively terminated by the book of Minsky and Papert [30] with the title *Perceptrons*. In order to get some concepts established, let us consider a multilayer feedforward network.

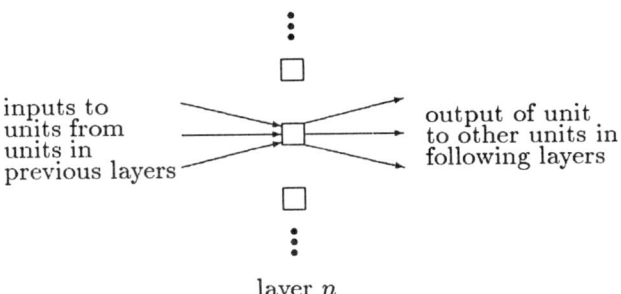

Figure E. A portion of a feedforward network.

Any network consists of *units* and *links* between units. Figure E shows a portion of such a network. Each layer consists of a string of units (which from the mathematical point of view can contain infinitely many such units). In a multilayer feedforward network, there are many such layers. The force of the term feedforward is that a unit in layer n can receive inputs only from layers 1 to $n-1$, and can transmit output only to layers $n+1$ and higher. No *backward* output is allowed and no *forward* input. Each link in the network is weighted numerically, and when a unit outputs along a given link, the value stored in the unit is multiplied by the weight and transmitted along the link.

Each unit has the capability to modify the input in some simple fashion (usually by a univariate function) before outputting to units *ahead* in the network. Figure F illustrates the typical action of a single unit. Inputs

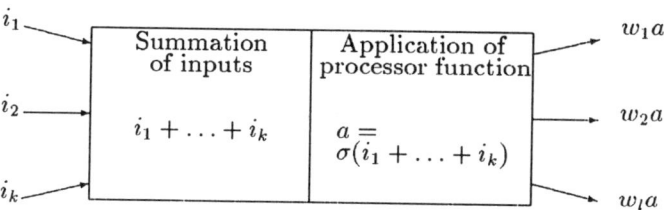

Figure F. A single unit in a neural network.

$i_1, \ldots, i_k$ are received, they are summed in the unit, the *processor function* is applied, and then *weighted* outputs are sent on. Usually, all units in the network have the same or similar processor functions and two very common choices are the sigmoidal functions

$$\sigma(t) = \begin{cases} 1 & \text{if } t \geq 0 \\ 0 & \text{if } t < 0 \end{cases} \quad \text{and} \quad \sigma(t) = \frac{e^t}{1 + e^t} \; .$$

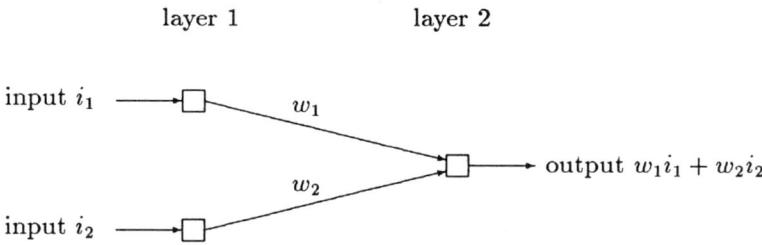

Figure G. A simple neural network.

The work of Minsky and Papert [30] in 1969 showed that a simple network could not recover a simple function. They took the network illustrated in Figure **G**, where the processor function for each unit is the identity function. Such a network cannot model the Boolean function given by

$$f(0,0) = f(1,1) = 0 \quad \text{and} \quad f(0,1) = f(1,0) = 1 \;,$$

since the second pair of conditions demands $\omega_1 = \omega_2 = 1$, while the condition $f(1,1) = 0$ demands $\omega_1 + \omega_2 = 0$. This example effectively stopped neural network research in its tracks until it was realised that continuous activation functions and additional layers in the network improved its performance dramatically.

The normal situation now is that a neural network has at least 3 layers — an input layer, an output layer and a *hidden* layer. Each unit is restricted to the simple tasks of summation of the input, followed by the application of a univariate function. However, units are often given individuality by allowing the processor function to depend on a single real parameter – usually involving the shifting of the processor function. Thus unit i might have the processor function $x \longmapsto \sigma(x - \theta_i)$ where $\sigma \in C(\mathbb{R})$ is a sigmoidal function and $\theta_i \in \mathbb{R}$.

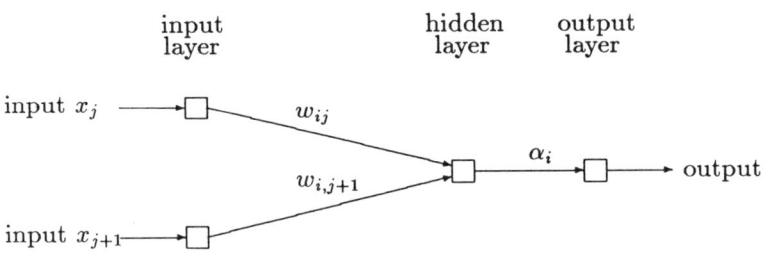

Figure H. Part of a three layer network.

A three layer network is illustrated in Figure H. It is a feedforward network with one hidden layer. As is conventional, the input layer units have the identity processor function, so they pass data straight through to the hidden layer. The output unit also has identity processor function. If there are n input units and N hidden units each with activation function $x \longmapsto \sigma(x - \theta_i)$, then the output of the network is

$$OUT(x_1, \ldots, x_n) = \sum_{i=1}^{N} \alpha_i \sigma \left(\sum_{j=1}^{n} \omega_{ij} x_j - \theta_i \right).$$

If the n inputs are the elements of a vector in $\mathbb{R}^n$, then OUT can be used to model a real-valued function on $\mathbb{R}^n$. Setting $x = (x_1, \ldots, x_n)$ and $\omega_i = (\omega_{i1}, \ldots, \omega_{in})$, we can write

$$OUT(x) = \sum_{i=1}^{N} \alpha_i \sigma(\omega_i x - \theta_i),$$

where $\omega_i x$ is the usual inner product of the vectors ω_i and x in $\mathbb{R}^n$. Now the question of the existence of weights ω_i and α_i, $1 \leq i \leq N$, plus shifts θ_i, $1 \leq i \leq N$, such that OUT represents a given function to a required accuracy is precisely the material dealt with in Section 3.

In practical applications, the idea is that the neural network "learns" the given function by being "trained." Suppose one has a good estimate of the number of units N needed in the hidden layer. Then the network is initially "dumb" in the sense that no weights w_i or α_i are prescribed. The network is then given a set of "training data," that is, a set of inputs and the corresponding outputs. The weights are then chosen to minimise the error between the actual output and the intended output. Two difficulties arise here. Firstly, the number of hidden units needed for a required accuracy may be very large. This in turn generates a large number of weights. Secondly, a large "training set" may be needed to determine the values of the weights, generating in turn a huge optimization problem with perhaps many local minima, slowly descending valleys and stationary features.

Some of the above difficulties may be alleviated if the topology of the network is allowed to become more complex. In addition, other techniques may be borrowed from optimization (for example, simulated annealing).

We conclude by briefly describing the work of Lenze [24]. Let x be the vector $(u_1, \ldots, u_n) \in \mathbb{R}^n$ and $a = (\alpha_1, \ldots, \alpha_n) \in \mathbb{R}^n$. Instead of the usual ridge function approximation

$$x \mapsto \sigma\left(\sum_{i=1}^{n} \alpha_i u_i\right),$$

Lenze uses the 'hyperbolic' function

$$x \mapsto \sigma\left(\prod_{i=1}^{n}(u_i - \alpha_i)\right).$$

Then, by using convolutions involving Lebesgue-Stieltjes integrals and quadrature, the following rather complicated form of approximation is produced:

$$\sum_{i=1}^{N} \beta_i \sigma\left(\sum_{k=1}^{n}\sum w_{j_1 \ldots j_k} u_{j_1} \ldots u_{j_k} - \theta_j\right)$$

where $x = (u_1, \ldots, u_n) \in \mathbb{R}^n$, β_i, θ_j, $w_{j_1 \ldots j_k}$ are in $\mathbb{R}$. This is in fact the sort of output generated by a neural network of the type illustrated in Figure I.

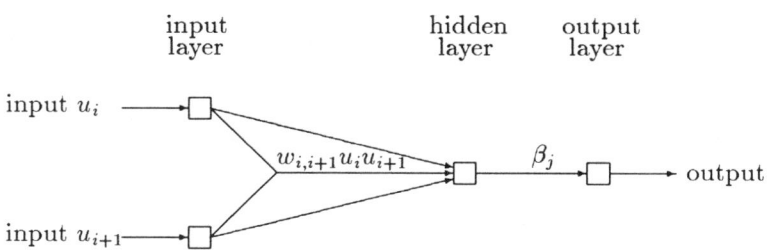

Figure I. Neural network for Lenze's approximation.

The novelty of this network is that input units are allowed to make weighted combinations of their outputs before passing these to the hidden layer. This is still a feedforward network, but the topology is considerably more complex than that previously considered.

Acknowledgements. It is a pleasure to acknowledge the help of various people in the compilation of this paper. Firstly, I had many useful conversations with my collaborators E.W. Cheney, X. Sun and Y. Xu. H. Mhaskar and C. Chui provided me with preprints of their work, as did A. Pinkus, T. Chen and B. Lenze. Secondly, Mrs. S. Carrington nobly slaved away with my unfamiliar handwriting and foibles, and is responsible for the production of this beautiful manuscript. Thirdly, I have profited from stimulating conversations with my colleague at Leicester, Dr. J. Levesley.

References

1. Barron, A.R., Universal approximation bounds for superpositions of a sigmoidal function, Technical Report No. 58, Dept. of Statistics, University of Illinois, Urbana-Champaign, 1991.
2. Braess, D. and A. Pinkus, Interpolation by ridge functions, J. Approx. Theory, (to appear).
3. Bressloff, P.C. and D.J. Weir, Neural Networks, GEC Journal of Research, **8** (1991), 151–169.
4. Broomhead, D. and D. Lowe, Multivariable functional interpolation and adaptive networks, Complex Systems **2** (1988), 321–355.
5. Chen, T., H. Chen and R. Liu, A constructive proof and extension of Cybenko's approximation theorem in *Computing Science and Statistics: Proceedings of the 22nd Symposium on the Interface* Springer-Verlag, (1991), 163-168.
6. Cheney, E.W. *Introduction to Approximation Theory*, McGraw-Hill, New York, 1966.
7. Chui, C. and X. Li, Approximation by ridge functions and neural networks with one hidden layer, J. Approx. Theory (to appear).
8. Chui, C. and X. Li, Realization of neural networks with one hidden layer, to appear in *Georg Fredrich Bernhard Riemann: a mathematical legacy* Rassias, T.M. and H.M. Srivastava (eds.), World Sci. Publ. Corp., 1992.
9. Cybenko, G., Approximation by superpositions of a sigmoidal function, Math. Control Signals Systems, **2** (1989), 303–314.
10. Cybenko, G., Continuous valued neural networks: approximation theoretic results, in *Proceedings of the 1988 Interface of Statistics and Computer Science, Reston, Virginia, 1988*, pp. 174–184.
11. Cybenko, G., Continuous valued neural networks with two hidden layers are sufficient, Center for Supercomputing Research Report No. 935, University of Illinios, 1988.

12. Cybenko, G., Mathematical problems in neural computing, in *Signal Processing, Scattering, Operator Theory, and Numerical Methods*, Birkhauser, Boston, 1990, pp. 47–63.
13. Dahmen, W. and C. Micchelli, Some remarks on ridge functions, Approx. Theory and its Appl. **3** (1987), 139–143.
14. Diaconis, P. and M. Shashahani, On nonlinear functions of linear combinations, SIAM J. Sci. Statist. Comput. **5** (1984), 175–191.
15. Dyn, N., W.A. Light and E.W. Cheney, Interpolation by piecewise-linear radial basis functions, J. Approx. Theory **59** (1989), 202–223.
16. Dyn, N. and C. Micchelli, Interpolation by sums of radial functions, Numer. Math. **58** (1990), 1–9.
17. Funahashi, K., On the approximate realisation of continuous mappings by neural networks, Neural Networks **2**, (1989), 183–192.
18. Gorman, R.P. and T.J. Sejnowski, Analysis of hidden units in a layered network trained to classify sonar targets, Neural Networks **1**, (1988), 75.
19. Hartman, E.J., J.D. Keeler and J.M. Kowalski, Layered neural networks with gaussian hidden units as universal approximations, Neural Computation **2** (1990), 210–215.
20. Hopfield, J.J. and D.W. Tank, Neural computation of decisions in optimisation problems, Biological Cybernetics **52** (1985), 141.
21. Hornik, K., M. Stinchcombe and H. White, Multilayer feedforward networks are universal approximators, Neural Networks **2** (1989), 359–366.
22. Jackson, I.R.H., Radial basis function methods for multivariable approximation, Ph.D. Dissertation, University of Cambridge, 1988.
23. John, F. *Plane waves and spherical means applied to partial differential equations*, Wiley-Interscience, New York, 1955.
24. Lenze, B., Constructive multivariate approximation via sigmoidal functions, Preprint, 1992.
25. Light, W.A. Some aspects of radial basis function approximation, to appear in *Splines and approximation theory*, S. Singh (ed.), North-Holland, 1992.
26. Lin, V. Ya. and A. Pinkus, Fundamentality of ridge functions, Math. Dept. Technical Report, Technion, I.I.T., Haifa, Israel, 1991.
27. Madych, W.R. Summability and approximate reconstruction from Radon transform data, Contemp. Math. **113** (1990), 189–219.
28. McCullough, W.S. and W. Pitts, Logical calculus of the ideas immanent in nervous activity, Bull. Math. Biophys., **3** (1943), 115–133.
29. Mhaskar, M. and C. Micchelli, Approximation by superposition of a sigmoidal function, Preprint, 1991.
30. Minsky, M. and S. Papert, *Perceptrons*, MIT Press, 1969.
31. Moody, J. and C.J. Darken, Fast learning in networks of locally-tuned processing units, Neural Computation **1** (1989), 281–294.
32. Roberts, S. and L. Tarassenko, A new method of automated sleep quantification, Oxford University Engineering Laboratory Report No. 1875, 1991.
33. Rudin, W. *Functional Analysis*, (2nd ed), McGraw-Hill, 1973.

34. Rumelhart, D.E., G.E. Hinton and J.L. McCelland, *Parallel distributed processing: explorations in the microstructure of cognition*, Vol I: Foundations, MIT Press, 1986.
35. Saarinen, S., R. Bramley and G. Cybenko, The numerical solution of neural network training problems, Center for Supercomputing Research Report No. 1089, University of Illinios, 1991.
36. Schoenberg, I.J., Metric spaces and completely monotone functions, Ann. Math. **39** (1938), 811–841.
37. Schoenberg, I.J., On certain metric spaces arising from Euclidean spaces..., Ann. Math. **36** (1937), 787–793.
38. Schumaker, L.L., *Spline Functions: Basic Theory*, Wiley, New York, 1981.
39. Stein, E. and G. Weiss, *Introduction to Fourier analysis on Euclidean spaces*, Princeton University Press, Princeton, 1971.
40. Stinchcombe, M. and H. White, Universal approximation using feedforward networks with non-sigmoid hidden layer activation functions, Proceedings of International Conference on Neural Networks, (1989), 613–617.
41. Stinchcombe, M. and H. White, Approximating and learning unknown mappings using multilayer feedforward networks with bounded weights, IEEE International Conference on Neural Networks, **3** (1990), 7–16.
42. Sun, X. Ridge function spaces and their interpolation property, J. Math. Anal. Appl. , (to appear).
43. Sun, X. and E.W. Cheney, The fundamentality of sets of ridge functions, Aequationes Math. (to appear), 1992.
44. Xu, Y., W.A. Light and E.W. Cheney, On kernels and approximation orders, in *Approximation Theory*, G. Anastassious (ed.), Marcel Dekker, New York, 1992, 227–242.
45. Xu, Y., W.A. Light and E.W. Cheney, Constructive methods of approximation by ridge functions and radial functions, Preprint 1991.

Will Light
Mathematics Department
University of Leicester
Leicester LE1 7RH
England
pwl@leicester.ac.uk

Knot Removal for Spline Curves and Surfaces

Tom Lyche

Abstract. We give a survey of techniques for removing knots from a spline without perturbing the spline more than a given tolerance. Knot removal can be used for data compression and for adaptive approximation of functions and data.

§1. Introduction

In areas like cartography [6], seismic processing [9,19], image processing [12, 30], and CAGD [20] one often encounters curves and surfaces defined using a huge amount of data. Evidently, there is a need for approximation methods which can handle such problems and reduce the number of parameters involved. In this paper we discuss a data compression technique known as knot removal. Briefly the knot removal methods work as follows. Given a spline function with many knots, the object is to remove as many knots as possible without perturbing the spline more than a given tolerance. This is different from the wavelet decomposition approach used for example in [10,11]. Knot removal methods were introduced for univariate functions in [26], and later extended to parametric curves and tensor product surfaces in [27]. An extension to piecewise polynomials on triangulations is given in [23,24].

The methods based on knot removal can also be used for adaptive approximation. Knot removal differs from conventional methods, found for example in [15], in one important aspect. Instead of starting with an approximation with few parameters and adaptively introducing more, we start with lots of parameters and remove those that are not needed for a satisfactory approximation.

The outline of the paper is as follows: After collecting some background material on splines in the next section we present in Section 3 the discrete approach to L^2 approximation used in this paper. The following sections then consider knot removal for various classes of splines; for functions in Section

4, parametric curves in Section 5, tensor product- and parametric surfaces in Section 6, and piecewise polynomials on triangulations in Section 7. Section 8 discusses a discrete approach to constrained spline approximation. Finally we discuss constrained knot removal in Section 9.

A Fortran implementation of the knot removal algorithms in [26,27] will be available on Netlib.

§2. Splines and Coefficient Norms

We will consider both univariate and multivariate piecewise polynomials. In the univariate case we will work with the usual space of spline functions of order k on a knot vector $t = (t_1, \ldots, t_{n+k})$

$$S_{k,t} = \left\{ \sum_{i=1}^{n} c_i B_i, \quad c_i \in \mathbb{R} \right\}. \tag{2.1}$$

Here $B_i = B_{i,k} = B_{i,k,t}$ are the usual B-splines on t of order k normalized so that

$$\sum_i B_i(x) = 1, \quad \text{for all} \quad x \in [t_k, t_{n+1}]. \tag{2.2}$$

(The B-splines are assumed to be right continuous on $[t_k, t_{n+1})$ and left continuous at t_{n+1}.)

To simplify the discussion we will in this paper restrict ourselves to knot vectors $t = (t_1, \ldots, t_{n+k})$ satisfying

1. The components of t are nondecreasing,
2. $n \geq k$, $t_1 = \cdots = t_k = a$, and $b = t_{n+1} = \cdots = t_{n+k}$,
3. $t_{i+k} > t_i$, $i = 1, 2, \ldots, n$.

We say that t is a k-extended knot vector on $[a, b]$ if 1.,2., and 3. hold. The knots $t_{k+1}, \ldots, t_n$ are called interior knots. We will work extensively with multiple knots, and use the symbol $m_t(x)$ to denote the total number of times the real number x occurs as a component in the vector t.

We need to combine vectors in various ways. The following definition makes this precise.

Definition 2.1. *If τ and t are any knot vectors we say that $\tau \subset t$ if $m_\tau(x) \leq m_t(x)$ for all real numbers x. We define vectors $\tau \cup t, t \setminus \tau, \tau \wedge t$ and $\tau \vee t$ to have nondecreasing components and satisfy*

$$m_{\tau \cup t}(x) = m_\tau(x) + m_t(x), \quad m_{t \setminus \tau}(x) = \max\{m_t(x) - m_\tau(x), 0\},$$
$$m_{\tau \wedge t}(x) = m_t(x) \wedge m_\tau(x) = \min\{m_\tau(x), m_t(x)\},$$
$$m_{\tau \vee t}(x) = m_\tau(x) \vee m_t(x) = \max\{m_\tau(x), m_t(x)\}.$$

We note that if $\tau \subset t$ then $S_{k,\tau} \subset S_{k,t}$. In [28] it is shown that for knot vectors τ, t

$$S_{k,\tau \wedge t} = S_{k,\tau} \cap S_{k,t}, \quad S_{k,\tau \vee t} = S_{k,\tau} + S_{k,t},$$

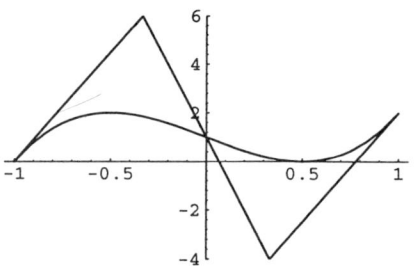

Figure 2.1. Spline with control polygon.

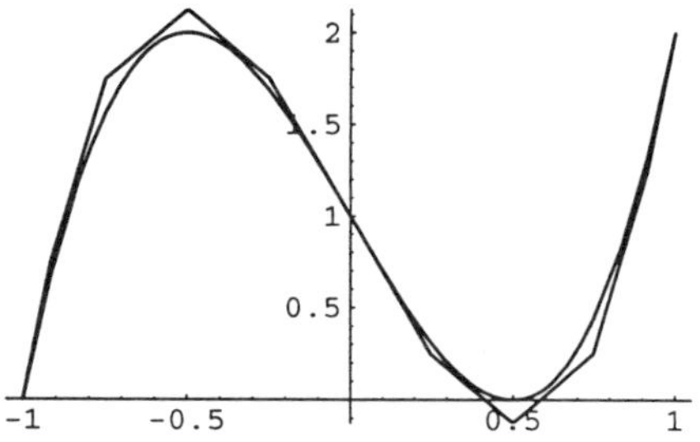

Figure 2.2. Spline with refined control polygon.

with

$$S_{k,\tau} + S_{k,t} = \{f + g : f \in S_{k,\tau} \text{ and } g \in S_{k,t}\}$$

denoting the sum of the two spaces. In this paper k, the interval $[a, b]$, and the end knots will be fixed. Only the interior knots will vary from knot vector to knot vector.

With $f \in S_{k,\tau}$ and $\tau \subset t$ we use the notation $\mathbf{c}(f, S)$ to denote the vector of B-spline coefficients of f in $S = S_{k,t}$. The *control polygon* of f in S, is the broken line connecting the points $(t_i^*, c_i), i = 1, \ldots, n$, where

$$t_i^* = (t_{i+1} + \cdots + t_{i+k-1})/(k-1) \tag{2.3}$$

are the knot averages, and the c_i are the B-spline coefficients of f in S.

"The B-spline coefficients model the function they represent." This quote from [3] could serve as a motto for the curve and surface fitting method presented in this paper. The quote is illustrated for the polynomial $f(x) = 4x^3 - 3x + 1$ in Figures 2.1 and 2.2. In Figure 2.1 the control polygon on the knot vector $\tau = (-1, -1, -1, -1, 1, 1, 1, 1)$ does not model the behavior of f very well. The points $(-1/3, 6)$ and $(1/3, -4)$ on the control polygon are quite far from the curve. In Figure 2.2 we have represented the same f together with its control polygon on a knot vector t obtained from τ by adding knots at $-3/4, -2/4, 0, 1/4, 2/4, 3/4$. The control polygon now gives a more accurate approximation of f. We conclude that the quote above must be taken with a grain of salt. It might be necessary to insert some knots in order to obtain a satisfactory model. We return to this example again in Section 9 when discussing constraints.

The following lemma shows that the coefficients of a cubic spline are always close to the spline if the knot spacing is small.

Lemma 2.1. *For any cubic spline $f = \sum_j c_j B_{j,4,t}$ we have*

$$|f(t_j^*) - c_j| \leq \frac{h^2}{6} \|D^2 f\|_{L^\infty(t_{j+1}, t_{j+3})},$$

where

$$h^2 = \frac{(\Delta t_{j+1})^2 + (\Delta t_{j+1})(\Delta t_{j+2}) + (\Delta t_{j+2})^2}{3}, \qquad \Delta t_i = t_{i+1} - t_i.$$

This result is proved in [21]. The constant $1/6$ in front of h^2 is best possible. For general k we also have $|c_i - f(t_i^*)| = O(h^2)$, where $h = \max(t_{j+1} - t_j)$, see [3,8,13].

For any function $f \in S = S_{k,\mathbf{t}}$ we recall the stability estimate

$$D_k^{-1}\|\mathbf{c}(f,S)\|_{p,\mathbf{t}} \leq \|f\|_{L^p} \leq \|\mathbf{c}(f,S)\|_{p,\mathbf{t}}, \tag{2.4}$$

for some positive constant D_k depending only on k ([3]). Here $\|\cdot\|_{L^p}$ is the classical L^p-norm on $[a,b]$ and

$$\|\mathbf{c}\|_{p,\mathbf{t}} = \|f\|_{\ell^p,\mathbf{t}} = \begin{cases} \left(\sum_i |c_i|^p (t_{i+k} - t_i)/k\right)^{1/p}, & 1 \leq p < \infty \\ \max_i |c_i|, & p = \infty, \end{cases} \tag{2.5}$$

is the weighted discrete ℓ^p-norm for splines. Thus, the L^p and $\ell^p, \mathbf{t}$ norms are equivalent. It was shown in [26] that

$$\|f\|_{L^p} \leq \|f\|_{\ell^p,\mathbf{t}} \leq \|f\|_{\ell^p,\boldsymbol{\tau}}, \quad \text{if} \quad \boldsymbol{\tau} \subset \mathbf{t}, \quad 1 \leq p \leq \infty, \tag{2.6}$$

and that

$$\|f\|_{\ell^2,\mathbf{t}} - \|f\|_{L^2} \leq \sqrt{2(t_{n+k} - t_1)}D_k\omega(f,h), \tag{2.7}$$

where $h = \max_j(t_{j+k-1} - t_{j-k+1})$ and $\omega(f,h) = \max\{|f(x) - f(y)| : x, y \in [t_1, t_{n+k}], |x-y| \leq h\}$ is the usual modulus of continuity. Thus the discrete norms approximate the continuous norms in a monotone way.

§3. Discrete ℓ^2 Approximation for Splines

In this section we give a method to approximate a spline with many knots by one with possibly fewer knots. The method was first considered in [25]. Rather that using the L^2 norm we propose in view of (2.7) to compute a best approximation in the $\ell^2, \mathbf{t}$ norm for a suitable $\mathbf{t}$. Suppose we are given two k-extended knot vectors $\boldsymbol{\rho}, \boldsymbol{\tau}$ on $[a,b]$, and an $f \in S_{k,\boldsymbol{\rho}}$. We want to find an approximation g to f from $S_{k,\boldsymbol{\tau}}$. Let $\mathbf{t}$ be a k-extended knot vector on $[a,b]$ containing $\boldsymbol{\rho} \vee \boldsymbol{\tau}$. We define $g = Gf = G(\boldsymbol{\tau}, \mathbf{t})f \in S_{k,\boldsymbol{\tau}}$ as the solution of the least squares problem

$$\min\{\|h - f\|_{\ell^2,\mathbf{t}} : h \in S_{k,\boldsymbol{\tau}}\}. \tag{3.1}$$

To discuss computational aspects of this problem we need to introduce the matrix $\mathbf{A} = \mathbf{A}_{k,\boldsymbol{\tau},\mathbf{t}}$ which transforms for $\boldsymbol{\tau} \subset \mathbf{t}$ the B-spline coefficients of a function h in $S_{k,\boldsymbol{\tau}}$ into the B-spline coefficients of h in $S_{k,\mathbf{t}}$. $\mathbf{A}$ is a rectangular matrix with $\dim(S_{k,\mathbf{t}})$ rows and $\dim(S_{k,\boldsymbol{\tau}})$ columns. It is known as the *knot insertion matrix of order k from $\boldsymbol{\tau}$ to $\mathbf{t}$*. In symbols

$$\mathbf{c}(h, S_{k,\mathbf{t}}) = \mathbf{A}\mathbf{c}(h, S_{k,\boldsymbol{\tau}}).$$

Since this is a change of basis transformation, $\mathbf{A}$ has full rank. We have

$$\|g - f\|_{\ell^2,\mathbf{t}} = \|\mathbf{E}_\mathbf{t}^{1/2}(\mathbf{A}\mathbf{c} - \mathbf{b})\|_2, \tag{3.2}$$

where $\mathbf{c} = c(g, S_{k,\tau})$ are the B-spline coefficients of the unknown best approximation g, $\mathbf{A}$ is the knot insertion matrix, $\mathbf{b} = c(f, S_{k,t})$, and $\mathbf{E}_t^{1/p}$ is a diagonal scaling matrix with diagonal elements given by

$$e_{i,i} = \begin{cases} [(t_{i+k} - t_i)/k]^{1/p}, & \text{for } 1 \le p < \infty, \\ 1, & \text{for } p = \infty. \end{cases} \quad (3.3)$$

From (3.2) it follows that (3.1) is a full rank linear least squares problem. The scaled matrix

$$\mathbf{F} = \mathbf{E}_t^{1/2} \mathbf{A} \mathbf{E}_t^{-1/2}$$

is well-conditioned, at least for moderate k. Indeed, with $\mathbf{F}^\dagger = (\mathbf{F}^T \mathbf{F})^{-1} \mathbf{F}^T$ we have ([26])

$$\|\mathbf{F}\|_2 \|\mathbf{F}^\dagger\|_2 \le D_k, \quad (3.4)$$

the constant in (2.4). Another virtue of the least squares problem (3.1) is that the solution is more local than best L^2 approximation. These and further aspects of the problem are discussed in [25,26]. Computationally (3.1) is easily solved. The matrix $\mathbf{A}$ is banded and the elements are conveniently generated row-wise using the Oslo algorithm [7]. Since $D_4 \approx 10$ we have, in view of (3.4), solved problem (3.1) for moderate k by Cholesky factorization of the normal equations.

As an example of this least squares process consider data $(x_i, y_i)_{i=1}^m$, where $x_1 < x_2 < \cdots < x_m$. As the spline f with many knots to be approximated we take the piecewise linear interpolant to the data written as a cubic spline with triple interior knots. Thus $f \in S_{4,\rho}$ where

$$\rho = (x_1, x_1, x_1, x_1, x_2, x_2, x_2, \ldots, x_{m-1}, x_{m-1}, x_{m-1}, x_m, x_m, x_m, x_m).$$

We can now choose τ and t and compute a possibly smooth approximation $g = G(\tau, t) f$. The B-spline representation of the piecewise linear f on t is given by

$$f(x) = \sum_{j=1}^n f(t_j^*) B_{j,4,t}(x), \quad t_j^* = \frac{t_{j+1} + t_{j+2} + t_{j+3}}{3}. \quad (3.5)$$

As a concrete example we consider the titanium heat data [3,p. 222], with $m = 49$. Inspired by the knot locations determined in [5] we choose τ to have 5 interior knots located at 830, 875, 905, 910, and 980. For t we use triple knots at every interior data point and an extra simple knot halfway between each of the data points. Adding 4-tuple knots at each end we obtain in this way a knot vector t with $4m + 1$ elements containing $\tau \vee \rho$. The solution of (3.1) is shown by a solid line in Figure 3.1. The τ knot locations have been indicated by dots under the top horizontal line. The doted line is the piecewise linear error curve $y_i - g(x_i), i = 1, 2, \ldots, 49$. The scale of the error curve is given along the vertical line to the right in the figure.

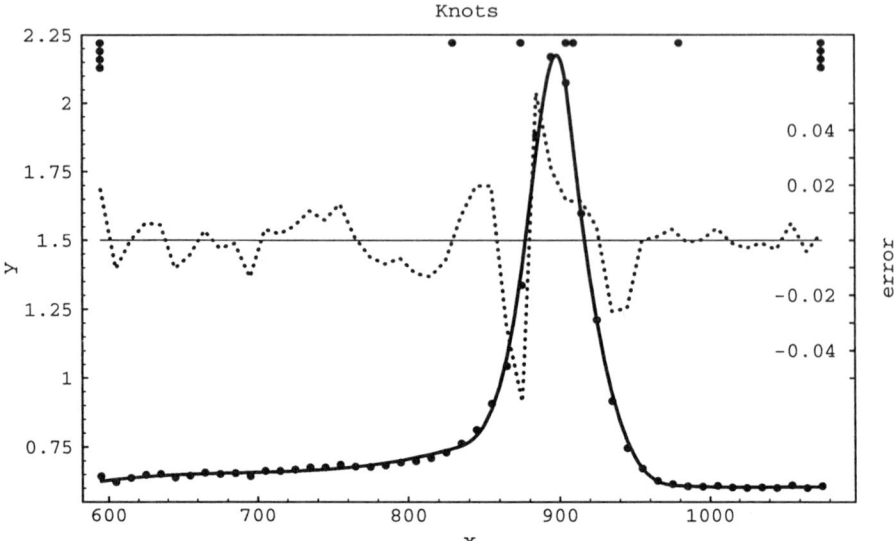

Figure 3.1. Least squares fit with 5 interior knots to titanium heat data.

This illustrates a process which can be used for fitting of possibly noisy data $(x_i, y_i)_{i=1}^m$, where m is large. The process consists of two main stages

1. Make a spline, say f, of the data. Examples are
   a) The piecewise linear interpolant to the data written as a higher order spline with multiple knots. (Cf. (3.5)).
   b) The variation diminishing spline approximation of Schoenberg $f = \sum_{i=1}^m y_i B_{i,4,\rho}$ with

   $$\rho = (x_1, x_1, x_1, x_1, x_3, x_4, \ldots, x_{m-2}, x_m, x_m, x_m, x_m).$$

   c) Cubic Hermite interpolation, $f = \sum_{i=1}^{2m} c_i B_{i,4,\rho}$ with

   $$\rho = (x_1, x_1, x_1, x_1, x_2, x_2, x_3, x_3, \ldots, x_{m-1}, x_{m-1}, x_m, x_m, x_m, x_m).$$

   $$c_{2i-1} = y_i - h_{i-1} d_i / 3 \quad c_{2i} = y_i + h_i d_i / 3,$$

   $$h_i = x_{i+1} - x_i, \quad x_0 = x_1, \quad x_{m+1} = x_m.$$

   The number d_i should be an approximation to the slope at x_i. For choices of d_i see [3].
   d) Local least squares or some other local approximation process.
2. Select τ and t and compute a best ℓ^2, t approximation to f from $S_{k,\tau}$.

§4. Knot Removal; Functions

In the previous section we computed an approximation g on a fixed knot vector τ. In this section we present a technique to determine a suitable knot vector automatically. We consider the following knot removal problem:

Problem 4.1. *Given a function $f \in S_{k,t}$ a norm $\|\ \|$ on $S_{k,t}$, and a tolerance $\epsilon > 0$. Find a subspace $S_{k,\tau}$ of $S_{k,t}$ of lowest possible dimension such that an element $g \in S_{k,\tau}$ can be found with $\|f - g\| \leq \epsilon$.*

In this section we describe the algorithm in [26] to find a subspace $S_{k,\tau}$ of low dimension and an element $g \in S_{k,\tau}$ so that $\|f - g\| \leq \epsilon$. The norm used in [26] is the ℓ^∞, t norm, but in practice any computable norm can be used.

In order to find τ we first assign a *ranking number* to each interior knot t_j of t. To find the ranking numbers we determine a *weight* w_j corresponding to t_j. The weight indicates the significance of the knot by measuring how well f can be approximated by a spline from $S_{k,t}$ without using the knot t_j. The weight is easy to define if $t_{j-1} < t_j$. We set

$$w_j = \min_{g \in S_j} \|f - g\|_{\ell^\infty, t} = \|f - f_j\|_{\ell^\infty, t}$$

where $S_j = S_{k, t \setminus t_j}$. If $t_{j-1} = t_j$ we do not define w_j in this way as it would lead to $w_j = w_{j-1}$. Assuming that w_{j-1} and f_{j-1} have been computed as just described and $t_{j-2} < t_{j-1}$, we set

$$w_j = w_{j-1} + \min_{g \in S_j} \|f_{j-1} - g\|_{\ell^\infty, t^{j-1}} \tag{4.1}$$

where $t^{j-1} = t \setminus \{t_j\}$ and $S_j = S_{k, t \setminus \{t_{j-1}, t_j\}}$. In general if

$$t^j = t \setminus \{t_{j-\mu+1}, \ldots, t_j\},$$

where

$$\mu = \max\{i \mid t_{j-i+1} = t_j\}$$

is the left multiplicity of t_j, we set $S_j = S_{k, t^j}$ and compute w_j by (4.1). To find the weight we have to solve an overdetermined linear system in the ℓ^∞ norm. A fast and stable algorithm for this purpose is given in [26]. Only $O(5k)$ number of arithmetic operations are required for each weight.

To find the ranking numbers we sort the knots into groups T_j given by

$$T_i = \begin{cases} \{t_j : 0 \leq w_j < \epsilon/2\}, & \text{if } i = 1, \\ \{t_j : 2^{i-3}\epsilon \leq w_j < 2^{i-2}\epsilon\}, & \text{if } i = 2, 3, \ldots. \end{cases}$$

The ranking number ν_j of t_j is now the unique integer such that $t_j \in T_{\nu_j}$. We also let $r_i = \sum_{j \leq i} \#T_j$ be the total number of knots in $T_1, T_2, \ldots, T_i, i = 1, 2, 3, \ldots$ (counting multiplicities).

Based on the ranking numbers and the r_i's we can now define reduced knot vectors u^j for $j = 0, 1, 2, \ldots$. We have $u^0 = t$, and u^j is defined from t

by removing j knots. More precisely, if $r_{i-1} < j \le r_i$ then we remove all the knots in $T_1, T_2, \ldots, T_{i-1}$, and if $v_1 \le v_2 \le \cdots \le v_q$ are the knots in T_i then we remove in addition $v_{i_1}, \ldots, v_{i_p}$ where $p = j - r_{i-1}$ and

$$i_\ell = \lfloor (q+1)(\ell - 1/2)/p + 1/2 \rfloor, \qquad \ell = 1, 2, \ldots, p.$$

Here $\lfloor x \rfloor$ is the largest integer $\le x$, and we see that i_ℓ is the number $(q+1)(\ell - 1/2)/p$ rounded to the nearest integer. In other words, we remove knots from T_i more or less uniformly on subscripts.

The reduced knot vector τ is now defined as $\tau = u^q$ where

$$q = \max\{j : \|f - G(u^j, t)f\|_{\ell^\infty, t} \le \epsilon\}. \tag{4.2}$$

(Recall that $G(\tau, t)f$ is the best ℓ^2, t approximation from $S_{k,\tau}$ to f.) The integer q in (4.2) can be found by a binary search. We try to remove half the knots in t and compute the best approximation in the ℓ^2, t norm. If the ℓ^∞, t error is less than ϵ then we try to remove 3/4th of the knots, otherwise we try to remove 1/4th of the knots. This process is continued until the exact number of knots that can be removed has been determined.

Suppose we have determined g and τ such that $\|f - g\| \le \epsilon$. If τ has no interior knots the knot removal was very successful. On the other extreme if $\tau = t$ we were not very successful. We can try again with a larger ϵ or stop (or maybe look for a better algorithm). In general we will be somewhere between these extremes. With the present ranking numbers we were unable to remove more knots. What we need is a new ranking strategy for the remaining knots. We base this ranking on g. Thus we compute new weights and ranking numbers for the reduced knot vector τ. This gives new reduced knot vectors $u^i = u^i(g)$ defined by removing i knots from τ, the knots of g. With these modified reduced knot vectors we try to remove more knots from f. Usually we can remove several more knots in this way. Applying this in an iterative fashion leads to the following algorithm:

1. $\tau^0 = t$; $g_0 = f$;   (Initialize)
2. **for** $j = 0, 1, 2, \ldots$
   1. **if** $|\tau^j| = 2k$ **then stop**;
   2. **determine** $u^i = u^i(g_j)$, for $i = 0, 1, 2, \ldots$;   (Rank)
   3. $\tau^{j+1} = u^q$ where $q = \max\{i : \|\|f - G(u^i, t)f\| \le \epsilon\}$;   (Remove)
   4. **if** $|\tau^{j+1}| = |\tau^j|$ **then stop**;
   5. $g_{j+1} = G(\tau^{j+1}, t)f$ (Approximate)

We can use any norm to measure the error in Statement 2.3 above. The use of one of the ℓ^p, t norms is convenient since this gives an upper bound on the

Knot Removal

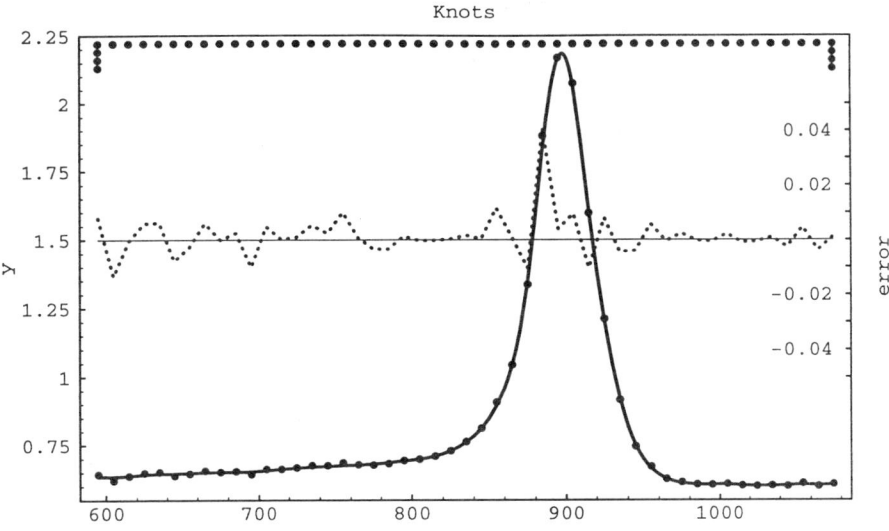

Figure 4.1. Smoothing spline fit to titanium heat data.

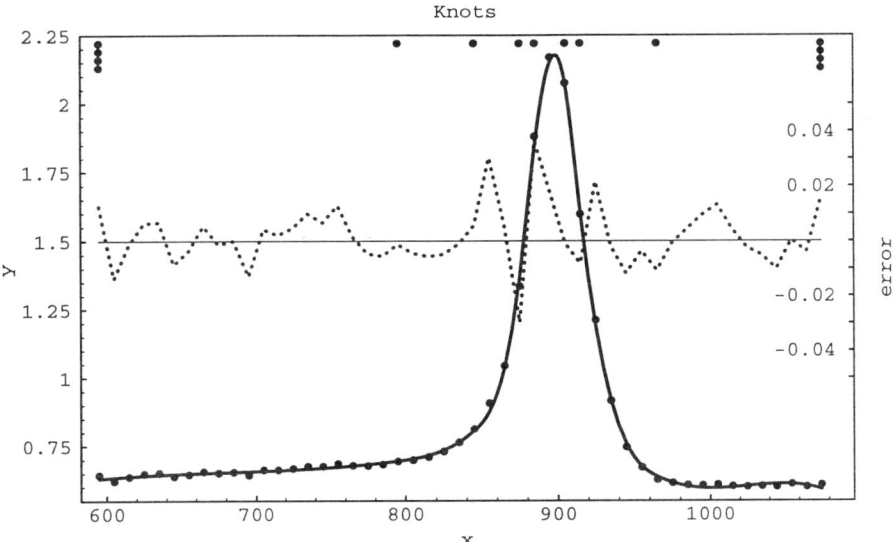

Figure 4.2. Knot removal on smoothing spline in Figure 4.1, $\epsilon = 0.04$.

L^p norm. Moreover, when the knots of t are quite dense this upper bound is not a bad overestimate.

As an example we consider again the titanium heat data. As the spline f with many knots we use the smoothing spline fit shown in Figure 4.1. This spline has a knot at every data point. The knot locations are shown as dots under the top horizontal line of the figure. The doted line is the piecewise linear error curve $y_i - f(x_i), i = 1, 2, \ldots, 49$. The scale of the error curve is given along the vertical line to the right. To compute this smoothing spline we used the Fortran package in [29].

Using the smoothing spline as input to our knot removal program with tolerance $\epsilon = 0.04$ we produced the spline in Figure 4.2. Seven interior knots remain. These are located at 795, 845, 875, 885, 905, 915, and 965.

Other examples on knot removal for functions can be found in [26].

§5. Knot Removal; Parametric Curves

We consider the following knot removal problem from [27]:

Definition 5.1. *Given a $d \geq 1$ dimensional parametric spline curve*

$$\mathbf{f} = (f^1, \ldots, f^d) = \sum_j (c_j^1, \ldots, c_j^d) B_{j,k,t} ,$$

and tolerances $\epsilon = (\epsilon^1, \ldots, \epsilon^d)$. Find a shortest possible $\tau \subset t$ such that $g^i \in S_{k,\tau}, i = 1, 2, \ldots, d$ exist with

$$\|f^i - g^i\|_{\ell^\infty, t} \leq \epsilon^i , \qquad i = 1, 2, \ldots, d . \tag{5.1}$$

The parametric curve $\mathbf{g} = (g_1, g_2, \ldots, g_d)$ will be an approximation to $\mathbf{f}$. To see how good this approximation is we need to introduce a measure for the distance between parametric curves. Suppose $\{\mathbf{f}(t) : t \in [a, b]\}$ and $\{\mathbf{g}(s) : s \in [c, d]\}$ are compact sets in $\mathbb{R}^d$. We can then define $d_H(\mathbf{f}, \mathbf{g})$ as the Hausdorff distance between these two pointsets

$$d_H(\mathbf{f}, \mathbf{g}) = \max \left\{ \max_{t \in [a,b]} \min_{s \in [c,d]} |\mathbf{f}(t) - \mathbf{g}(s)|, \max_{s \in [c,d]} \min_{t \in [a,b]} |\mathbf{f}(t) - \mathbf{g}(s)| \right\} .$$

Here $| \ |$ denotes the Euclidean distance in $\mathbb{R}^d$. A disadvantage with this measure is that irregular behavior like small loops and local change of orientation is not excluded. See [14] for a discussion and the introduction of an alternative distance called the *normal distance*.

Suppose now $\mathbf{f}$ and $\mathbf{g}$ are the parametric spline curves of the knot removal problem. Note that $\mathbf{g}$ inherits the parametrization of $\mathbf{f}$. If (5.1) holds then

$$d_H(\mathbf{f}, \mathbf{g}) \leq \max\{|\mathbf{f}(t) - \mathbf{g}(t)| : t \in [t_k, t_{n+1}]\} \leq |\epsilon| .$$

The approach to the knot removal problem taken in [27] only requires small modifications of the strategy explained in the previous section. We start by computing weights w_j^i and ranking numbers ν_j^i for all components f^i, and all interior knots t_j. We then let

$$\nu_j = \max_i \nu_j^i ,$$

be the ranking number for **f** at t_j. Using these ranking numbers we can find reduced knot vectors u^i precisely as in the function case. Once these vectors are determined the reduced knot vector τ is given by $\tau = u^q$ where

$$q = \min\{p : \|f^i - G(u^p, t)f^i\|_{\ell^\infty, t} \leq \epsilon^i, \quad i = 1, \ldots, d\}.$$

Finally we obtain the parametric curve **g** by

$$\mathbf{g} = G(\tau, t)\mathbf{f}.$$

As in the last section we apply this process iteratively by reranking the knots of **g** and computing new approximations to f until no more knots can be removed. Several examples of this process can be seen in [27].

§6. Knot Removal; Tensor Product Surfaces

For simplicity we consider only functions of two variables. A tensor product B-spline surface of order (k_1, k_2) with knot vectors s, t can be written

$$F(u, v) = \sum_{i=1}^{M} \sum_{j=1}^{N} c_{i,j} B_{i,k_1,s}(u) B_{j,k_2,t}(v). \tag{6.1}$$

We denote the space of such functions by $S_{k_1,k_2,s,t}$. More compactly,

$$F(u, v) = \mathbf{b}_s(u)^T \mathbf{C} \mathbf{b}_t(v),$$

where

$$\mathbf{b}_s = (B_{1,k_1,s}, \ldots, B_{M,k_1,s})^T,$$
$$\mathbf{b}_t = (B_{1,k_2,t}, \ldots, B_{N,k_2,t})^T,$$

and **C** is a matrix with M rows and N columns containing the B-spline coefficients $c_{i,j}$. The coefficient norm of a tensor product spline surface is given by

$$\|F\|_{\ell^p, s, t} = \|\mathbf{E}_s^{1/p} \mathbf{C} \mathbf{E}_t^{1/p}\|_{\ell^p},$$

where the matrix $\mathbf{E}_t^{1/p}$ is a diagonal scaling matrix with diagonal elements given by (3.3), and $\|\mathbf{A}\|_{\ell^p}$ is the ℓ^p vector norm of the $M \times N$ matrix **A** viewed as a vector in $\mathbb{R}^{MN}$. (For $p = 2$ this is the Frobenius norm). For any $F \in S_{k_1,k_2,s,t}$ the stability estimate (2.4) becomes

$$D_{k_1,p}^{-1} D_{k_2,p}^{-1} \|F\|_{\ell^p, s, t} \leq \|F\|_{L^p} \leq \|F\|_{\ell^p, s, t}. \tag{6.2}$$

The knot removal problem for tensor product surfaces can be stated as follows. Given a tolerance $\epsilon > 0$ and a spline $F(u, v) \in S_{k_1,k_2,s,t}$, find $\sigma \subset s$, $\tau \subset t$ and $\Phi(u, v) \in S_{k_1,k_2,\sigma,\tau}$ such that $\|F - \Phi\|_{\ell^\infty, s, t} \leq \epsilon$.

The process in [27] is easily described in terms of knot removal for parametric curves. We start by choosing one of the directions, say s. Then we apply knot removal to the parametric curve $\mathbf{f}_s = \mathbf{b}_s^T \mathbf{C}$ with $\epsilon = (\epsilon/2,\ldots,\epsilon/2)$. The result is a knot vector $\sigma \subset s$ and a parametric curve $\mathbf{f}_\sigma = \mathbf{b}_\sigma^T \mathbf{C}'$. We then obtain a surface $F'(u,v) = \mathbf{b}_\sigma(u)^T \mathbf{C}' \mathbf{b}_t(v)$ which satisfies $\|F - F'\|_{\ell^\infty,s,t} \le \epsilon/2$. Knot removal for parametric curves is next applied to the curve $\mathbf{f}_t = \mathbf{C}' \mathbf{b}_t$ again using the tolerance $\epsilon = (\epsilon/2,\ldots,\epsilon/2)$. This leads to a knot vector $\tau \subset t$ and coefficients Γ of a parametric curve, $\mathbf{f}_\tau = \Gamma \mathbf{b}_\tau$. The reduced surface is now given by

$$\Phi(u,v) = \mathbf{b}_\sigma(u)^T \Gamma \mathbf{b}_\tau(v) .$$

For the total error we obtain

$$\|F - \Phi\|_{L^\infty} \le \|F - F'\|_{\ell^\infty,s,t} + \|F' - \Phi\|_{\ell^\infty,s,t}$$
$$\le \|F - F'\|_{\ell^\infty,s,t} + \|F' - \Phi\|_{\ell^\infty,\sigma,t} \le \epsilon ,$$

where we have used the inequalities (6.2), and (2.6).

An advantage of this approach is that it requires little programming effort. A disadvantage is the lack of symmetry between the two parameter directions. An alternative approach is to compute ranking numbers for each of the parametric curves $\mathbf{f}_s = \mathbf{b}_s^T \mathbf{C}$ and $\mathbf{f}_t = \mathbf{C} \mathbf{b}_t$. We can then sort the union of these ranking numbers. In this case we have to solve slightly more complicated least squares problems. We refer to [27] for more details.

We have described knot removal for explicit surfaces. The method generalizes quite naturally to parametric surfaces. By a parametric surface we mean a surface of the form (6.1) where each $c_{i,j}$ is a vector in $\mathbb{R}^d$ for some positive integer d. Typically we have $d = 3$. To obtain ranking numbers for the two knot vectors we consider curves of the type $\mathbf{f}_s = \mathbf{b}_s^T \mathbf{C}$ which we think of as d parametric curves $(f_s^1,\ldots,f_s^d)$ in $\mathbb{R}^M$. Given a tolerance vector $\epsilon \in \mathbb{R}^d$ we can then (for each interior knot s_j of s) compute ranking numbers ν_j^i for each of the d parametric curves $(f_s^1,\ldots,f_s^d)$. To obtain a single ranking number ν_j we set $\nu_j = \max_i \nu_j^i$ just as we did for parametric curves. We can now compute approximations to each of the d components of the surface.

There are numerous applications of knot removal for surfaces. One important application is to convert a surface given in any format to a tensor product B-spline format. Our approach to this problem is resampling and knot removal. We evaluate the surface at a dense rectangular grid. We can then proceed as described earlier in the univariate case. First we compute say a bicubic spline approximation to this gridded data by some local method. Examples of such methods are bilinear interpolation, bicubic interpolation with gradient estimation, or a variation diminishing approximation. We then remove knots from this tensor product spline. Several numerical examples can be found in [27].

Tensor product knot removal can be used to convert a surface defined on triangles into tensor product form. As an example we consider the 100 point data set sampled from the test function used by Franke, see [17], and [18] equation (3.1) and Figure 3. Using the package in [31] we found the Delauny triangulation of this data set and a C^1 piecewise quintic approximation. To convert this surface to tensor product format we sampled points on a 33×33

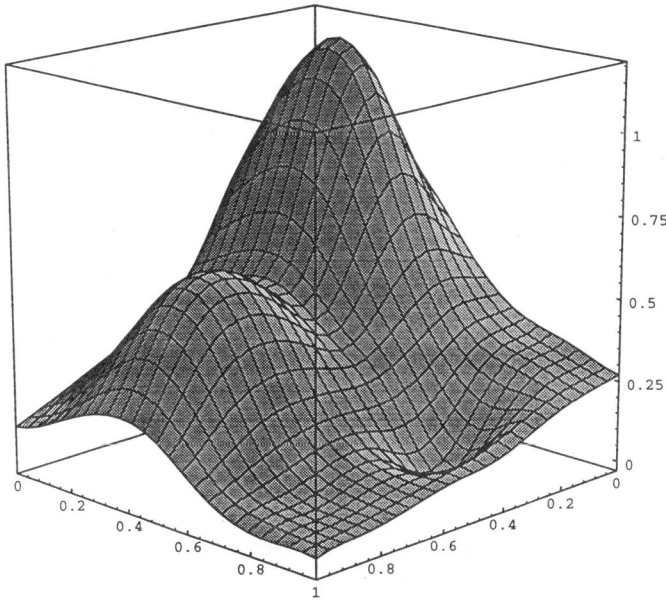

Figure 6.1. A bicubic spline surface constructed from 100 data points.

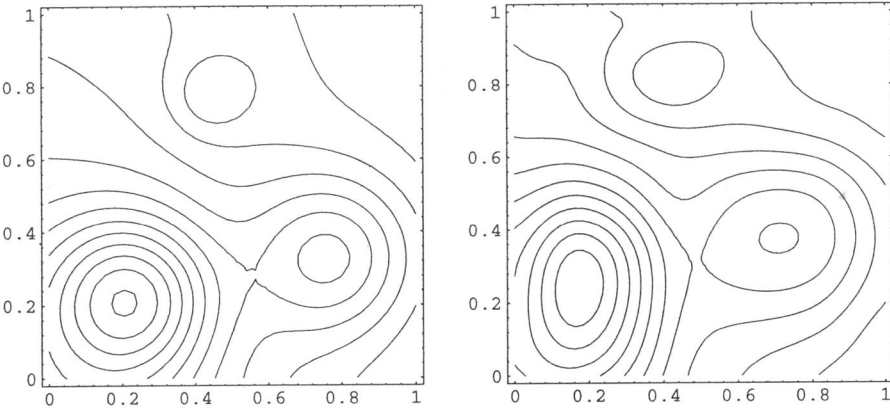

Figure 6.2a. Contours of the Franke function.

Figure 6.2b. Contours of the surface in Figure 6.1.

grid. We then wrote the bilinear interpolant on this grid as a bicubic surface using triple knotlines. Finally we applied knot removal to this surface using a tolerance of 0.05. We then obtained a bicubic tensor product surface with 5×5 interior knotlines. This surface is seen from the (1,1) direction in Figure 6.1. Contour plots of the original Franke function and the final tensor product approximation are shown in Figures 6.2a and 6.2b.

§7. Knot Removal; Scattered Data

Corresponding to a given triangulation Δ of a polygonal domain $\Omega \subset \mathbb{R}^2$ let $S_k^r(\Delta)$ denote the space of real valued functions $f \in C^r(\Omega)$ which reduce to a polynomial of degree k on each subtriangle of Δ. We denote the vertices of the triangulation by

$$\mathcal{A} \cup \mathcal{B} = \{A_1, A_2, \ldots, A_N\} \cup \{B_1, B_2, \ldots, B_p\}, \quad (7.1)$$

where the A's and B's are interior and boundary points of Ω, respectively.

We consider the following knot removal problem on $S_k^r(\Delta)$:

Problem 7.1. *Given $f \in S_k^r(\Delta)$, a norm $\|\ \|$ on $S_k^r(\Delta)$, and a tolerance $\epsilon > 0$, find a smallest possible subset $\mathcal{A}'$ of the interior vertices $\mathcal{A}$ of Δ, such that a triangulation Δ' of $\mathcal{A}' \cup \mathcal{B}$, and a function $g \in S_k^r(\Delta')$ can be found with $\|f - g\| < \epsilon$.*

We have only removed interior vertices. Sometimes we might have more than two boundary points lying on a straight line and then we can also possibly remove some boundary points without changing the region Ω.

The basic philosophy we have described for removing knots can in principle be used for Problem 7.1. First we assign a weight to each vertex of the triangulation. Then we use these weights to rank the vertices in order of removal. Given an approximation scheme for scattered data we could then try (within tolerance) to remove as many vertices as possible. This approach might be costly. A number of triangulations and approximations will have to be found. We describe the approach taken in [23,24] where a local approximation scheme of finite element type is used. Before describing this approach we recall some facts about the B(ernstein-Bézier)-form of piecewise polynomials on triangles ([4]).

Suppose $\mathbf{V} = (V_0, V_1, \ldots, V_s)$ is a collection of $s+1$ points in $\mathbb{R}^s$ spanning a nontrivial simplex $[\mathbf{V}]$ in $\mathbb{R}^s$. Let $\lambda = \lambda(x) = (\lambda_0, \lambda_1, \ldots, \lambda_s)$ given by $\sum \lambda_i V_i = x$ and $\sum \lambda_i = 1$ be the usual barycentric coordinates of $x \in \mathbb{R}^s$ with respect to $\mathbf{V}$. In standard multiindex notation the B-form of a polynomial f of degree k in s variables takes the form

$$f(x) = \sum_{|\alpha|=k} c_\alpha B_\alpha(\lambda(x)), \quad B_\alpha(\lambda) = \frac{|\alpha|!}{\alpha!}\lambda^\alpha, \quad (7.2)$$

where $\alpha = (\alpha_0, \ldots, \alpha_s)$ is a collection of nonnegative integers and $|\alpha| = \sum_{i=1}^s \alpha_i$. We call $\mathbf{c} = \mathbf{c}(f, \mathbf{V}) = (c_\alpha)$ the B-coefficients of f with respect to $\mathbf{V}$. The norm

$$\|\mathbf{c}(f, \mathbf{V})\|_{\ell^\infty} = \max_{|\alpha|=k} |c_\alpha|$$

is a norm on Π_k the class of polynomials of degree $\leq k$ in s variables. For $f = \sum c_\alpha B_\alpha$ we have the stability estimate

$$C^{-1}\|\mathbf{c}\|_{\ell^\infty} \leq \|f\|_{L^\infty[\mathbf{V}]} \leq \|\mathbf{c}\|_{\ell^\infty}$$

Knot Removal

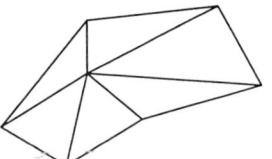
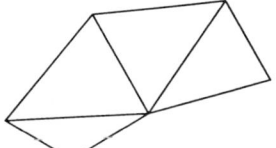
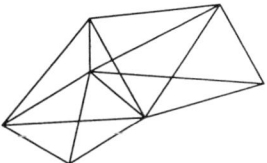

Figure 7.1. The triangulation using A_j.

Figure 7.2. The triangulation not using A_j.

Figure 7.3. The union of the two triangulations on the left.

for some positive constant C depending only on k and s, (see (4.16) in [13]). From this it follows ([13] Theorem 4.1) that

$$|f(v_\alpha) - c_\alpha| \leq C(\operatorname{diam}[\mathbf{V}])^2 \sum_{|\alpha|=2} \|D^\alpha f\|_{L^\infty[\mathbf{V}]}, \quad v_\alpha = \sum \alpha_i V_i], \quad (7.3)$$

where the constant C only depends on s and k. The collection of points $(v_\alpha, c_\alpha)_{|\alpha|=k}$ in $\mathbb{R}^{s+1}$ is called a *B-net* for f.

For a function $f \in S_k^r(\Delta)$ we can use the B-form on each triangle of Δ to represent f. On $S_k^r(\Delta)$ we have the norm

$$\|f\|_{\ell^\infty(\Delta)} = \max_{T \in \Delta} \|\mathbf{c}(f, T)\|_{\ell^\infty} \quad (7.4)$$

where $\mathbf{c}(f, T)$ denotes the B-coefficients of f on the triangle T.

We now describe the knot removal strategy in [23,24]. It is based on a quintic C^1 finite element, but as remarked by the authors one could try other composite C^1 elements like Clough & Tocher or Powell & Sabin. We refer to [18] for a survey of other data fitting methods for scattered data. Suppose we know all the data ϕ_0 needed to construct a quintic C^1 element on Δ. We denote this interpolant by $f = G_5^1(\Delta)\phi_0$. This process will be our basic approximation scheme used in the knot removal process. We note that this process is local. The value of f on a triangle T depends only on data on T. We refer to [22] for more details on this approximation.

To remove knots from f we assign weights to the interior vertices A_j. We set

$$\mathcal{A}_j = \mathcal{A} \setminus A_j, \quad j = 1, 2, \ldots, N.$$

Let P_j be the union of all the triangles that share A_j as a common vertex. We consider 3 triangulations, $\Delta_j^1, \Delta_j^2, \Delta_j^3$ of P_j. Δ_j^1 is the restriction of Δ to P_j. This triangulation is shown in Figure 7.1. Δ_j^2 shown in Figure 7.2 is a triangulation of P_j using only boundary points. The third triangulation (Figure 7.3) is the union of the first two triangulations, $\Delta_j^3 = \Delta_j^1 \cup \Delta_j^2$. With $f_j = G_5^1(\Delta_j^2)\phi_0$ the weight w_j of A_j is now determined in [24] as

$$w_j = \max_{T \in \Delta_j^3} \|\mathbf{c}(f - f_j, T)\|_{\ell^\infty}. \quad (7.5)$$

This is a measure of how well we can do without the vertex A_j.

Once the weights are computed the interior vertices are ranked in increasing order of weights, $w_{r_1} \leq w_{r_2} \leq \cdots \leq w_{r_N}$. Suppose s of these weights are $\leq \epsilon$. We start by removing A_{r_1}. We remove A_{r_2} provided $s > 1$ and A_{r_2} does not belong to P_{r_1}. Continuing in increasing order of weights we remove A_{r_i} for $i = 1, 2, \ldots, s$ provided it is not a boundary point of any of the regions P_j corresponding to vertices A_j already removed. In this way we remove all vertices with weight less that the tolerance and none of the removed points will lie on a common edge of Δ. Thus the interior of the regions P_j corresponding to removed points will be pairwise disjoint. For the reduced vertex set we can now define the following triangulation Δ^1. For the regions P_j corresponding to removed points A_j we use the triangulation showed in Figure 7.2. Outside this region we keep the triangulation Δ. Now let $g = G_5^1(\Delta^1)\phi_0$ be our approximation to f on the new triangulation. We then have

$$\|f - g\|_{L^\infty(\Delta)} \leq \epsilon .$$

This follows from the locality of degree 5 C^1 finite element approximation. For on a region P_j corresponding to a removed vertex A_j we have $g = f_j$ where f_j was used to compute the weight in (7.5). Since $w_j \leq \epsilon$ and no two P_j's corresponding to removed knots overlap we have $\|g - f\|_{L^\infty} = \|f_j - f\|_{L^\infty} \leq w_j \leq \epsilon$. Since $g = f$ outside such regions P_j we conclude that $\|f - g\|_{L^\infty(\Delta)} \leq \epsilon$.

The process can be iterated. We can try to remove points which were neighbors of already removed points. But to stay within tolerance we have to measure the error with respect to the original function.

§8. Constrained Spline Fitting

In data fitting we often want curves and surfaces which satisfy constraints. We will here restrict our discussion to explicit curves $f = f(x), x \in [a, b]$. (cf. Section 3). Typical constraints which might have to be satisfied one one or more subintervals of $[a, b]$ could be

1. Nonnegativity.
2. Monotonicity.
3. Convexity.
4. Curved constraints.
5. Integral constraints.

Other constraints only applies to one or more points of $[a, b]$.

6. Interpolation.
7. Smoothness.
8. Discontinuities.

The curved constraints in Item 4 can be of the general form

$$\hat{\ell}_r(x) \leq f^{(i_r)}(x) \leq \hat{u}_r(x) , \qquad x \in [a_r, b_r] . \tag{8.1}$$

Here $\hat{\ell}_r$ and $\hat{u}_r$ are given functions, i_r is a nonnegative integer, and $[a_r, b_r]$ is a subinterval of $[a, b]$.

To get finite dimensional optimization problems we need discrete constraints. Following our general principle we replace functional constraints by corresponding constraints on the B-spline coefficients. For example, we replace nonnegativity of a spline function by nonnegativity of the B-spline coefficients. For the constraint list above this is described in detail in [2].

In general we will consider linear constraints of the form

$$\mathbf{E}_1\mathbf{b} \geq \mathbf{v}_1, \qquad \mathbf{E}_2\mathbf{b} = \mathbf{v}_2, \qquad (8.2)$$

where $\mathbf{E}_1$ and $\mathbf{E}_2$ are rectangular matrices and $\mathbf{v}_1$ and $\mathbf{v}_2$ are vectors. (The notation $\mathbf{u} \geq \mathbf{w}$ for vectors $\mathbf{u}$ and $\mathbf{w}$ denotes the component-wise inequalities $u_i \geq w_i$ for $i = 1, 2, \ldots$.) The constraints characterize the set of permissible spline functions

$$F_{k,t} = \left\{ \psi = \sum_{i=1}^m b_i B_{i,k,t} \,\Big|\, \mathbf{E}_1\mathbf{b} \geq \mathbf{v}_1 \text{ and } \mathbf{E}_2\mathbf{b} = \mathbf{v}_2 \right\}. \qquad (8.3)$$

Our general approach to constrained spline approximation will follow the same pattern as for unconstrained approximation. First we compute an initial approximation satisfying the constraints, but in general requiring a large number of parameters for its representation. Then we remove knots from this spline. In general we also determine the initial approximation in two stages. First, the data are converted into a spline ϕ which does not necessarily satisfy the constraints. Then, on a sufficiently fine knot vector t we compute a spline approximation f to ϕ satisfying the constraints. For the construction of ϕ we can typically use one of the local schemes described at the end of Section 3. To find f we can solve the problem

$$\min_{f \in F_{k,t}} \|\phi - f\|_{\ell^2, t}, \qquad (8.4)$$

where $F_{k,t}$ is given by (8.3). This is equivalent to the quadratic minimization problem

$$\min_{\mathbf{z} \in \mathbb{R}^m} \|\mathbf{E}_t^{1/2}(\mathbf{z} - \mathbf{a})\|_2$$

subject to

$$\mathbf{E}_1\mathbf{z} \geq \mathbf{v}_1, \quad \mathbf{E}_2\mathbf{z} = \mathbf{v}_2.$$

This problem can be solved using a general purpose routine for quadratic optimization.

It should be stressed that some care and judgement may be needed when introducing the discrete constraints. In particular we have to make sure that we use a sufficiently fine knot vector to define the constraints.

§9. Knot Removal; Constraints

We consider the following constrained knot removal problem.

Problem 9.1. *Given $f \in F_{k,t}$ (cf. (8.3)), a tolerance $\epsilon > 0$, and a norm $\|\ \|$. Find a smallest possible τ, with $\tau \subset t$ for which we can find $g \in S_{k,\tau} \cap F_{k,t}$ such that $\|f - g\| \leq \epsilon$.*

Note that the constraints are always given with respect to $S_{k,t}$. It is important that the knot vector t is such that the discrete constraints approximate continuous constraints well (cf. Example 8.1).

It would in fact have been possible to define problem 9.1 in terms of three knot vectors. In addition to the unknown τ we would have one where f is defined and from which knots are removed, and then a refined one used for the constraints and the discrete norm in the approximation process.

The general strategy used in the unconstrained case in Section 4 can also be applied to the constrained case. The most important change is that the fixed knot approximation scheme has to preserve the constraints. In the unconstrained case we computed a best approximation in the ℓ^2, t norm. More generally we can consider one of the ℓ^p, t norms. Thus, given $1 \leq p \leq \infty$, $\tau \subset t$ and a constrained spline space $F_{k,t}$ we consider the minimization problem

$$\min_{g \in F_{k,t} \cap S_{k,\tau}} \|f - g\|_{\ell^p, t}.$$

With $F_{k,t}$ given by (8.3) and with $\mathbf{A}$ the knot insertion matrix from τ to t this problem can be written

$$\min_{z \in \mathbb{R}^n} \|\mathbf{E}_t^{1/p}(\mathbf{Az} - \mathbf{d})\|_p \quad \text{s.t.} \quad \mathbf{E}_1 \mathbf{Az} \geq \mathbf{v}_1, \quad \mathbf{E}_2 \mathbf{Az} = \mathbf{v}_2. \tag{9.1}$$

Here n is the dimension of $S_{k,\tau}$. For $p = 1$ and $p = \infty$ (9.1) can be written as a linear optimization problem, while for $p = 2$ we have a quadratic problem with linear constraints. If the constraint set is nonempty the latter problem has a unique solution. This follows since the quadratic term is always positive definite, see [16]. We denote the solution of (9.1) by $G_F(\tau, t)f$.

The constrained knot removal implementation in [2] is obtained if we replace G by G_F in the algorithm described in Section 4. No other changes are made. In particular we compute the weights and do the ranking as in the unconstrained case. Several univariate examples, with monotonicity, convexity, and curved constraints are given in [2].

It might be argued that the importance of a knot might change when constraints are introduced. Thus maybe the weights should be computed differently now. Computing weights more accurately from constrained problems was considered in [1].

Acknowledgements. Supported in part by the Royal Norwegian Council for Scientific and Industrial Research - NTNF and Center for Industrial Research - SI. I would like to thank SI for making their knot removal software available for general use.

References

1. Abraham, T., Constrained data reduction of constrained splines, master thesis, Univ. of Oslo, 1990.
2. Arge, E., M. Dæhlen, T. Lyche, and K. Mørken, Constrained spline approximation of functions and data based on constrained knot removal, in *Algorithms for Approximation II*, J. C. Mason and M. G. Cox (eds.), Chapman and Hall, London, 1990, 4–20.
3. de Boor, C., *A Practical Guide to Splines*, Springer-Verlag, New York, 1978.
4. de Boor, C., B-form basics, in *Geometric Modeling: Algorithms and New Trends*, G. E. Farin (ed.), SIAM Philadelphia, 1987, 131–148.
5. de Boor, C. and J. R. Rice, Least squares cubic spline approximation, II - Variable knots, CSD TR 21, Purdue Univ, 1968.
6. Buttenfield, B. P. and R. B. McMaster (eds.), *Map Generalization: Making Rules for Knowledge Representation*, Longman Scientific and Technical, New York, 1991.
7. Cohen, E., T. Lyche, and R. Riesenfeld, Discrete B-splines and subdivision techniques in computer-aided geometric design and computer graphics, Computer Graphics Image Proc. **14** (1980), 87–111.
8. Cohen, E. and L. L. Schumaker, Rates of convergence of control polygons, Computer Aided Geom. Design **2** (1985), 229–235.
9. Dæhlen, M. and Ø. Hjelle, Compact representation of seismic sections, preprint.
10. Dæhlen, M. and T. Lyche, Decomposition of splines, in *Mathematical Methods in Computer Aided Geometric Design II*, T. Lyche and L. L. Schumaker (eds.), Academic Press, Boston, 1992, 135–160.
11. DeVore, R.A., B. Jawerth, and B. Lucier, Surface compression, preprint.
12. DeVore, R.A., B. Jawerth, and B. Lucier, Image compression through wavelet transform coding, preprint.
13. Dahmen, W., Subdivision algorithms converge quadratically, J. Comp. Appl. Math. **16** (1986), 145–158.
14. Degen, W. L. F., Best approximation of parametric curves by splines, in *Mathematical Methods in Computer Aided Geometric Design II*, T. Lyche and L. L. Schumaker (eds.), Academic Press, Boston, 1992, 171–184.
15. Dierckx, P., A fast algorithm for smoothing data on a rectangular grid while using spline functions, SIAM J. Numer. Anal. **19** (1982), 1286–1304.
16. Fletcher, R., *Practical Methods of Optimization*, Second Edition, John Wiley & Sons, Chichester, 1987.
17. Franke, R., Scattered data interpolation: tests of some methods, Math. Comp. **38** (1982), 181–200.
18. Franke, R. and G. M. Nielson, Scattered data interpolation and applications: A tutorial and survey, in *Geometric Modeling*, H. Hagen and D. Roller (eds.), Springer Verlag, Berlin, 1991, 131–160.

19. Hatten, L., M. H. Worthington, and J. Makin, *Seismic Data Processing: Theory and Practice*, Blackwell Scientific Publication, Oxford, 1986.
20. Hoschek, J. and D. Lasser, *Grundlagen der geometrischen Datenverarbeitung*, B. G. Teubner, Stuttgart, 1989.
21. Koch, P. E. and T. Lyche, Calculating with exponential B-splines in tension, preprint.
22. Le Méhauté, A. J. Y., A finite element approach to surface reconstruction, in *Computation of curves and surfaces*, W. Dahmen, M. Gasca, and C. A. Micchelli (eds), Kluwer Academic Publishers, Dordrecht, 1990, 237–274.
23. Le Méhauté, A. J. Y. and Y. Lafranche, A knot removal strategy for scattered data in $\mathbb{R}^2$, in *Mathematical Methods in Computer Aided Geometric Design*, Tom Lyche and L. L. Schumaker (eds.), Academic Press, New York, 1989, 419–426.
24. Le Méhauté, A. J. Y. and Y. Lafranche, Knot removal for scattered data, preprint.
25. Lyche, T. and K. Mørken, A discrete approach to knot removal and degree reduction algorithms for splines, in *Algorithms for Approximation*, J. C. Mason and M. G. Cox (eds.), Clarendon Press, Oxford, 1987, 67–82.
26. Lyche, T. and K. Mørken, A data reduction strategy for splines with applications to the approximation of functions and data, IMA J. of Num. Anal. **8** (1988), 185–208.
27. Lyche, T. and K. Mørken, Knot removal for parametric B-spline curves and surfaces, Computer Aided Geom. Design **4** (1987), 217–230.
28. Lyche, T., K. Mørken, and K. Strøm, Conversion between B-spline bases using the generalized Oslo Algorithm, Preprint 1990-12, Inst. for informatikk, Univ. of Oslo.
29. Lyche, T., L. L. Schumaker, and K. Sepehrnoori, Fortran subroutines for computing smoothing and interpolating natural splines, Adv. Eng. Software **5** (1983), 2–5.
30. Rao, K. R. and P Yip, *Discrete Cosine Transform: Algorithms, Advantages, Applications*, Academic Press, New York, 1990.
31. Renka, R. J., Algorithm 624, Triangulation and interpolation at arbitrary distributed points in the plane, ACM Trans. Math. Software **10** (1984), 437–439.

Tom Lyche
Institutt for Informatikk, University of Oslo
P.O.Box 1080, Blindern
0316 Oslo 3, Norway
EMAIL: tom@ifi.uio.no

Approximation by Algebraic Polynomials

Vilmos Totik

Abstract. We review some recent developments in (real) polynomial approximation of a single variable. Some open problems are also mentioned.

Table of Contents

1. Bernstein Polynomials   227
2. Best Approximation   231
3. Weighted Polynomial Approximation with Varying Weights   233
    3.1. A Quick Course on Potential Theory   236
    3.2. Solution of the Approximation Problem for Varying Weights   237
    3.3. Examples and Problems   239
4. Fast Decreasing Polynomials   242
5. Locally Geometric Rates   244
References   246

§1. Bernstein Polynomials

We start with Weierstrass' theorem, which asserts that every continuous function on $[0,1]$ can be uniformly approximated by polynomials. This theorem, along with its first proofs, is by no means constructive; that is, if f is given in $C[0,1]$, then it does not tell *how* to construct the polynomials that uniformly approximate f on $[0,1]$.

The most well-known constructive variant is due to Bernstein, who used the polynomials

$$B_n(f;x) = \sum_{k=0}^{n} f\left(\frac{k}{n}\right) \binom{n}{k} x^k (1-x)^{n-k}$$

to approximate f. These so-called Bernstein polynomials have many additional properties (like shape preservation and variation diminution), besides uniformly converging to f, and they have penetrated several disciplines of mathematics. It is no wonder that their literature is vast and that there are several hundred papers devoted just to the study of their approximation properties.

In this connection, we first note that the sequence of Bernstein polynomials is saturated with order 1 as proved in [25]. This means that no functions other than linear ones can be approximated with order better than $O(n^{-1})$. Thus, even though Bernstein polynomials uniformly approximate every continuous function, the convergence cannot be too fast (this is a general principle: if a method is fast, then it fails for some objects).

Despite their intense study, one basic question concerning the approximation properties of B_n has eluded every effort until very recently, and this is *how well* they approximate in the supremum norm. In other words, we ask for a description of the error

$$\|B_n f - f\|_{[0,1]},$$

where $\|\ \|$ denotes the uniform norm. We would like to characterize this error in terms of structural properties of the function f. By such a characterization we provide an equivalent quantity that can be explicitly computed.

We can do that by using the modulus of smoothness

$$\begin{aligned}\omega_\varphi(f;\delta) &= \sup_{0\le t\le\delta} \|\Delta^2_{t\varphi}f\|_{[0,1]} \\ &= \sup_{0\le t\le\delta}\sup_x |f(x-t\varphi(x)) - 2f(x) + f(x+t\varphi(x))|,\end{aligned} \qquad (1.1)$$

where $\varphi(x) = \sqrt{x(1-x)}$, and where the second supremum is taken over those values x for which every argument belongs to $[-1,1]$. Note that in this second order modulus of smoothness, the increment in the second symmetric difference $\Delta^2_{t\varphi}f$ changes according to the position of x in $[0,1]$.

Theorem 1.1. (Totik, [39, Theorem 1]) *For $f \in C[0,1]$ we have*

$$\|B_n f - f\| \sim \omega_\varphi\left(f; \frac{1}{\sqrt{n}}\right), \qquad (1.2)$$

where $\sim$ means that the ratio of the two sides lies between two positive constants independent of f and n.

We emphasize that, in addition to telling us exactly how good the approximation via Bernstein polynomials is, this theorem allows us to calculate explicitly the order. For example, if $f(x) = x^\delta(\log 2/x)^\gamma$, $1 \ge \delta > 0$, then (see [6, p.35] for the exact calculation of the rate of the right hand side in (1.2))

$$\|B_n f - f\| \sim \begin{cases} n^{-\delta}(\log n)^\gamma & \text{for } \delta \ne 1 \\ n^{-1}(\log n)^{\gamma-1} & \text{for } \delta = 1 \text{ and } \gamma \ge 1 \end{cases}$$

Approximation by Polynomials

(when $\delta > 1$ or $\delta = 1$ and $\gamma < 1$, then the rate of approximation is the saturation rate n^{-1}).

As for the history leading to Theorem 1.1, it is impossible to list all contributions in a survey paper. Let me just mention that before the eighties, the people who came closest to characterizing the error were Berens and Lorentz [1]: they established the equivalence of $f \in \text{Lip } \alpha$ and pointwise rate of approximation. So let us jump now to the late seventies and early eighties, when several Bulgarian mathematicians (Andreev, Ivanov, Popov, Sendov), Ditzian and the present author independently started to use the modulus of smoothness ω_φ or some variation of it. Eventually, as an analogue of the Jackson theorem, the inequality

$$\|B_n f - f\| \le C \omega_\varphi \left(f; \frac{1}{\sqrt{n}}\right)$$

was proved in [6, Chapter 9]. However, the converse result was not expected until very recently. In fact, the so-called Stechkin-type weak inequalities were the strongest known converse estimates. The inequality

$$\omega_\varphi \left(f; \frac{1}{\sqrt{n}}\right) \le \frac{C}{n^s} \sum_{k=1}^{n} k^{s-1} \|B_k f - f\|$$

was proved in [42] for $s < 1$; then in [45] and almost simultaneously in [6, Theorem 9.3.6] for $s = 1$. The analogy with best polynomial approximation (see the next section) suggested that such estimates are perfectly natural and perhaps cannot be improved. Only very recently did Ditzian and Ivanov [5] prove a converse result involving only two terms. More precisely, they showed that for some K ($K = 20$ would work)

$$\omega_\varphi \left(f; \frac{1}{\sqrt{n}}\right) \le C \frac{m}{n} (\|B_n f - f\| + \|B_m f - f\|) \tag{1.3}$$

holds for every $m \ge Kn$, $n = 1, 2, \ldots$. Thus, by choosing $m = Kn$, they got the equivalence

$$\omega_\varphi \left(f; \frac{1}{\sqrt{n}}\right) \sim \|B_n f - f\| + \|B_{Kn} f - f\|.$$

They also conjectured that the second term can be dropped; i.e., that Theorem 1.1 is true.

In [41] we showed by an argument that is very close to the classical parabola technique that such two-term converse results are very natural, and they actually hold for a large family of operators.

Even though this survey is about polynomial approximation, it is appropriate at this point to mention that the analogue of Theorem 1.1 holds for some other approximation processes. For many commutative ones this was done by Ditzian and Ivanov in [5]. For noncommutative operators, we know it for the most well-known ones: Szász-Mirakjan and Baskakov operators. The Szász-Mirakjan operators are defined as

$$S_n(f;x) = \sum_{k=0}^{\infty} f\left(\frac{k}{n}\right) e^{-nx} \frac{(nx)^k}{k!}, \qquad f \in C[0,\infty).$$

The corresponding φ is given by $\varphi(x) = \sqrt{x}$, and the modulus ω_φ is defined as in (1.1), except that now the second supremum is taken for those x for which the arguments belong to $[0,\infty)$. It was shown in [39] that

$$\|S_n f - f\| \sim \omega_\varphi\left(f; \frac{1}{\sqrt{n}}\right) \tag{1.4}$$

for all $f \in C[0,\infty)$ and n. The so-called *Baskakov* operators are defined by

$$V_n(f;x) = \sum_{k=0}^{\infty} f\left(\frac{k}{n}\right) \binom{n+k-1}{k} x^k (1+x)^{-n-k}, \qquad f \in C[0,\infty),$$

For them we use the function $\varphi(x) = \sqrt{x(1+x)}$, and obtain the equivalence

$$\|V_n f - f\| \sim \omega_\varphi\left(f; \frac{1}{\sqrt{n}}\right) \tag{1.5}$$

for all $f \in C[0,\infty)$ and n.

In the proofs of these three theorems I made essential use of some special properties of the operators (it is known that the structure of the coefficient functions for the Bernstein, Szász-Mirakjan and Baskakov operators is very similar because of the Moivre-Laplace formula; see e.g., [12, p. 201–203]). We do not have a *general theorem* on positive operators which would imply the above equivalences (1.2), (1.4) and (1.5).

Probably the most important open problem in the subject is the case of L^p approximation. In $L^p[0,1]$ we use a modification of the Bernstein polynomials due to L. Kantorovich [21]:

$$K_n(f;x) = \sum_{k=0}^{n} \left((n+1) \int_{k/(n+1)}^{(k+1)/(n+1)} f(u)\, du\right) \binom{n}{k} x^k (1-x)^{n-k}.$$

Here the function values appearing in the Bernstein operator have been replaced by averages over a uniform partition of the interval $[0,1]$. It is known (see [6, Chapter 9]) that for every p in the interval $[1,\infty)$

$$\|K_n f - f\|_{L^p[0,1]} \leq C\omega_\varphi\left(f; \frac{1}{\sqrt{n}}\right)_p + \frac{1}{n}\|f\|_p,$$

where the definition of the L^p modulus is similar to (1.1), namely

$$\omega_\varphi(f;\delta)_p = \sup_{0\le t\le\delta} \|\Delta^2_{t\varphi}f\|_{L^p[0,1]}.$$

In the converse direction we also know (see [6, Chapter 9]) that

$$\omega_\varphi\left(f;\frac{1}{\sqrt{n}}\right)_p \le \frac{C}{n}\sum_{k=1}^{n}\|K_k f - f\|_{L^p[0,1]},$$

and with some effort one can probably show even a two term converse in the sense of (1.3). But the question whether

$$\|K_n f - f\|_{L^p[0,1]} \ge c\omega_\varphi\left(f;\frac{1}{\sqrt{n}}\right)_p$$

remains open. The difficulty is that in [39] our method was more or less geometric (in one part of the proof we used a modification of the classical parabola argument), while in L^p no similar technique is known.

§2. Best Approximation

The moduli of smoothness presented in the first section have many other applications. In this section, we shall discuss how to get direct and converse estimates for best polynomial approximation. For other applications, see the monograph [6].

Let the best approximation of f by polynomials of degree at most n in L^p be defined by

$$E_n(f)_p = \inf_{\deg P_n \le n} \|f - P_n\|_{L^p[0,1]}.$$

As usual, when $p = \infty$ we think of uniform approximation. Weierstrass' theorem asserts that $E_n(f)_\infty$ tends to zero as $n \to \infty$ for every continuous f. In 1912, Jackson [19] noticed that the rate with which $E_n(f)_p$ tends to zero is closely related to smoothness properties of f. To provide a background for the type of results we can expect, let us discuss first the case of trigonometric approximation, for which the results are classical and can be found in virtually any book on approximation theory.

For trigonometric approximation, we shall attach an asterisk to $E_n(f)_p$ and related quantities. The general form of Jackson's estimate is

$$E_n(f)^*_p \le C\omega^r(f;n^{-1})^*_p. \tag{2.1}$$

Here for a positive integer r, the ordinary r-th modulus of smoothness ω^r is

$$\omega^r(f;\delta)^*_p = \sup_{0\le t\le\delta} \|\Delta^r_t f\|_{L^p_{2\pi}},$$

where
$$\Delta_t^r f(x) = \sum_{k=0}^{r}(-1)^k \binom{r}{k} f\big(x+(r/2-k)t\big)$$

is the r-th symmetric difference of the (now) 2π-periodic function f. The converse is given in a weak form:

$$\omega^r(f;n^{-1})_p^* \leq \frac{C_r}{n^r}\sum_{k=1}^{n} k^{r-1} E_k(f)_p^*, \qquad (2.2)$$

and this is an easy consequence of Bernstein's inequality for the norm of the derivative of a trigonometric polynomial (stating that it is at most the degree times the original norm).

For algebraic approximation the analogue of the Jackson estimate is

$$E_n(f)_p \leq C\omega^r(f;n^{-1})_p,$$

and has been known for a long time. However, this is not exact because the analogue of a Stechkin-type weak converse like (2.2) *is not true*. The problem is that in the algebraic case the analogue of Bernstein's inequality is Markoff's inequality, but the Markoff constant is the square of the degree of the polynomial, which results in a nonmatching converse. For example, until the early eighties it was not known for what functions we have the Lipschitz rate $E_n(f)_p \leq Cn^{-\alpha}$, even in the continuous case. In the literature this deficiency was partially resolved by the so-called Timan-Gopengauz type theorems that relate the regular moduli of smoothness to pointwise estimates by polynomials. However, this is more or less the transformation of the trigonometric case, and the original problem of characterizing the rate of $E_n(f)_p$ in terms of structural properties remained open.

The φ moduli from the preceding section offered a solution in the early eighties. The problem was first resolved by Ivanov [15,16] with the so-called Bulgarian modulus of smoothness, which is closely related to our ω_φ^r (we also mention that the so-called Jacobi transform had been utilized before by Butzer, Stens and Wehrens [4] to obtain matching direct and converse results). For a positive integer, r the r-th modulus of smoothness ω_φ^r is defined as a perfect analogue of the above ω^r; the only change is that we replace again the increment in the symmetric difference by $h\varphi(x)$:

$$\omega_\varphi^r(f;\delta)_p := \sup_{0\leq t\leq \delta} \|\Delta_{t\varphi}^r f\|_{L^p[0,1]}.$$

Here φ can be arbitrary, but for algebraic approximation we use again $\varphi(x) = \sqrt{x(1-x)}$. Thus, the ω_φ used in connection with Bernstein polynomials is actually ω_φ^2 with this notation. In terms of these moduli, the exact equivalent of the aforementioned trigonometric approximation results can be formulated as

Theorem 2.1. (Ditzian-Totik, [6, Theorem 7.2.1])

$$E_n(f)_p \leq C\omega_\varphi^r(f; n^{-1})_p,$$

and

$$\omega_\varphi^r(f; n^{-1})_p \leq \frac{C_r}{n^r} \sum_{k=1}^n k^{r-1} E_k(f)_p. \qquad (2.3)$$

In general, $\omega_\varphi^r(f;\delta)_p \leq C\omega^r(f;\delta)_p$ (which is not a trivial fact, but immediately follows from the K-functional characterization that we shall mention below), and these two moduli can have completely different orders; thus Jackson's estimate in the algebraic case is actually very far from the correct one.

In general, no better converse result can be given than (2.3). But if $1 < p < \infty$, then a sharper converse holds, which is the analogue of a result of Zygmund [46] for trigonometric approximation. Namely, we have

Theorem 2.2. (Totik, [44, Theorem 1]) *Let* $1 < p < \infty$. *Then*

$$\omega_\varphi^r(f; n^{-1})_p \leq C_{r,p} \begin{cases} n^{-r} \left(\sum_{k=1}^n k^{pr-1} E_k(f)_p^p \right)^{1/p} & \text{if } 1 < p \leq 2 \\ n^{-r} \left(\sum_{k=1}^n k^{2r-1} E_k(f)_p^2 \right)^{1/2} & \text{if } 2 < p < \infty. \end{cases}$$

The wide applicability of ω_φ^r (see [6]) is explained by the fact that it is equivalent to a K-functional between a weighted Sobolev space and an L^p space: if

$$K_{r,\varphi}(f; t^r)_p = \inf_g \left(\|f - g\|_p + t^r \|\varphi^r g^{(r)}\|_p \right),$$

then [6, Theorem 2.1.1]

$$K_{r,\varphi}(f; t^r)_p \sim \omega_\varphi^r(f; t).$$

Analogous theorems hold for weighted approximation and for approximation on polytopes in several variables. For details, see the monograph [6]. The complete analogue of Theorem 2.1 for approximation in several variables on compact sets other than polytopes is not known.

§3. Weighted Polynomial Approximation with Varying Weights

Recently much attention has been given to approximation with weighted polynomials of the form $w^n P_n$, where w is some fixed weight and the degree of P_n is at most n. We emphasize that the exponent of the weight w^n changes together with n, so this is a different (and more difficult) type of approximation from what is usually called "weighted approximation." In fact, in the present case the polynomial P_n must balance exponential oscillations in w^n.

The applications of this type of approximation are very diverse; here we only mention its relation to Freud type orthogonal polynomials. Another application will be discussed in the next section.

Let $w_\alpha(x) = \exp(-c|x|^\alpha)$, $c > 0$, be a so-called Freud weight, and consider the orthonormal polynomials

$$p_n(x) = \gamma_n(w_\alpha)x^n + \cdots$$

associated with w_α^2:

$$\int p_n p_m w_\alpha^2 = \delta_{n,m} .$$

When $\alpha = 2$, these are the classical Hermite polynomials; for other α's Freud [7] initiated their investigation, and they appeared in some physical problems as well. It is well known that the monic polynomials

$$\frac{1}{\gamma_n(w_\alpha)} p_n(x)$$

are the extremal polynomials in the minimum problem

$$\inf \int (x^n + \cdots)^2 w_\alpha^2(x)\, dx$$

with the minimum itself being equal to $\gamma_n^{-2}(w_\alpha)$, where the infimum is taken for all polynomials of the form indicated. If we carry out the substitution $x \to n^{1/\alpha}x$, then we see that the polynomials

$$P_n(x) = \frac{1}{\gamma_n(w_\alpha)} n^{-n\alpha} p_n(n^\alpha x)$$

are the extremal polynomials in the minimum problem

$$\inf \int (x^n + \cdots)^2 w_\alpha^{2n}(x)\, dx .$$

Now it turns out that with appropriate normalization (that is, with appropriate choice of c, whose exact value is discussed below) in this infimum the integral can be restricted to integrals on $[-1, 1]$ because the weight $w_\alpha^n(x) = \exp(-nc|x|^\alpha)$ decreases very fast outside $[-1, 1]$ compared to the increase of polynomials of degree at most n (see next subsection). Now, by a formula of Bernstein we know the explicit solution of the minimum problem

$$\inf \int_0^1 (x^n + \cdots)^2 \frac{\sqrt{1 - x^2}}{R_{2n}(x)}\, dx$$

if R_{2n} is a polynomial of degree at most $2n$. Hence, if we succeed in approximating $\sqrt{1-x^2}$ by weighted polynomials of the form $w_\alpha^{2n}(x) R_{2n}(x)$ on $[-1, 1]$, then we may be able to achieve sharp asymptotics on $\gamma_n(w_\alpha)$, which is enough

Approximation by Polynomials

to derive sharp asymptotics on the orthogonal polynomials themselves. This approach, which is due to Lubinsky and Saff [27] (see also Rahmanov [33]), eventually produced the so-called power type asymptotics on $\gamma_n(w_\alpha)$.

The problem of approximation with weighted polynomials of the form $w^n P_n$ originated in Lorentz's incomplete polynomial approximation (see Section 3.3 below). It was generalized by Mhaskar and Saff [29]. They found that the norm of these weighted polynomials is attained on a compact set S_w; i.e., for every such weighted polynomial we have

$$\|w^n P_n\|_{\mathbf{R}} = \|w^n P_n\|_{S_w} ,$$

and furthermore, $w^n P_n$ tends to zero outside S_w as $n \to \infty$. They also explicitly determined S_w for the Freud weight $w = w_\alpha$:

$$S_{w_\alpha} = [-\gamma_\alpha^{1/\alpha} c^{-1/\alpha}, \gamma_\alpha^{1/\alpha} c^{-1/\alpha}] , \qquad (3.1)$$

where

$$\gamma_\alpha := \int_0^1 \frac{v^{\alpha-1}}{\sqrt{1-v^2}} dv = \Gamma\left(\frac{\alpha}{2}\right) \Gamma\left(\frac{1}{2}\right) / \left(2\Gamma\left(\frac{\alpha}{2}+\frac{1}{2}\right)\right) .$$

This was the start. The next impulse came from orthogonal polynomials. One of the most challenging problems in the eighties in the theory of orthogonal polynomials was Freud's conjecture about the asymptotic behavior of the recurrence coefficients for Freud type orthogonal polynomials (with respect to the weights $w_\alpha(x) = \exp(-c|x|^\alpha)$), which asserted that the ratios $n^{-1/\alpha} \gamma_n(w_\alpha)/\gamma_{n+1}(w_\alpha)$ tend to a limit as $n \to \infty$. The solution came in three papers [22], [28] and [26] by Lubinsky, Knopfmacher, Nevai, Mhaskar and Saff. The most difficult part of the proof was the following approximation theorem of Lubinsky and Saff [28].

Theorem. *If $w_\alpha(t)$ is a Freud weight $\exp(-\gamma_\alpha |t|^\alpha)$, $\alpha > 1$, normalized so that $S_{w_\alpha} = [-1,1]$, then for every continuous f which vanishes outside $[-1,1]$ there are polynomials P_n of degree at most n, $n = 1, 2, \ldots$ such that $w_\alpha^n P_n$ tends uniformly to f.*

By what we have said about $w_\alpha^n P_n$ tending to zero outside $[-1,1]$, if f can be uniformly approximated by $w_\alpha^n P_n$, then it must vanish outside $[-1,1]$.

Besides this result and a few other particular ones for concrete weight functions, nothing has been known for the general approximation problem with varying weights, that is, about what functions can be approximated by weighted polynomials of the form $w^n P_n$ for a given weight w. In what follows we will describe a few new results that solve this problem for a large class of weights that contain as special cases all the concrete results known so far. To formulate our main theorem, we shall need to take a quick excursion into potential theory.

§3.1. A Quick Course on Potential Theory

Let Σ be a closed subset of the real line. For simplicity we shall assume that Σ is the union of finitely many (finite or infinite) intervals.

A weight function w on Σ is said to be *admissible* if it satisfies the following three conditions

(i) w is continuous;
(ii) $\Sigma_0 := \{x \in \Sigma \,|\, w(x) > 0\}$ has positive capacity;
(iii) if Σ is unbounded, then $|x|w(x) \to 0$ as $|x| \to \infty$, $x \in \Sigma$.

We define $Q = Q_w$ by

$$w(x) =: \exp(-Q(x)) \,. \tag{3.3}$$

The theory of logarithmic potentials can be based on the notion of energy. In the case when we have a weight present (in our case Q), then the same can be done. Let $\mathcal{M}(\Sigma)$ be the set of all positive unit Borel measures μ with $\operatorname{supp}(\mu) \subseteq \Sigma$, and define the weighted energy integral

$$I_w(\mu) := \iint \log[|z-t|w(z)w(t)]^{-1}\, d\mu(z)\, d\mu(t) \tag{3.4}$$

$$= \iint \left[\log\frac{1}{|z-t|} + Q(z) + Q(t)\right] d\mu(z)\, d\mu(t) \,.$$

The next theorem is the "weighted" analogue of a classical theorem of Frostman [9] in potential theory, and was essentially proved in [29] by Mhaskar and Saff.

Theorem A. *Let w be an admissible weight on the set Σ, and let*

$$V_w := \inf\{I_w(\mu) \,|\, \mu \in \mathcal{M}(\Sigma)\} \,. \tag{3.5}$$

Then the following assertions are true.
(a) There exists a unique $\mu_w \in \mathcal{M}(\Sigma)$ such that

$$I_w(\mu_w) = V_w \,.$$

(b) $S_w := \operatorname{supp}(\mu_w)$ is compact.
(c) For the potential

$$U^{\mu_w}(z) := \int \log\frac{1}{|z-t|}\, d\mu_w(t)$$

the following inequality holds on Σ:

$$U^{\mu_w}(z) \geq -Q(z) + V_w - \int Q\, d\mu_w =: -Q(z) + F_w \,.$$

(d) *The inequality*
$$U^{\mu_w}(z) \leq -Q(z) + F_w$$
holds for all $z \in \mathcal{S}_w$.

In fact, in [29] property (c) was proved to hold for quasi-every $z \in \Sigma$. But the regularity of Σ implies that then the set of points where
$$U^{\mu_w}(z) + Q(z) \geq F_w$$
is dense at every point of Σ in the fine topology (see [14, Chapter 10] or [23, Chapter III]), hence the inequality in question is true at every $z \in \Sigma$ by the continuity of Q (where it is finite) and the continuity of logarithmic potentials in the fine topology.

The measure μ_w is called the *equilibrium* or *extremal measure* associated with w.

We cite another theorem of Mhaskar and Saff [29, Theorem 2.1], which is an easy consequence of Theorem A and the principle of domination, and which says that the supremum norm of weighted polynomials $w^n P_n$ is attained on $\mathcal{S}_w$.

Theorem B. *Let w be an admissible weight on $\Sigma \subseteq \mathbf{R}$. If P_n is a polynomial of degree at most n and if*
$$|w(z)^n P_n(z)| \leq M \quad \text{for} \quad z \in \mathcal{S}_w , \tag{3.6}$$
then for all $z \in \mathbf{C}$
$$|P_n(z)| \leq M \exp\bigl(n(-U^{\mu_w}(z) + F_w)\bigr) . \tag{3.7}$$
Furthermore, (3.6) implies
$$|w(z)^n P_n(z)| \leq M \quad \text{for} \quad z \in \Sigma . \tag{3.8}$$

This theorem asserts that every weighted polynomial must assume its maximum modulus on $\mathcal{S}_w$ (it can, however, assume it outside $\mathcal{S}_w$, as well). It can be shown that $\mathcal{S}_w$ is the smallest compact set with this property.

§3.2. Solution of the Approximation Problem for Varying Weights

Let Σ be a closed subset of the real line consisting of finitely many intervals, and let w be an admissible weight on Σ. As we have indicated in the beginning of this section, we consider the problem of approximating a continuous function f by weighted polynomials $w^n P_n$. Since, in general, the exponent in (3.7) is negative outside $\mathcal{S}_w$ (see Theorem A(c)), such weighted polynomials tend to zero outside $\mathcal{S}_w$, hence it is reasonable to assume that f vanishes outside $\mathcal{S}_w$. Strictly speaking, this is not a necessary condition, but a convenient working assumption which is necessary in most of the interesting cases (like the Freud case discussed above).

Let O be an open subset of $\mathbf{R}$, and assume that $O \subset \Sigma$. The space of continuous real functions that vanish outside O will be denoted by $C_0(O)$.

Definition. *We say that w has the approximation property on the open set O if for every $f \in C_0(O)$ there is a sequence of polynomials $\{P_n\}_{n=1}^{\infty}$ of corresponding degree $n = 1, 2, \ldots$ such that $w^n P_n$ converges to f on Σ.*

Thus, what we have said above implies that, in general, we can hope for the approximation property on an open set O only if $O \subseteq S_w$, that is O should be part of the interior $\text{Int}(S_w)$ of S_w. Our main result is that on the other hand, if μ_w has continuous and positive density function on the interior of S_w, then w does have the approximation property on $\text{Int}(S_w)$.

To formulate the main result of this section we introduce the following definition:

Definition. *Let S^w denote the set of those points x_0 where the equilibrium measure μ_w has continuous and positive density, that is*

$$d\mu_w(t) = v_w(t)\,dt,$$

and the density function v_w is continuous and positive in a neighborhood of x_0. This S^w is called the restricted support of μ_w.

Thus, if μ has positive and continuous density on $\text{Int}(S_w)$, then $S^w = \text{Int}(S_w)$. On the other hand, if at x_0 we have $v_w(x_0) = 0$, then this x_0 does *not* belong to the restricted support.

Theorem 3.1. (Totik, [40, Theorem 4.1]) *Let w be an admissible weight on Σ. Then w has the approximation property on the restricted support S^w. In particular, if μ_w has continuous and positive density on $\text{Int}(S_w)$, then every continuous function that vanishes outside S_w can be uniformly approximated on Σ by weighted polynomials of the form $w^n P_n$, where the degree of P_n is at most n.*

Let us add to the second part that if $\Sigma = \mathbf{R}$ and strict inequality holds outside S_w in Theorem A(c), then no other function can be uniformly approximated (see Theorem B).

Theorem 3.2. (Totik, [40, Theorem 4.2]) *Suppose that Σ is the union of finitely many disjoint intervals I_j, and w is an admissible weight of class $C^{1+\epsilon}$ for some $\epsilon > 0$ such that $Q = \log 1/w$ is convex on every I_j. Then w has the approximation property on the interior of the support S^w.*

A similar result holds if the convexity is replaced by the property that $(x - a_j)Q'(x)$ increases on I_j, where a_j is the left endpoint of the interval I_j.

In the case considered in Theorem 3.2, if S_w lies in the interior of Σ, then the interior of the extremal support S_w is the largest set where the approximation property can hold. In fact, if f is a real function that is uniformly approximable by weighted polynomials $w^n P_n$, then f must vanish outside S_w. This follows from the fact that the equilibrium potential U^{μ_w} is strictly convex outside the support S_w of its generating measure, hence the strict inequality has to be true in Theorem A(c) at every $z \notin S_w$. Now

Approximation by Polynomials 239

apply Theorem B to conclude that $w^n(z)P_n(z) \to 0$ (in fact, exponentially) if $z \notin \mathcal{S}_w$.

Now we show that Theorem 3.1 is sharp in a certain sense. To illustrate Theorem 3.1, let us consider the case when $\mathcal{S}_w = [-1,1]$, and μ_w has continuous density v_w in $(-1,1)$. We have seen that if this density is positive in $(-1,1)$, then on $\text{Int}(\mathcal{S}_w) = (-1,1)$ w has the approximation property, and in general this is the largest set where approximation is possible. Now what happens if v_w vanishes at a single point, say at $x = 0$? We have constructed an example where this single zero prohibits approximation in a very strong sense, namely approximation is possible only for functions that vanish at the origin. In fact we have

Theorem 3.3. (Totik, [40, Example 1]) *There exists a weight w such that the support of the corresponding extremal measure is $[-1,1]$, this measure has continuous density in $(-1,1)$ which is positive everywhere except at 0, and still no function that is nonzero at 0 can be approximated by weighted polynomials of the form $w^n P_n$.*

Hence, in this case the largest set for the approximation problem of the present section is the restricted support $(-1,0) \cup (0,1)$. This shows that, in general, on no larger set than the restricted support can w have the approximation property; in other words, Theorem 3.1 cannot be improved.

Theorem 3.1 does not tell us whether approximation is possible on $\text{Int}(\mathcal{S}_w)$ if the density v_w vanishes there. Theorem 3.3 shows that such internal zeros may prevent approximation, but this does not rule out the possibility of the approximation property on the whole $\text{Int}(\mathcal{S}_w)$ for concrete weights w. The next example together with Theorem 3.3 shows that the situation is indeed very delicate: approximation in the presence of internal zeros depends on the weight in a subtle way.

Theorem 3.4. (Totik, [40, Example 2]) *There exists a weight w on $[-1,1]$ such that the support of the corresponding extremal measure μ_w is $[-1,1]$, μ_w has continuous density in $(-1,1)$ which vanishes at the origin, and still every continuous f that is zero at ± 1 can be uniformly approximated by weighted polynomials of the form $w^n P_n$.*

§3.3. Examples and Problems

The type of approximation we are discussing has evolved from Lorentz's incomplete polynomials. Lorentz [24] studied polynomials on $[0,1]$ that vanish at zero with high order. That is, he considered polynomials of the form

$$P_n(x) = \sum_{k=s_n}^{n} a_k x^k, \qquad (3.9)$$

and he verified that if $s_n/n \to \theta$ and the P_n's are bounded on $[0,1]$, then $P_n(x)$ tends to zero uniformly on compact subsets of $[0, \theta^2)$. In our terminology this result means that the support of the extremal measure for the weight $w(x) = x^{\theta/(1-\theta)}$, $\Sigma = [0,1]$, is $[\theta^2, 1]$ (to get this weight write $P_n(x)$ in the form $(x^{s_n/(n-s_n)})^{n-s_n} R_{n-s_n}(x)$). The corresponding approximation theorem, namely that every $f \in C[0,1]$ that vanishes on $[0, \theta^2)$ is the uniform limit of polynomials of the form (3.9), was independently proved by Golitschek [10] and Saff and Varga [36].

In [34] Saff generalized the problem to exponential weights of the form $w_\alpha = \exp(-c|x|^\alpha)$, $\alpha > 1$, $\Sigma = (-\infty, \infty)$. He and Mhaskar proved in [30] that in this case the extremal support is (3.1). In [34] Saff conjectured that every continuous function that vanishes outside (3.1) can be uniformly approximated by weighted polynomials $w_\alpha^n P_n$. This was shown to be true in the special case $\alpha = 2$ in [31] by Mhaskar and Saff, and, as we have already mentioned, by Lubinsky and Saff [28] for all $\alpha > 1$.

In [31] the conjecture was made that even for general continuous weights w, approximation by weighted polynomials $w^n P_n$ is possible for an f if and only if f vanishes outside S_w. Theorem 3.3 shows that this is not the case. In [3] the weaker conjecture was stated that at least for the case when $Q = \log 1/w$ is convex, a necessary and sufficient condition for approximation is the same as before, that is that the function vanishes outside S_w (in that case the support S_w is an interval). Theorem 3.2 verifies this under the smoothness assumption that Q is a $C^{1+\epsilon}$, $\epsilon > 0$ function on the support S_w (note that if Q is convex on an interval I, then it is automatically Lip 1 inside I). The conjecture for general convex Q's remains open.

We have already mentioned that if $w(x) = x^{\theta/(1-\theta)}$, then $S_w = [\theta^2, 1]$ ([24]). The generalization to Jacobi weights was given in [35], where it was shown that if $w(x) = (1-x)^\alpha (1+x)^\beta$, $\alpha, \beta \geq 0$, $\Sigma = [-1,1]$, then the support of the extremal measure is

$$[\theta_2^2 - \theta_1^2 - \sqrt{\Delta}, \theta_2^2 - \theta_1^2 + \sqrt{\Delta}],$$

where $\theta_1 := \alpha/(1+\alpha+\beta)$, $\theta_2 := \beta/(1+\alpha+\beta)$ and

$$\Delta := \{1 - (\theta_1 + \theta_2)^2\}\{1 - (\theta_1 - \theta_2)^2\}.$$

In this case the approximation problem was settled by He and Li [13].

A "midway" case between Jacobi weights and Freud-type exponential weights is given by the Laguerre weights $w(x) = x^\alpha e^{-x}$, $\alpha \geq 0$, $\Sigma = [0, \infty)$, for which (see [32])

$$S_w = \left[1 + \alpha - \sqrt{(1+\alpha^2) - \alpha^2}, 1 + \alpha + \sqrt{(1+\alpha^2) - \alpha^2}\right].$$

In all these cases $Q = \log 1/w$ is convex; hence Theorem 3.2 can be applied, and we can deduce the approximation property of the corresponding w on the interior of the support S_w. This is even true for the Golitschek-Saff-Varga theorem (note that in that theorem the function need not vanish at the right endpoint of $S_w = [\theta^2, 1]$), for it is enough to approximate $f \in C[-1, 1]$

that vanish on some $[0, \theta_1]$ with $\theta_1 > \theta^2$, and for such functions the claim follows from the fact that the function $f((\theta_1/\theta^2)\cdot)$ can be uniformly approximated on $[0, \theta^2/\theta_1]$ by weighted polynomials of the form $x^{n\theta/(1-\theta)}P_n(x)$ (extend this function to $[0,1]$ continuously so that it vanishes at 1).

In our theorems we required that the density of the extremal measure be continuous (and positive). It is possible that finitely many logarithmic type singularities in the density (these arise for example at the origin if one considers $w(x) = \exp(-|x|)$) can be handled. The situation is much worse if the infinite singularity is not of logarithmic type. For example, if $w_\alpha(x) = \exp(-c|x|^\alpha)$ with $0 < \alpha < 1$, then the density of the extremal measure has a singularity of the form $\sim t^{\alpha-1}$ at the origin. In fact, it turns out that out that for Freud weights $w_\alpha(x)$ with $0 < \alpha \le 1$ approximation is possible if and only if $\alpha = 1$.

Internal zeros in the density function constitute another problem. We have seen in our first example that even a single zero may rule out the possibility of approximation in the sense of Theorem 3.1. On the other hand, Theorem 3.4 shows that in some cases approximation is possible even in the presence of an internal zero, and it seems to be a very delicate problem to clarify the effect of internal zeros on the approximation problem for given individual weights.

A typical example of this kind of difficulty is encountered if we consider the weight $w(x) = e^{x^2}$ (note the positive coefficient in the exponent) considered on $\Sigma := [-1, 1]$. It can be shown that $S_w = [-1, 1]$, the density v of μ_w is given by

$$v(t) = \frac{2t^2}{\pi\sqrt{1-t^2}},$$

which has a zero at the origin, and has a $(1-x^2)^{-1/2}$ type singularity at ± 1.

This example raises another problem that I was not able to solve. Namely, in this case the weight is not defined outside of S_w, and there is no a priori reason why we should restrict our attention to functions f that vanish at ± 1. In particular, is it true that for every function $f \in C[-1, 1]$ there is a sequence of polynomials P_n of degree at most n such that

$$e^{nx^2}P_n(x) \to f(x)$$

uniformly on $[-1, 1]$? The answer would be interesting even for $f \equiv 1$.

Recently some efforts have been made to find a "soft" approach to the approximation problem considered in this paper. In some restricted cases such an approach is possible, for example in [3] and [11] simple sign change counting was used to prove such theorems (this works for example if $w(x) = e^{-x^2}$). Although I do not believe that this approach can produce very general results, the simplicity of the method warrants further research in this direction.

§4. Fast Decreasing Polynomials

In this section, we discuss another application of weighted polynomial approximation with varying weights. This will be in connection with so-called fast decreasing polynomials. We call polynomials P_n, $\deg P_n \leq n$, fast decreasing on $[-1,1]$ if they attain the value 1 at $x = 0$ and decrease rapidly away from the origin:

$$P_n(0) = 1, \qquad |P_n(x)| \leq e^{-\Phi(x)} \quad x \in [-1,1]. \tag{4.1}$$

We shall discuss the problem of what Φ and n are possible.

Obviously, the significance of such fast decreasing polynomials lies in the fact that they approximate the "Dirac delta function." Hence these are good polynomial kernels for convolution operators to reproduce the identity. By integration, we can get from the above polynomials good polynomial approximants Q_n of the signum function in the sense

$$|\text{sign } x - Q_m(x)| \leq e^{-\Phi(x)} \quad x \in [-1,1], \tag{4.2}$$

which in turn can be used to construct well localized "partitions of unity" (cf. the construction on p. 156 in [18]) consisting of polynomials of a given degree n.

The problem can be formulated in two different ways: one can ask what possible decrease (i.e., what Φ) is possible for a given degree, or, alternatively, for a given Φ what is the smallest degree n for which there are polynomials with properties (4.1). Let n_Φ denote this degree. For symmetric Φ's that are increasing on $[0,1]$ this problem was completely solved modulo a constant in [17]:

Theorem 4.1. (Ivanov-Totik, [17, Theorem 1]) *Let Φ be an even function, right continuous and increasing on $[0,1]$. Then*

$$\frac{1}{5} N_\Phi \leq n_\Phi \leq 12 N_\Phi,$$

where $N_\Phi = 0$ if $\Phi(1) \leq 0$, and

$$N_\Phi = 2 \sup_{\Phi^{-1}(0) \leq x < b} \sqrt{\frac{\Phi(x)}{x^2}} + \int_b^{1/2} \frac{\Phi(x)}{x^2} dx + \sup_{1/2 \leq x < 1} \frac{\Phi(x)}{-\log(1-x)} + 1,$$

$b = \min(\Phi^{-1}(1), 1/2)$, *otherwise*.

We emphasize that the estimate is given for all Φ, in particular, Φ can depend on n. As an easy corollary we get

Corollary 4.2. *Let φ be even and increasing on $[0,1]$. Then there are polynomials P_n of degree at most n satisfying*

$$P_n(0) = 1, \qquad |P_n(x)| \leq Ce^{-cn\varphi(x)} \quad x \in [-1,1], \; n = 0,1,\ldots$$

for some constants $C, c > 0$ if and only if

$$\int_0^1 \frac{\varphi(u)}{u^2} du < \infty.$$

As a second consequence we mention

Corollary 4.3. *Let φ be even and increasing on $[0,\infty]$. Then there are polynomials P_n of degree at most n satisfying*

$$P_n(0) = 1, \qquad |P_n(x)| \leq Ce^{-c\varphi(nx)} \quad x \in [-1,1], \quad n = 0, 1, \ldots$$

for some constants $C, c > 0$ if and only if

$$\int_{-\infty}^{\infty} \frac{\varphi(u)}{1+u^2} \, du < \infty.$$

We have also determined the size of the minimal m for which (4.2) is possible. If this is denoted by m_Φ, then m_Φ has the order given below.

Theorem 4.4. (Ivanov-Totik, [17, Theorem 1]) *Let Φ be an even function, right continuous and increasing on $[0,1]$. Then*

$$\frac{1}{C} M_\Phi \leq m_\Phi \leq C M_\Phi,$$

where $M_\Phi = 0$ if $\Phi(1) \leq 0$, and

$$M_\Phi = 2 \sup_{\Phi^{-1}(0) \leq x < b} \frac{\Phi(x)}{x^2} + \int_b^{1/2} \frac{\Phi(x)}{x^2} \, dx + \sup_{1/2 \leq x < 1} \frac{\Phi(x)}{-\log(1-x)} + 1,$$

otherwise. Here $b = \min(\Phi^{-1}(1), 1/2)$.

Similar estimates hold if we require further properties of the polynomials, e.g., if we also require that the Q_m in (4.2) be monotone increasing.

The above results completely solve our problem modulo a constant. The method of their proof was not suitable to obtain sharp constants. Such results can be obtained, at least for the special case considered in Corollary 4.2, by potential theoretic methods.

Theorem 4.5. (Totik, [43, Theorem 3.3]) *Suppose that φ is even, increasing on $[0,1]$, and $\varphi(\sqrt{x})$ is concave on $[0,1]$. Then there are polynomials satisfying*

$$P_n(0) = 1, \qquad |P_n(x)| \leq e^{-n(\varphi(x)+o(1))} \quad x \in [-1,1], \tag{4.3}$$

if and only if

$$\frac{2}{\pi} \int_0^1 \frac{\varphi(t)}{t^2 \sqrt{1-t^2}} \, dt \leq 1$$

holds.

This can be applied to any $\varphi(t) = c|t|^\lambda$, $\lambda \leq 2$, and we conclude that there are polynomials P_n with

$$P_n(0) = 1, \qquad |P_n(x)| \leq e^{-n(c|x|^\lambda + o(1))}, \quad x \in [-1,1],$$

if and only if $\lambda > 1$ and $c \leq \sqrt{\pi}\Gamma\left(\frac{\lambda}{2}\right)/\Gamma\left(\frac{\lambda-1}{2}\right)$. We also mention that the latter result is no longer true for $\lambda > 2$. It is an open problem to determine the exact conditions on φ that allow (4.3) when $\varphi(\sqrt{x})$ is not concave, e.g., if $\varphi(t) = c|t|^\lambda$ with $\lambda > 2$.

How is this problem connected to the approximation problem discussed in the preceding section? If we multiply through in (4.3) by $w(x) := \exp(n\varphi(x))$, then (assuming $\varphi(0) = 0$) we find that the weighted polynomials $w^n(x)P_n(x)$ satisfy

$$|w^n(x)P_n(x)| \leq e^{o(n)} \qquad x \in [-1,1]$$

while $w^n(0)P_n(0) = 1$; this means that they essentially "take their maximum" at $x = 0$ on $[-1,1]$ (note that typical behaviour of such polynomials is exponential, i.e., in general, they are exponentially small away from the support S_w, by the results mentioned in the preceding section). But we know that this can happen if and only if 0 belongs to the support S_w of the extremal measure associated with w, and we have effective ways of telling when a point belongs to the support S_w.

§5. Locally Geometric Rates

Until now we have discussed rates for polynomial approximation that were not geometric. The case of geometric rate has been completely settled by Bernstein, who proved that for $f \in C[-1,1]$, $E_n(f)$ tends to zero exponentially fast if and only if f is analytic on $[-1,1]$. In fact, a more precise formula holds: if

$$\rho = 1/\limsup_{n\to\infty} E_n(f)^{1/n} \ ,$$

then ρ equals the sum of the half axes of the largest ellipse G with foci at ± 1 which has the property that f can be analytically extended to the interior of G.

Now let us assume that f is continuous but not analytic on $[-1,1]$, and is analytic on some parts of $[-1,1]$. (As a typical such f, think of the absolute value function). The local problem concerning geometric approximation on intervals of analyticity was first considered by Bochner [2], but he obtained the false result that geometric approximation is not possible in general (we also want the approximating polynomials to converge uniformly to f on $[-1,1]$). Later results of Frey [8] and Temlyakov [38] showed that a geometric rate is possible, and these authors obtained very precise information on this rate away from the boundary of the region where f is analytic. We shall briefly describe what happens if we do not exclude a neighborhood of this boundary.

Our results were motivated by the fact that best polynomial approximants are very far from giving good approximation on *subsets* of the original set. In fact, let $Q_n = Q_n(f)$ be the best uniform approximant to $f \in C[-1,1]$ among all polynomials of degree at most n. A celebrated result of Kadec [20] says that the extremal points of $|f - Q_n(f)|$, $n = 1, 2, \ldots$, are dense on $[-1,1]$, and so on any subinterval $I \subset [-1,1]$ the approximation given by

$\{Q_n(f)\}_{n=0}^{\infty}$ (considering the whole sequence) is not better than on the whole interval $[-1, 1]$, no matter how smooth f is on I.

Now what about other polynomials? To be more precise, we ask the following: let $f \in C[-1, 1]$ be analytic on the (relative to $[-1, 1]$) open subset D of $[-1, 1]$. Is it possible to find polynomials P_n of degree at most n, such that $\|f - P_n\|_{[-1,1]} \to 0$ as $n \to \infty$, and at every point of D we have geometric convergence, i.e.,

$$\limsup_{n \to \infty} |f(x) - P_n(x)|^{1/n} < 1, \qquad x \in D. \tag{5.1}$$

In [18] we showed that this is always possible and described the behavior of the left-hand side of (5.1) which is, in a certain sense, best possible.

In fact, assume that D is the exact set of analyticity, i.e., D contains every regular point of f. For $x \in [-1, 1]$ let $d(x)$ be the distance from x to the nearest singularity of f, where f is considered to be extended to the complex plane and we also count the singularities outside $[-1, 1]$. In other words, $d(x)$ is the largest radius such that the Taylor expansion of f about x converges in $\{z \in \mathbf{C} : |z - x| < d(x)\}$. Of course, if x is not a regular point of f, then $d(x) = 0$.

If $d(x) > 0$ for every $x \in [-1, 1]$, i.e., if f is analytic on $[-1, 1]$, then the best uniform approximants converge geometrically to f, and so in what follows we assume that f has a singularity somewhere on $[-1, 1]$.

Theorem 5.1. (Ivanov-Saff-Totik, [18, Theorem 1]) *Suppose that $\beta > 1$, $f \in C[-1, 1]$, and f has a singularity on $[-1, 1]$. There are polynomials P_n of degree n, $n = 0, 1, \ldots$ such that*

$$\|f - P_n\|_{[-1,1]} \to 0 \quad \text{as} \quad n \to \infty,$$

and for every $x \in [-1, 1]$

$$|f(x) - P_n(x)| \leq C_{f,x} \exp(-cn[d(x)]^{\beta}), \tag{5.2}$$

where $c > 0$ is an absolute constant and the constant $C_{f,x}$ is bounded in x in any compact subset of D.

This is sharp in the sense that, in general, it is impossible to set $\beta = 1$.

Theorem 5.2. (Ivanov-Saff-Totik, [18, Theorem 2]) *There are no positive constants C_x, $x \in [-1, 1]$, and $c > 0$ such that the C_x's are bounded for x belonging to any compact subset of $[-1, 1]$ not containing the origin, and for every n there are polynomials P_n of degree at most n with*

$$||x| - P_n(x)| \leq C_x \exp(-cn|x|), \qquad x \in [-1, 1].$$

Note that for $f(x) = |x|$, we have $d(x) = |x|$.

Finally, we mention that Theorem 5.1 cannot be sharpened by using a constant C_f — in other words, the constant in (5.2) must depend on x, even allowing c and β to depend on f.

Theorem 5.3. (Ivanov-Saff-Totik, [18, Theorem 3]) *There exists an f in $C[-1, 1]$ such that for no constants β, C, $c > 0$ can one find polynomials $P_n \in \Pi_n$, $n = 1, 2, ...$, with the property*

$$|f(x) - P_n(x)| \leq C_f \exp(-cn[d(x)]^\beta) , \qquad n = 1, 2, ..., \ x \in [-1, 1] .$$

As an open problem let us mention that in Theorem 5.1 we did not address the rate with which the polynomials P_n approximate f. It can be shown that Jackson type rates (see Section 2) can be achieved. We have not been able to determine whether the following is true: If f is as in Theorem 5.1, then there are polynomials P_n of degree n, $n = 0, 1, ...$ such that

$$\|f - P_n\|_{[-1,1]} \leq CE_n(f)$$

and

$$\limsup_{n \to \infty} |f(x) - P_n(x)|^{1/n} < 1 , \qquad x \in D .$$

Acknowledgements. This work was supported by NSF grant DMS 9101380 and by the Hungarian Science Foundation for Research, grant No. 1990/3.

References

1. Berens, H., and G. G. Lorentz, Inverse theorems for Bernstein polynomials, Indiana Univ. Math. J. **21** (1972), 693–708.
2. Bochner, S., Localization of best polynomial approximation, in *Contributions to Fourier Analysis*, Annals of Mathematics Studies **25**, 3–23, Princeton University Press, 1950.
3. Borwein, P., and E. B. Saff, On the denseness of weighted incomplete approximations, in *Proceedings of the First US–Soviet Conference on Approx. Theory*, Tampa 1990, Springer–Verlag, to appear.
4. Butzer, P. L., R. L. Stens, and M. Wehrens, Approximation by algebraic convolution integrals, in *Approximation Theory and Functional Analysis*, J. B. Prolla (ed.), North Holland Publ. Co., 1979, 71–120.
5. Ditzian, Z., and K. G. Ivanov, Strong converse inequalities, J. Analyse Math., to appear.
6. Ditzian, Z., and V. Totik, *Moduli of Smoothness*, Springer Series for Computational Mathematics, **9**, Springer-Verlag, New York, 1987.
7. Freud, G., On the coefficients in the recursion formulae of orthogonal polynomials, Proc. Roy. Irish Acad. Sect. A **76** (1976), 1–6.
8. Frey, T., An the localization of best polynomial approximation I, II (in Hungarian), Magyar Tud. Akad. Mat. Fiz. Oszt. Közl. **7** (1957), 403–412; **8** (1958), 89–112.
9. Frostman, O., *Potentiel d'Équilibre et Capacité des Ensembles avec Quelques Application là Théorie des Fonctions*, Thesis, Lunds Univ. Mat. Sem. **3** (1935), 1–118.

10. v. Golitschek, M., Approximation by incomplete polynomials, J. Approx. Theory **28** (1980), 155–160.
11. v. Golitschek, M., G. G. Lorentz and Y. Makovoz, Asymptotics of weighted polynomials, in *Proceedings of the First US-Soviet Conference on Approx. Theory*, Tampa 1990, Springer-Verlag, to appear.
12. Hardy, G. H., *Divergent Series*, Clarendon Press, Oxford, 1949.
13. He X., and X. Li, Uniform convergence of polynomials associated with varying weights, Rocky Mountain J. Math. **21** (1991), 281–300.
14. Helms, L. L., *Introduction to Potential Theory*, Wiley-Interscience, New York, 1969.
15. Ivanov, K. G., Direct and converse theorem for best algebraic approximation in $C[-1,1]$ and $L_p[-1,1]$, C. R. Acad. Bulgare Sci. **33** (1980), 1309–1312.
16. Ivanov, K. G., On a new characterization of functions II, direct and converse theorems for the best algebraic approximation in $C[-1,1]$ and $L_p[-1,1]$, Pliska Stud. Math. Bulgar. **5** (1983), 151–163.
17. Ivanov, K. G., and V. Totik, Fast decreasing polynomials, Constr. Approx. **6** (1990), 1–20.
18. Ivanov, K. G., E. B. Saff and V. Totik, Approximation by polynomials with locally geometric rates, Proc. Amer. Math. Soc. **106** (1989), 153–161.
19. Jackson, D., On the approximation by trigonometric sums and polynomials, Trans. Amer. Math. Soc **13** (1912), 491–515.
20. Kadec, M. I., On the distribution of maximum deviation in the approximation of continuous functions, Amer. Math. Soc. Transl. **26** (1963), 231–234.
21. Kantorovich, L., Sur certains développements suivant les polynômes de la forme de S. Bernstein, I, II, C. R. Acad. Sci. URSS, 1930, 563–568, 595–600.
22. Knopfmacher, A., D. S. Lubinsky and P. Nevai, Freud's conjecture and approximation of reciprocals of weights by polynomials, Constr. Approx. **4** (1988), 9–20.
23. Landkof, N. S., *Foundations of Modern Potential Theory*, Grundlehren der Mathematischen Wissenschaften, **190**, Springer-Verlag, New York, 1972.
24. Lorentz, G. G., Approximation by incomplete polynomials, in *Padé and Rational Approximation, Theory and Applications*, E. B. Saff and R. S. Varga (eds.), Academic Press, New York 1977, 289–302.
25. Lorentz, G. G., *Approximation of Function*, Holt, Rinehart and Winston, Athena series, New York, 1966. Reprinted by Chelsea Publ., New York, 1991.
26. Lubinsky, D. S., H. N. Mhaskar and E. B. Saff, A proof of Freud's conjecture for exponential weights, Constr. Approx. **4** (1988), 65–83.
27. Lubinsky, D. S., and E. B. Saff, *Strong Asymptotics for Extremal Polynomials Associated with Weights on* $\mathbb{R}$, Lecture Notes in Mathematics **1305**, Springer-Verlag, New York, 1988.

28. Lubinsky, D. S., and E. B. Saff, Uniform and mean approximation by certain weighted polynomials, with applications, Constr. Approx. **4** (1988), 21–64, 239–282.
29. Mhaskar, H. N., and E. B. Saff, Where does the sup norm of a weighted polynomial live?, Constr. Approx. **1** (1985), 71–91.
30. Mhaskar, H. N., and E. B. Saff, Extremal problems for polynomials with exponential weights, Trans. Amer. Math. Soc. **285** (1984), 203–234.
31. Mhaskar, H. N., and E. B. Saff, A Weierstrass-type approximation theorem for certain weighted polynomials, in *Approximation Theory and Applications*, S. P. Singh (ed.), Pitman Publ. Ltd., 1985, 115–123.
32. Mhaskar, H. N., and E. B. Saff, Polynomials with Laguerre weights in L^p, in *Rational Approximation and Interpolation*, P. R. Graves-Morris, E. B. Saff and R. S. Varga (eds.), Lecture Notes in Mathematics, **1105**, Springer-Verlag, Berlin, 1984, 511–523.
33. Rahmanov, E. A., On asymptotic properties of polynomials orthogonal on the real axis, Mat. Sb. **119** (161)(1982), 163–203. English transl., Math. USSR-Sb. **47** (1984), 155–193.
34. Saff, E. B., Incomplete and orthogonal polynomials, in *Approximation Theory IV*, C. K. Chui, L. L. Schumaker and J. D. Ward (eds.), Academic Press, New York, 1983, 219–256.
35. Saff, E. B., J. L Ullman and R. S. Varga, Incomplete polynomials, an electrostatic approach, in *Approximation Theory IV*, C. K. Chui, L. L. Schumaker and J. D. Ward (eds.), Academic Press, New York, 1983, 769–782.
36. Saff, E. B., and R. S. Varga, On incomplete polynomials, in *Numerische Methoden der Approximationstheorie*, L. Collatz, G. Meinardus and H. Werner (eds.), ISNM **42**, Birkhäuser-Verlag, Basel, 1978, 281–298.
37. Saff, E. B., and R. S. Varga, The sharpness of Lorentz's Theorem on incomplete polynomials, Trans. Amer. Math. Soc. **249** (1979), 159–162.
38. Temlyakov, V., On the localization of the trigonometric approximation, Anal. Math. **3** (1977), 151–169.
39. Totik, V., Approximation by Bernstein polynomials, (manuscript).
40. Totik, V., Approximation by weighted polynomials with varying weights (manuscript).
41. Totik, V., Strong converse inequalities, (manuscript).
42. Totik, V., The necessity of a new kind of modulus of smoothness, in *Anniversary Volume on Approximation Theory and Functional Analysis*, Birkhäuser, Basel, 1984, 233–337.
43. Totik, V., Fast decreasing polynomials via potentials, J. Analyse Math., (to appear).
44. Totik, V., Sharp converse theorem of L^p-polynomial approximation, Constr. Approx. **4** (1988), 419–433.
45. van Wickeren, E., Stechkin-Marchaud-type inequalities in connection with Bernstein polynomials, Constr. Approx. **2** (1986), 331–337.
46. Zygmund, A., A remark on the integral modulus of continuity, Univ. Nac. Tucumán rev., Ser. A **7** (1950), 259–269.

Vilmos Totik
Bolyai Institute
Aradi V. tere 1.
Szeged, 6720 Hungary
h871@ella.hu

and

Department of Mathematics
University of South Florida
Tampa, FL 33620, USA
totik@gauss.math.usf.edu

ISBN 0-12-174589-9